Sustainability and Climate Services: Critique, Integration, and Reimagination

Sustainability and Climate Services: Critique, Integration, and Reimagination

Editors

Charles Herrick
Jason Vogel
Glen Anderson

MDPI • Basel • Beijing • Wuhan • Barcelona • Belgrade • Manchester • Tokyo • Cluj • Tianjin

MDPI

Editors
Charles Herrick
New York University
Washington DC Center
Washington DC USA

Jason Vogel
Climate Impacts Group
University of Washington
Seattle
WA
USA

Glen Anderson
Independent Consultant on
International Development
and Climate Adaptation
Assignments
Raleigh
NC
USA

Editorial Office
MDPI
St. Alban-Anlage 66
4052 Basel, Switzerland

This is a reprint of articles from the Special Issue published online in the open access journal *Sustainability* (ISSN 2071-1050) (available at: https://www.mdpi.com/journal/sustainability/special_issues/climate_service).

For citation purposes, cite each article independently as indicated on the article page online and as indicated below:

LastName, A.A.; LastName, B.B.; LastName, C.C. Article Title. *Journal Name* **Year**, *Volume Number*, Page Range.

ISBN 978-3-0365-7836-1 (Hbk)
ISBN 978-3-0365-7837-8 (PDF)

Cover image courtesy of Sarah Le Sueur

Contents

About the Editors vii

Preface to "Sustainability and Climate Services: Critique, Integration, and Reimagination" . ix

Charles Herrick and Jason Vogel
Climate Services: Critique, Integration, and Reimagination
Reprinted from: *Sustainability* **2023**, *15*, 6789, doi:10.3390/su15086789 1

Robert J. Lempert, Lisa Busch, Ryan Brown, Annette Patton, Sara Turner, Jacyn Schmidt and Tammy Young
Community-Level, Participatory Co-Design for Landslide Warning with Implications for Climate Services
Reprinted from: *Sustainability* **2023**, *15*, 4294, doi:10.3390/su15054294 9

Elena Lioubimtseva and Charlotte da Cunha
The Role of Non-Climate Data in Equitable Climate Adaptation Planning: Lessons from Small French and American Cities
Reprinted from: *Sustainability* **2023**, *15*, 1556, doi:10.3390/su15021556 35

Aparna Bamzai-Dodson and Renee A. McPherson
When Do Climate Services Achieve Societal Impact? Evaluations of Actionable Climate Adaptation Science
Reprinted from: *Sustainability* **2022**, *14*, 14026, doi:10.3390/su142114026 57

Edward R. Carr
Climate Services and Transformational Adaptation
Reprinted from: *Sustainability* **2023**, *15*, 289, doi:10.3390/su15010289 71

Arsum Pathak, Laura E. Hilberg, Lara J. Hansen and Bruce A. Stein
Key Considerations for the Use of Nature-Based Solutions in Climate Services and Adaptation
Reprinted from: *Sustainability* **2022**, *14*, 16817, doi:10.3390/su142416817 81

Alfredo Pena-Vega, Marianne Cohen, Luis Manuel Flores, Hervé Le Treut, Marcelo Lagos, Juan Carlos Castilla, et al.
Young People Are Changing Their Socio-Ecological Reality to Face Climate Change: Contrasting Transformative Youth Commitment with Division and Inertia of Governments
Reprinted from: *Sustainability* **2022**, *14*, 15116, doi:10.3390/su142215116 93

Anna Boqué Ciurana, Melisa Ménendez, María Suárez Bilbao and Enric Aguilar
Exploring the Climatic Potential of Somo's Surf Spot for Tourist Destination Management
Reprinted from: *Sustainability* **2022**, *14*, 8496, doi:10.3390/su14148496 111

Ann Grodnik-Nagle, Ashima Sukhdev, Jason Vogel and Charles Herrick
Beyond Climate Ready? A History of Seattle Public Utilities' Ongoing Evolution from Environmental and Climate Risk Management to Integrated Sustainability
Reprinted from: *Sustainability* **2023**, *15*, 4977, doi:10.3390/su15064977 133

Werner Krauß
Slowing Down Climate Services: Climate Change as a Matter of Concern
Reprinted from: *Sustainability* **2023**, *15*, 6458, doi:10.3390/su15086458 159

Charles Herrick and Jason Vogel
Climate Adaptation at the Local Scale: Using Federal Climate Adaptation Policy Regimes to Enhance Climate Services
Reprinted from: *Sustainability* **2022**, *14*, 8135, doi:10.3390/su14138135 173

About the Editors

Charles Herrick

Charles Herrick: Before retiring in December 2016, Dr. Herrick consulted with a wide range of U.S. government agencies, providing analytical input and strategic direction on a variety of environmental and science policy issues, including climate change adaptation, sustainable fisheries management, acid rain, environmental indicators and indices, the valuation of ecosystem services, and control of invasive species. A leading expert in program evaluation, he conducted third-party evaluations and designed program theories of change and logic models to characterize diverse program delivery mechanisms, process outputs, and outcomes for major foundations, government agencies, and non-profit service-delivery organizations, including the National Fish and Wildlife Foundation, Gordon and Betty Moore Foundation, David and Lucile Packard Foundation, Pew Charitable Trusts, Sloan Foundation, Health Canada, and the U.S. EPA, DOE, NOAA, and CDC. As a consultant to municipal and water sector research foundations, he partnered with municipal and utility executives and managers to address topics including stormwater management, watershed management, organizational cultural change to support sustainable operations, climate change adaptation, emergency response planning, the protection of critical information, and implementation and evaluation of environmental regulatory programs. Herrick is a seasoned executive with oversight experience across all corporate functions. Throughout his career, Dr. Herrick has maintained an active association with academia, teaching regularly and publishing frequently in the peer-reviewed literature. He is currently an adjunct professor at the New York University, Washington DC Center.

Jason Vogel

Jason Vogel: As the Interim Director and Deputy Director of the University of Washington Climate Impacts Group, Dr. Vogel provides leadership and continuity of operations for the organization, including external relations, fundraising, government and University relations, budgeting, strategic planning, and technical excellence in applied interdisciplinary climate impacts and adaptation research and engagement. He currently serves as the lead principal investigator for the Northwest Climate Resilience Collaborative, a five-year multi-institution effort funded by the NOAA. The Resilience Collaborative was envisioned as an effort to center tribes and frontline communities in the implementation of a climate resilience research and engagement program. Vogel is committed to systemic change in social–environmental systems to ensure a sustainable and equitable future for human and non-human beings. He is a fierce skeptic, always questioning how to achieve the greatest impact, and he is committed to his team and their success. Before this, he worked as a consultant specializing in climate change impacts and adaptation at Stratus Consulting and Abt Associates, working both domestically and internationally. He conducted early work with U.S. water utilities, who were among the first institutions to grapple with how climate change might affect their ability to deliver on their mission. Later, he worked more with communities and decision makers in developing and emerging nations; observing how climate and international finance impacts these nations reinforced his commitment to being an agent of change.

Glen Anderson

Glen Anderson: Over the last 23 years, Dr. Anderson has worked in 41 countries in the areas of climate adaptation, benefit–cost analysis, environmental and economic policy, water resources management, and environmental finance. In addition to his project management duties as a USAID Chief of Party, Anderson served as editor and contributing author of a primer on the economics of weather, climate, and hydrological services, a joint effort of the WMO, the World Bank, and the CSP (with financial support from USAID through CCRD). He received his BA from the University of Washington in Economics and holds MA and Ph.D. degrees in Agricultural Economics from the University of Wisconsin–Madison.

Preface to "Sustainability and Climate Services: Critique, Integration, and Reimagination"

This volume explores the relationship between scientific information and its application in social, operational, and policy environments, a domain of research and analysis with which we have grappled for a collective 45 years. Although this volume addresses climate change adaptation, we have puzzled over this issue in numerous contexts, including acid rain regulation, sustainable fisheries management, security and water utility operations, governmental agency efforts to understand and bridge digital divides, regulation of invasive plant and animal species, management of agricultural systems in the Caribbean Basin, high-level radioactive waste disposal, U.S. chemical regulation, and others.

We have observed that the common denominator across all these topics is the irrefutable reality that scientific outputs—no matter how sound, accurate, and precise, or in the parlance of science and technology policy studies, no matter how salient, credible, and legitimate—do not and cannot drive or compel any particular action on the part of so-called users. Our professional and policy discourse is rife with admonitions to 'follow the science' or that decisions be 'evidence-driven' or 'fact-based'. Such urgings are symptomatic of deeply held assumptions that center science in decision-making in a way that does not comport with the realities we have observed. As suggested, or maybe even demonstrated, in this volume's essays, factual inputs nearly always need to be steeped within a brew of practice, value orientation, circumstantial limitations, entrenched operational environments, and obscure but deeply held affectations of place, self, and group identity.

To say this is not to hand the keys of rationale deliberation over to some sloppy, come-what-may relativism. It is instead to recognize and embrace the full complexity of the human experience and insist that all involved in public decisions, including the producers of scientific information, must be broadly thoughtful across a much broader range of considerations than what may be 'in our wheelhouse'. It is our hope that this volume and its constituent essays are viewed by readers as a gateway to understanding this complexity and an opportunity to de-center science in understanding how public decisions are made.

Charles Herrick and Jason Vogel

Editors

sustainability MDPI

Editorial

Climate Services: Critique, Integration, and Reimagination

Charles Herrick [1,*] and Jason Vogel [2]

1 Washington, DC Center, New York University, Washington, DC 20005, USA
2 Climate Impacts Group, College of the Environment, University of Washington, Seattle, WA 98195, USA; jmvogel@uw.edu
* Correspondence: ch133@nyu.edu

As an ideal, sustainable development—or sustainability—integrates economic growth, social equity, and enduring environmental quality. Studies of sustainability are multi-disciplinary, cross-sectoral, and reflexive—some would claim transdisciplinary. More broadly, sustainable development has been characterized as a societal process of learning, adaptation, and creation. Climate change represents a clear and pressing challenge toward efforts to forge a sustainable future. This Special Issue is premised on the proposition that the effort to develop and apply climate services is—or at least ought to be—part and parcel of the larger enterprise of sustainable development.

For this project, we sought critical input from scholars and practitioners outside the climate services community, and especially pieces co-authored by information users and their technical/scientific partners. We insisted that contributions reflect the voice, perspective, and existential situation of climate service users. Entitled "*Sustainability and Climate Services: Critique, Integration, and Reimagination*", the purpose of this Special Issue is to empower unconventional thinking in the hopes of accelerating the relevance of climate services at a time when many communities, public and private organizations, tribes, and all level of government agencies are pursuing programs of climate adaptation and resilience. It is our hope that this body of work will provide outside-the-box critique and help to identify and facilitate a 'next generation' of weather- and climate-related services, both more relevant information products, but also services that may go beyond the conventional scientific orientation of providing information products.

Summary of Contributed Papers: The Special Issue is a collection of ten research articles, reviews, and viewpoints covering a wide variety of circumstantial, experiential, and geographic variation. It includes the work of 33 authors from three sovereign tribal nations (Tlingit, Haida, Sitka), Chile, France, Germany, Spain, and the United States. The authors are academic and institutional researchers, federal agency program managers, water utility employees, consultants, scientists affiliated with environmental and social advocacy groups, and local and tribal government officials. The articles cover research and interpretive analyses focused on large- and medium-sized U.S. municipalities; small communities in France and the U.S.; a remote village in southeastern Alaska; student-led educational activities in developed, emerging, and developing countries world-wide; agrarian communities in East Africa; federal resource management programs in the U.S.; the operations of a large U.S. water and solid waste management utility; regional and community level activists in Northern Germany; and geographically non-specific research dealing with the potential for nature-based solutions in a wide range of climate change and sustainability challenges.

Short summaries of each essay follow below:

A paper by **Robert Lempert**, **Lisa Busch,** and colleagues describes a community-level co-design process among academic, state, and federal scientists; citizens and local officials; and potentially impacted tribal council members to develop a landslide warning system for Sitka, Alaska—a small, remote coastal town (Contribution 1). The decentralized system features an online dashboard which displays current and forecasted risk levels to help

Citation: Herrick, C.; Vogel, J. Climate Services: Critique, Integration, and Reimagination. *Sustainability* **2023**, *15*, 6789. https://doi.org/10.3390/su15086789

Received: 10 April 2023
Accepted: 13 April 2023
Published: 18 April 2023

residents make their own risk management decisions. This case study addresses questions including: what activities did the project team conduct, what did these activities intend to accomplish, and did these activities accomplish what they intended? The paper describes the co-design process, the associated changes in system design and research activities, and formal and informal evaluations of the system and process. Overall, the co-design process appears to have generated a warning system the Sitka community finds valuable, helped to align system design with local knowledge and community values, significantly modified the scientists' research agendas, and helped navigate sensitivities such as the effect of landslide exposure maps on property values. Other communities in southeastern Alaska are now adopting this engagement approach. The paper concludes with broader implications for the role of community-level, participatory co-design, and risk governance for climate services.

There is a growing consensus that to effectively adapt to climate change, cities need user-friendly tools and reliable high-resolution biophysical and socio-economic data for analysis, mapping, modeling, and visualization. A study by **Elena Lioubimtseva** and **Charlotte da Cunha** examines the availability of various types of information used in climate adaptation plans of 40 municipalities with populations of less than 300,000 people in the United States and France (Contribution 2). The authors argue that non-climatic spatial data, such as population demographic and socio-economic patterns, urban infrastructure, and environmental data, must be integrated with climate tools and datasets to inform effective vulnerability assessments and equitable adaptation planning goals. Most climate adaptation plans examined in this study fail to address the existing structural inequalities and environmental injustices in urban infrastructure and land use, with challenges such as methodological and ideological barriers, data quality issues, and a lack of meaningful community connections. Adaptation methodological approaches should be reassessed in the context of much-needed societal transformation. Lessons learned from this and associated studies offer valuable insights for the potential development of national and state-level climate adaptation information services for cities.

Aparna Bamzai-Dodson and **Renee McPherson** explore how the discipline and professional practice of project and program evaluation might be engaged to help assure the applicability, relevance, and overall usability of climate services (Contribution 3). To achieve the intended societal impact, scientists are using climate services to engage directly with stakeholders to better understand their needs and inform knowledge production. However, the wide variety of climate-services outcomes—ranging from establishing collegial relationships with stakeholders to obtaining specific information for inclusion into a pre-existing decision process—do not directly connect to traditional methods of measuring scientific impact (e.g., publication citations, journal impact factor). In this paper, Bamzai-Dodson and McPherson describe how concepts and methods from project-program evaluation can be used to examine the societal impacts of climate services. Working with desired outcomes in mind, those who conduct and fund applied climate research would benefit from the inclusion and execution of evaluation activities at the beginning of project development.

Drawing on the author's research in East Africa, **Edward Carr's** article explores the potential for climate services to catalyze and foster transformational adaptation (Contribution 4). Carr argues that weather and climate information are not, in and of themselves, tools for transformation. When designed and delivered without careful identification of the intended users of the service and the needs that service addresses, they can fail to catalyze change among the users of that information. At worst, they can reinforce the status quo and drive maladaptive outcomes. He goes on to argue that for climate services to serve as agents of transformational adaptation, the climate services community will have to change how it understands the users of these services and their needs. Building climate services around contemporary understandings of how people make decisions about their lives and livelihoods offers designers and implementers of climate services opportunities to create services that catalyze transformational adaptation.

Arsum Pathak, Laura Hilberg and others explore how the application of nature-based systems (NbS) can enhance community resilience by providing both climate adaptation and mitigation outcomes (Contribution 5). While NbS do not necessarily represent new "technology" or methods, the planned incorporation of these approaches into climate adaptation efforts is often considered novel, particularly within the climate services sector where interventions have historically prioritized structural infrastructure approaches. Pathak, Hilberg and colleagues argue that NbS can offer an effective replacement for or complement many traditional infrastructure applications. Additionally, natural and nature-based systems can respond to climate change in a manner that engineered solutions often cannot, providing long-term holistic adaptation and mitigation success with additional sustainability benefits to ecosystem services such as improved air and water quality, carbon sequestration, outdoor recreation, and biodiversity protection. This article supports implementation of NbS through a set of seven "key considerations" for their use in community-based adaptation.

A paper by **Alfredo Pena-Vega**, **Marianne Cohen** and colleagues contributes to a critical re-reading of the notion of climate services by problematizing the discontinuity between young people's commitment to climate change and the lack of a unified, action-oriented vision regarding climate policy among governments (Contribution 6). This essay reminds us that the activities of young people can help to build civic awareness and drive action to arrest climate change. In this sense, climate services, directed to young people, could contribute to the design of a sustainable future. To help actualize this vision, the authors propose a 'dialogical link' between the enterprise of climate services and the ways in which distinct groups of young people come to visualize and develop relationships with their environments, organize themselves, and then take action to transform reality.

Anna Boqué Ciurana, Melisa Ménendez and others describe an interesting episode of co-production and explore climate service provision through a recreational lens (Contribution 7). Surfing is one of the most popular activities in coastal tourism resorts; but the sport depends strongly on met-ocean weather conditions, particularly on surface wind-generated waves that reach the coast. This study provides examples of how users' needs, and user perspectives can be captured and operationalized by climate data specialists to develop useful information addressing human and social needs. This paper describes the research team's collaborations with the surfing community to co-define a series of indices to quantify surfing days, surfing days stratified by surfers' levels of skill, and other useful types of information. A hindcasting exercise was undertaken to illustrate the potential applicability of the indices in a real-world context, specifically Somo Beach near Cantabria, Spain.

Ann Grodnik-Nagle, **Ashima Sukhdev**, and others chronicle the evolution of Seattle Public Utilities (SPU) as a 'beyond climate-ready' organization (Contribution 8). This utility has explored the impacts of climate change and supported climate adaptation work since 1997. Faced with threats such as sea level rise, drought, wildfires, and extreme precipitation events, SPU has worked to "mainstream" climate science throughout its strategic and capital investment planning, management, operations, staffing, institutional culture, and more. This paper provides a descriptive, chronologically ordered account of how SPU's climate-change-related work has evolved to become an aspect of a broader social and environmental sustainability orientation, aimed at resilience against climate impacts, but also addressing a diverse palate of services, including greenhouse gas emissions reduction, carbon sequestration, water and waste circularity, green infrastructure, ecosystem and species stewardship, green and blue workforce development, service affordability, an intergenerational perspective, and environmental justice. The authors frame this transition as a movement from a core focus on risk management toward a proactive and integrated mode of sustainable operations. Acknowledging that SPU's journey has been enabled by a co-productive approach to climate services, the authors end the essay with questions and speculations about how the climate services enterprise can be broadened and diversified to help SPU and other progressive utilities to pursue their goal of attaining sustainable operations.

For whom is climate change a matter of concern, how does climate change come to matter, and what do the answers to questions such as these mean for the practice of climate services? The essay by **Werner Krauß** builds upon the proposition that climate change always happens somewhere, in some place, and that there is a difference between results gained from a model or a dataset and the actual changes caused by climate in real world places (Contribution 9). This distinction draws upon Bruno Latour's conception of political ecology, in which he challenged the separation of science on the one side and society, law, culture, or politics on the other, and in so doing, shifted attention from data and methods to the assemblies of actors that are involved in any given issue where climate data come to matter. In this article, Krauß argues that climate services should imagine new forms of provider/user encounters to develop more effective forms of climate protection. Adopting the moniker of 'slow science,' Krauß points toward a more inclusive and participatory approach, based on long-term ethnographic and participatory research in coastal landscapes of Northern Germany.

Finally, the piece by **Charles Herrick** and **Jason Vogel** provides findings from an interpretive reanalysis of a series of case studies of community-based climate adaptation sponsored by the Kresge Foundation between 2014 and 2016 (Contribution 10). The essay draws on the political science and international relations literature to identify and characterize a "regime" of U.S. federal policies that drive and enable climate change adaptation programs and activities at the local level. The authors find that a wide variety of federal policies are used by localities to either compel and/or support adaptation objectives and propose that the enterprise of climate services may need to move beyond existing models of co-production to embrace an 'apprenticeship' model, immersing technical information providers in the milieu of policy and governance so that they might learn to recognize factors that influence the applicability, usefulness, and uptake of climate products and services.

Building off the title of this Special Issue of *Sustainability*, we briefly summarize and synthesize this collection of papers in terms of critique, integration, and reimagination.

1. Critique

For a long time, the entry point for most communities and decision makers concerned about climate change was to ask "what does the science tell us is going to happen"? Consequently, even though some effort was made to bring social and policy sciences to bearin the climate services enterprise, the demand for technical advances, such as downscaled climate model projections and climate impacts science, drove much of the climate services agenda in its first decade and a half. By the 2000s, "users" of climate information were becoming more sophisticated, starting with a handful of utilities and municipalities before expanding to other sectors. While this demand-driven innovation is being recognized in peer-reviewed discussions of climate services, many—perhaps most—authors continue to presume, even privilege a top-down, science-first flow of knowledge production and innovation. As explored in the Krauss essay, this privileging of science and information as a precursor to action is consistent with and may inadvertently reinforce the conventional norms of modern capitalistic societies driven by a scientific management paradigm, particularly the reliance on science as the foundation for policies made through central administrative authorities.

While the parlance and methodological orientation of co-production seems well-established within the climate services community, this collection makes it clear that opportunities remain to reexamine approaches and basic, orienting assumptions. As Herrick and Vogel write, "The literature on climate services takes it as an article of faith that local scale adaptation is being impeded, constrained, or blocked entirely by mismatches and incongruities between available information and the perceived needs of local decision makers and stakeholders. Careful review suggests the opposite. Indeed, none of the case materials reviewed in our analysis indicate the kind of stark bifurcation between knowledge users and knowledge producers that has become a fixture of the climate services literature. Stakeholder interviews provide little evidence of a debilitating distraction due to policy actors insisting upon answers to questions that the scientific community is unable to provide". As papers by Boqué Ciurana et al., Carr, Grodnik-Nagel et al., Herrick and Vogel, Krauss, and Lempert et al. illustrate, a hard and fast distinction between knowledge producers and knowledge users seems neither realistic nor especially helpful. Indeed, such a distinction may serve to shield climate services providers from 'users' demanding a reasonable level of accountability from the climate services community. A compelling example of this comes from Boqué Ciurana et al., in which we learn that some members of the research team are avid surfers—both conducting rigorous science, but also acting in the interest of a closely held personal value. The lack of meaningful epistemic separation between the user and producer is especially marked when discussions advance beyond climate readiness to address sustainability. Sustainability is always a composite of values, knowledge, natural conditions, technological capabilities and constraints, and stakeholder lifeworld experience. Within such a milieu, there is no one who is simply and purely a knowledge user or a knowledge producer.

Perhaps the most significant aspect of this bifurcation between information users and producers is to isolate normative considerations into the realm of the user and thus insulate the producer from value-based critiques. It is clear, for example, from Grodnik-Nagle et al., that a clear value orientation is driving Seattle Public Utilities' efforts toward sustainability and that they are looking for partners in the climate services community to assist them in achieving the sustainable future they envision. However, can the climate services enterprise be responsive to such a challenge if we remain steadfast in our commitment to being 'honest brokers' of scientific information and thus hold normative commitments at arms-length? What does such an arms-length relationship with normative considerations mean in light of Carr's considerations about climate services—without a normative grounding—perhaps resulting in maladaptation for something as fundamental as food production in developing and food-stressed nations? The emerging prevalence of climate justice considerations in climate services brings such normative issues to the fore as well. The contributions to this Special Issue suggest that this is an topic whose time has come.

2. Integration

As Lioubimtseva and da Cunha observe, "The integration of climate and weather data with social, economic, cultural, and environmental data is paramount [for the characterization and evaluation of] present and future human vulnerability to climate change, addressing disproportionate socioeconomic risk to climate impacts, and engaging overburdened communities in the planning process". As examples outlined in this issue illustrate, sustainability can be site- or situation-specific, making it nearly impossible to stipulate in advance how much of any given knowledge domain will be necessary to inform a particular effort to pursue sustainable operations. As Grodnik-Nagle et al. illustrate, the quest for sustainable modes of operation entails a multi-generational perspective and integrates economic vitality, social equity, and environmental stewardship. Sustainability initiatives involve the combination of scientific characterizations and projections, technological and engineering applications, professional standards and expectations, and clearly articulated commitments to value-based objectives. Within Seattle Public Utilities, sustainability initiatives involve dozens of disciplines and topical domains, including physiological factors that influence the population-level dynamics of endangered species; capillary theory and compost engineering design; socio-cultural determinants for equitable, generational planning; hydro-geological variables that affect watershed functions; principles of sustainable landscape design that emphasize native species; and financial forecasting and modeling capabilities that can help to actualize concepts such as intergenerational planning. The quest for sustainable or resilient operations, then, requires a co-productive enterprise that accommodates, and indeed draws in, scientific and technical inputs across a far broader range of knowledge and competencies than can be provided by the 'traditional' disciplines of climate science—i.e., climate model projections.

Ultimately, as suggested by the 'apprenticeship model' of Herrick and Vogel, climate service practitioners have as much to learn from the people they work with as those people have to learn from climate services practitioners. The longest-lived climate services organizations have been around for a quarter century or more. In their early years, these organizations served important—mostly scientific—purposes, as illustrated in the historical account provided by Grodnik-Nagle and colleagues. However, as suggested by that same history, those original purposes—largely scientific translation and awareness raising—have been either accomplished, internalized by the service 'user', or transcended by more practical considerations of implementing solutions to ever more pressing climate-related challenges. In a meaningful sense, the original climate services enterprise has "solved" its original challenge of getting people to take climate change seriously. Now, it may be time to listen more closely to long-standing collaborators to understand what they need to amplify and accelerate their efforts at addressing climate change and making intentional decisions about which of those needs can and should be taken on by the climate services community. This suggestion seems to be something more than what is conventionally implied by the term 'co-production', as it may move beyond simply agreeing to conduct science together and may include common value orientations, normative commitments, and/or explicit co-efforts at policy implementation.

3. Reimagination

Sustainability implies social, economic, and cultural transformation. To move beyond 'climate readiness', climate service providers need to expand their networks and prepare to interact with numerous other disciplines. Such a shift will likely have methodological implications, networking implications, organizational implications, and ontological implications. Climate services providers are the 'proprietors' of a critical set of resources in efforts to achieve sustainability. In other words, the climate service community can use its methods and outputs to help drive change. Climate service practitioners should imagine working as advocates for sustainable futures, or at least contemplate and anticipate how their unique work products can be effectively deployed in an advocacy context.

Diversity, equity, inclusiveness, and social justice are becoming increasingly important with respect to the development, availability, and delivery of government-sponsored infrastructure and services, including, and perhaps especially, sustainability initiatives and climate change adaptation planning. Papers in this collection make it clear that equity-related concerns are stark and lingering. As Pathak and colleagues emphasize, attention to NbS tends to benefit predominantly white and comparatively affluent communities. In a similar vein, Lioubimtseva and da Cunha emphasize that adaptation plans often lack "air quality monitoring in locations and at scales that can indicate the potential for systemic inequalities in climate adaptation and sustainability planning". Climate service providers should imagine how the enterprise might change if its leaders and practitioners came to see themselves as provocateurs in the battle of social justice. While some climate services providers may recoil at the presumed loss of so-called 'credibility' associated with such a suggestion, it is worth thinking explicitly about both what is gained as well as what is lost—especially given the alarming intensification of climate impacts.

In our view, nearly all the papers in this collection either articulate or implicitly support the notion that climate services need to 'know their place'. From the vantage point of sustainability, climate services are but one voice in a large choir. From a practical perspective, this means that would-be climate service providers—like any other good professional—need to understand their client's operational environment. This means that climate services providers need (i) to identify and seek to work through the professional staff and/or acculturated representatives of existing, action-oriented institutions; and (ii) become connoisseurs of existing knowledge networks and experiential lifeworld's to figure out how to 'fit-in,' especially if that means going out of their comfort zone to provide information that is less than cutting-edge, qualitative, or primarily narrative. As we suggest above, this may mean that there will be situations in which climate service providers will need to work from within an explicitly 'ideological' frame of reference. As Carr spells this out, "[b]uilding climate services around contemporary understandings of how people make decisions about their lives and livelihoods offers designers and implementers of climate services opportunities to create services that catalyze transformational adaptation". For example, helping underprivileged high school students to 'problematize' climate change in their communities (Pena-Vega et al.). In this case, metrics for good climate services would not merely include accuracy, precision, full exposition of uncertainty, and other of the old chestnuts of "good science", but also information that can be used within the context of full-throated narratives of place-based, culturally embedded advocacy for change.

4. List of Contributions

1. Lempert, R.; Busch, L.; Brown, R.; Patton, A.; Turner, S.; Schmidt, J.; Young T. Community-Level, Participatory Co-Design for Landslide Warning with Implications for Climate Services.
2. Lioubimtseva, E.; da Cunha, C. The Role of Non-Climate Data in Equitable Climate Adaptation Planning: Lessons from Small French and American Cities.
3. Bamzai-Dodson, A.; McPherson, R.A. When Do Climate Services Achieve Societal Impact? Evaluations of Actionable Climate Adaptation Science.
4. Carr, E.R. Climate Services and Transformational Adaptation.
5. Pathak, A.; Hilberg, L.; Hansen, L.J.; Stein, B.A. Key Considerations for the Use of Nature-Based Solutions in Climate Services and Adaptation.
6. Pena-Vega, A.; Cohen, M.; Flores, L.M.; Le Treut, H.; Lagos, M.; Castilla, J.C.; Gaxiola, A.; Marquet, P. Young People Are Changing Their Socio-Ecological Reality to Face Climate Change: Contrasting Transformative Youth Commitment with Division and Inertia of Governments.
7. Boqué Ciurana, A.; Menendez, M.; Suarez Bilbao, M.; Aguilar, E. Exploring the Climatic Potential of Somo's Surf Spot for Tourist Destination Management.
8. Grodnik-Nagle, A.; Sukhdev, A.; Vogel, J.; Herrick, C. Beyond Climate Ready? A History of Seattle Public Utilities' Ongoing Evolution from Environmental and Climate Risk Management to Integrated Sustainability.
9. Krauss, W. Slowing Down Climate Services: Climate Change as a Matter of Concern
10. Herrick, C.; Vogel, J. Climate Adaptation at the Local Scale: Using Federal Climate Adaptation Policy Regimes to Enhance Climate Services.

Author Contributions: C.H. and J.V. shared equally in all aspects of producing this editorial piece. All authors have read and agreed to the published version of the manuscript.

Acknowledgments: We thank the editorial staff at MDPI/*Sustainability* for their flexibility, attention to detail, and patient oversight throughout this process. We also thank the journal for providing an open-access platform as a venue for this important discussion. We have enjoyed getting to know all the contributors to this issue and greatly appreciate their rigor and willingness to discuss ideas and entertain revisions. Finally, we thank the numerous peer reviewers who shared their thoughts and time to help sharpen and improve this issue.

Conflicts of Interest: The authors declare no conflict of interest.

sustainability

MDPI

Article

Community-Level, Participatory Co-Design for Landslide Warning with Implications for Climate Services

Robert J. Lempert [1,*], Lisa Busch [2], Ryan Brown [1], Annette Patton [2], Sara Turner [3], Jacyn Schmidt [4] and Tammy Young [5]

1 RAND Corporation, Santa Monica, CA 90401, USA
2 Sitka Sound Science Center, Sitka, AK 99835, USA
3 Pardee RAND Graduate School, Santa Monica, CA 90401, USA
4 Central Council of the Tlingit & Haida Indian Tribes of Alaska, Juneau, AK 99801, USA
5 Sitka Tribe of Alaska, Sitka, AK 99835, USA
* Correspondence: lempert@rand.org

Abstract: Inclusive, participatory governance is a key enabler of effective responses to natural hazard risks exacerbated by climate change. This paper describes a community-level co-design process among academic, state, and federal scientists and the community of Sitka, Alaska to develop a novel landslide warning system for this small coastal town. The decentralized system features an online dashboard which displays current and forecast risk levels to help residents make their own risk management decisions. The system and associated risk communications are informed by new geoscience, social, and information science generated during the course of the project. This case study focuses on our project team's activities and addresses questions including: what activities did the project team conduct, what did these activities intend to accomplish, and did these activities accomplish what they intended? The paper describes the co-design process, the associated changes in system design and research activities, and formal and informal evaluations of the system and process. Overall, the co-design process appears to have generated a warning system the Sitka community finds valuable, helped to align system design with local knowledge and community values, significantly modified the scientists' research agendas, and helped navigate sensitivities such as the effect of landslide exposure maps on property values. Other communities in SE Alaska are now adopting this engagement approach. The paper concludes with broader implications for the role of community-level, participatory co-design and risk governance for climate services.

Keywords: landslide warning; warning systems; participatory co-design; risk governance; community engagement; Southeast Alaska

Citation: Lempert, R.J.; Busch, L.; Brown, R.; Patton, A.; Turner, S.; Schmidt, J.; Young, T. Community-Level, Participatory Co-Design for Landslide Warning with Implications for Climate Services. *Sustainability* **2023**, *15*, 4294. https://doi.org/10.3390/su15054294

Academic Editors: Charles Herrick, Jason Vogel and Glen Anderson

Received: 4 January 2023
Revised: 17 February 2023
Accepted: 21 February 2023
Published: 28 February 2023

1. Introduction

Inclusive, participatory governance is a key enabler of effective responses to climate change [1] (Sect C5). Such participatory processes, which directly involve members of the public in making decisions in matters that affect them [2], can make decisions more effective by engaging multiple sources of knowledge, such as formal science and local and Indigenous knowledge; can better align decisions with community values; and enhance community ownership and acceptance. Participation is also a normative good consistent with principles of procedural justice [3] (Sect 1.4.1.1). Addressing climate change will often require significant changes in lifestyles and daily routines informed and enabled by new and evolving science and technology. Under such conditions, inclusive, participatory governance may be particularly important in helping to achieve equitable outcomes and to reduce tensions in this era of polarization, inequality, and distrust of elites and science.

This paper describes an exercise in community level, inclusive, participatory governance focused on the co-design and deployment of a landslide early warning system (LWS) for the small town of Sitka, Alaska. In the aftermath of the fatal Kramer Ave landslide in 2015, (see Figure 1) members of the Sitka community became concerned about landslide risk. They quickly realized that this risk had always been present and was likely to grow in the future as climate change increases the incidence of extreme rainfall events. Such realization generated anxiety among community members about personal safety and concern about the town's economic future [4].

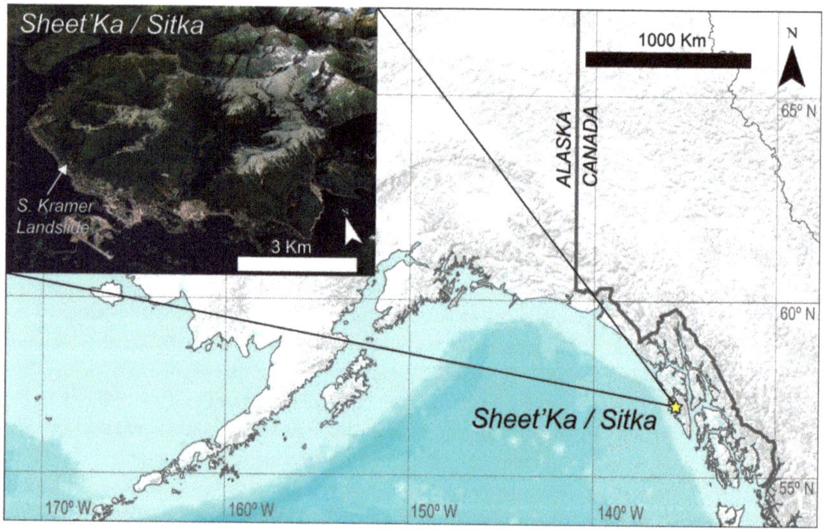

Figure 1. Map of the study area in Sitka, Alaska. Source: Service layer image credits: ESRI, USGS, NOAA. Inset map from Google Earth (2023), Maxar Technologies, Landsat/Copernicus, and Airbus (accessed on 6 February 2023).

With funding from the National Science Foundation, a team of community members and geoscience, information, and social scientists based at external institutions conducted a participatory co-design process which resulted in a novel, community-run, decentralized landslide warning system. Consistent with community values, the system employs a landslide risk dashboard that enables Sitkans to make their own evacuation decisions rather than rely on centralized evacuation warning. The warning system design and associated risk communications are informed by novel science and technology, including networks of low-cost sensors and social network analysis aimed at efficient and equitable dissemination of risk information throughout the community.

By co-design, we mean a creative, participatory process of fashioning solutions to public policy challenges in which community members and scientists collaborate as equals. In the context of early warning systems, such solutions can include information products, information services, and policies. The literature offers many versions of the co-design concept. The public administration literature adapts participatory processes of product design and design-thinking from the private sector and applies them to public sector solutions [5]. The sustainability literature often envisions co-design as a process in which academic and non-academic partners jointly develop research questions and research agendas as an initial phase in the co-production of knowledge [6].

Here we implement co-design as a participatory risk governance process embedded in a wide-ranging process of community engagement. The co-design, organized as a process of "deliberation with analysis" [7], engages multiple stakeholders; includes storytelling, serious games, and information provision by experts; facilitates deliberative dialogues; and follows an iterative process of framing and reframing [8].

This paper presents this effort to create a Sitka landslide warning system (LWS) as a case study of the type of community-level, participatory co-design process which may prove important to the effective provision of climate services in the United States and beyond. The case study focuses on our project team's activities and addresses questions including: what activities did the project team conduct, what did these activities intend to accomplish, and did these activities accomplish what they intended? To address these questions, the paper describes the co-design process, the associated changes in system design and research activities, and formal and informal evaluations of the system and the process.

Overall, the co-design process appears to have generated an LWS the Sitka community finds valuable. The system provides useful warning and reduces anxiety during periods of heavy rain. Community residents and officials report their satisfaction. The co-design process significantly influenced the design of the LWS consistent with local knowledge and community values, significantly modified the scientists' research agendas, and helped navigate sensitivities such as those related to the effect of landslide exposure maps on property values. Small towns across Southeast Alaska are now replicating this co-design and engagement approach to provide warning for climate-related risks.

Many U.S. communities face natural hazard risks exacerbated by climate change. This Sitka landslide early warning system and the co-design process that produced it have implications for U.S. climate services. The effort suggests that co-design processes previously used with expert stakeholders can be adapted for small town, lay audiences. The effort highlights the importance of strong local partners; active engagement by state and local agencies; an ability to engage with a broad range of information focused not only on the hazard but also exposure, vulnerability, and the responses to risk; the need for resources to support on-going operations of the system; and a cadre of personnel trained to facilitate the co-design process and supporting research. The Sitka system also provides insight into the contexts in which an individual-centered system is more appropriate than one based on centralized judgments regarding evacuation.

2. Context

Sitka, a town of 8700 residents, is located on Baranof Island, on the outer coast of Southeast Alaska. The Pacific Ocean borders one side of the town. Rising above the other side are the steep, forested mountains of the Tongass National Forest (see Figure 1), the largest temperate rainforest in the world. Accessible only by boat or plane, the community's road system is 14 miles.

The town of Sitka is located immediately below very steep post-glacial hillslopes on the western coast of Baranof Island. The landscape has extreme topographic relief. One local mountain rises from sea level to ~900 m in 2.2 km. The local geology and geomorphology is extremely spatially heterogeneous with a history of tectonic, volcanic, and glacial processes resulting in a complex topography and a patchwork of different soil types and bedrock [9]. The town and the surrounding area are exposed to persistent landslide hazards from multiple susceptible hillslopes [9–11]. On 18 August 2015, an extreme atmospheric river initiated over 40 landslides in and near Sitka, including the Kramer slide, a debris flow that resulted in three fatalities.

The Tlingit, the Indigenous people, have inhabited Sheet' Ka, the ancient name for the Sitka, area for more than ten thousand years. Today they comprise over 17% of the Sitka region's population [12]. Commercial fishing represents the largest economic sector, but the Southeast Alaska Regional Health Consortium, the region's hub hospital, is the biggest single employer. Every year during the summer, Sitka receives tens of thousands of visitors via its cruise ship dock, which helps fuel a large local tourism industry.

The Sitka community has two distinct forms of local government, the Sitka Tribe of Alaska (STA) and the City and Borough of Sitka (CBS). The latter has a city manager form of government, in which the elected CBS Assembly appoints the CBS Municipal Administrator, who is responsible for municipal operations including but not limited to the hiring and managing of department executive leadership, the oversight of Municipal property, public works, and budget [13]. The STA is the long-standing, federally recognized government for the more than 4000 Tribal Citizens who live in the CBS and traditional

Sheet'Ka area [14]. The STA provides programs including but not limited to addressing employment, natural and cultural resources protection, higher and vocational education, domestic violence prevention, and social services.

The CBS manages emergency services for the community. In support of this role, the city maintains a 19-member Local Emergency Planning Committee (LEPC), established by a State of Alaska statute and implemented through local resolution [15]. The LEPC is responsible for emergency planning to guide the community in preparing for, mitigating, and responding to disasters and emergencies. The committee further functions as a community convener determining the appropriateness of emergency response and the related use of public funds.

State and federal agencies, and their regional offices, also play an important role in Sitka's management of natural hazards. The National Weather Service (NWS) office in Juneau, AK provides forecasts for Sitka and maintains the weather station at the airport. The Alaska Division of Geological and Geophysical Surveys (DGGS) provides local geologic information, including collecting lidar data for the region and developing exposure maps of landslide runout zones. The U.S. Forest Service (USFS) maintains a Ranger District Office in Sitka and owns significant land in the Tongass National Forest area surrounding the town. Local USFS personnel also developed and maintain a landslide inventory for Baranof Island.

The project team, funded by the NSF Smart and Connected Cities Program, was led by the RAND Corporation, a non-profit, public policy research organization headquartered in California. RAND also led the social science and risk management project teams. The Sitka Sound Science Center (SSSC), a non-profit community science education and research organization located in Sitka, served as the project's main community partner. The STA served as the project's primary liaison with the Indigenous community. Other funded team members included the University of Oregon, which led the project's geosciences team, the University of Southern California, which led the project's information sciences team, and DGGS, which installed a weather station atop Harbor Mountain with project support. The web developer Azavea built the dashboard. Project partners included the regional NWS and USGS offices, researchers from USGS, local emergency services agencies, local schools, and the chamber of commerce.

As suggested by Figure 2, the Sitka LWS presents a challenge of polycentric governance [16], with many independent yet interdependent actors required to act in coordinated ways to create and operate it. Climate change adaptation increasingly faces this type of governance challenge [3] (Sect 1.4.2.2). In Sitka, the city and tribe have primary responsibility for the safety of Sitka residents and making it possible for Sitkans to respond effectively to any warning. Federal and state agencies are the primary sources of data on which the warning system depends. The system empowers Sitka residents to make their own decisions regarding landslide risk, so the system only functions if they use it appropriately. The warning system design also gives the SSSC significant responsibilities for maintaining the system and the community's awareness of it. Alongside these decision-making organizations, the project team functioned as a boundary organization [17] mediating among information relevant to landslide warning and the community actors whose decisions are integral to the warning system's function.

In this context, the project team faced a number of challenges. The team needed to engage a lay public and local officials with new and evolving science in a context of considerable anxiety about landslides, grief about previous loss of life, the potential effect of increasing the salience of landslides on life in the town, and distrust of outside experts. In addition, the project team also had balance among the related but different interests of the researchers and community. With its NSF funding, the research team had promised both to serve the community as well as to explore the utility of new and untested networks of low-cost moisture sensors to geoscience as well as novel applications of social networks and influence maximization to risk communications. While there was little contestation of values regarding landslides among Sitkans, some people were primarily concerned with personal safety and others worried that undue concern with landslides could harm property values and disrupt the community's daily life [4]. The project also needed to address a highly diverse community including blue collar workers in fishing and related industries, retirees, Indigenous people, Coast Guard personnel, and seasonal workers in retail, tourism, and related industries.

Figure 2. Organizations involved with Sitka landslide warning system. Source: authors.

3. Frameworks

Community engagement has played an increasingly important role in the theory and practice of disaster risk and environmental management in recent decades, as part of an overall turn towards including lay people in what had previously been seen as primarily technocratic decisions of experts [18–20]. Community engagement is a normative principle, seen as a key component of procedural justice, which examines who makes and participates in societal decisions [3] (Sect 1.4.1.1). Recognition, another component of justice, also requires engagement. Recognition entails acknowledgement, basic respect, and fair consideration of diverse cultures, values, perspectives, and worldviews. These principles are broadly recognized including, for instance, in Article 6 of the Framework Convention on Climate Change, which creates a binding commitment on parties to promote participation by civil society in climate-related decisions.

Principles of public consultation and participation have also become embedded in U.S. environmental policymaking. Public meetings and comment periods are required by law in many jurisdictions, and many U.S. and Alaska regulatory bodies have processes for public consultation and participation as part of the regulatory process. In Alaska specifically, public participation processes have been developed to address transportation policy and fisheries management. When well-executed, engagement makes disaster risk and environmental management more effective by enabling better integration of local knowledge, expertise, and values while also increasing the legitimacy and public support for the resulting risk management approaches.

This project conducted a community-level, participatory co-design process to fashion Sitka's landslide warning system. Several frameworks informed this process: *risk governance*, *warning systems* as defined by United Nations International Strategy for Disaster Risk Reduction (UNISDR) [21], and the concept of *co-design*.

Risk governance [22,23] provides a definition of risk along with a conceptual framework and normative principles for guiding multiple actors in identifying, assessing, managing, and communicating risks. Most broadly, risk is the effect of uncertainty on objectives [24]. In Sitka, landslide risk can be usefully estimated as the product of hazard, exposure, and vulnerability. Hazard is the potential occurrence of a landslide at a particular time and place, exposure is the presence of people or infrastructure in places that could be adversely affected by a landslide, and vulnerability is the propensity of a person or structure exposed to landslides to be adversely affected. It also proves vital to consider the psychological and social dimensions of risk. In Sitka, landsliding and warning systems interact with many features of daily life, including anxiety about personal safety, disruption of daily routines, personal independence, and choices of where to live.

Risk governance is not a method, but rather a set of principles that inform methodological choices in activities intended to understand, assess, and manage risk. Our co-design process aimed to embrace these principles which are: (1) *communication and inclusion*, in which information flows in multiple directions in an ongoing process of social learning that includes in a meaningful way all relevant parties; (2) *integration*, which recognizes the need to collect and synthesize all relevant knowledge and experience, including scientific, Indigenous, and local knowledge; and (3) *reflection*, which avoids reliance on potentially familiar but inappropriate understandings through a continuous process of revisiting assumptions and frames in light of new knowledge and interactions among the parties.

Risk governance is most appropriate when risks are systematic, characterized by deep uncertainty, and interpreted through multiple, legitimate points of view [23]. For Sitka's landslide warning, deep uncertainty and multiple points of view were most salient. In addition, risk governance is also most appropriate when, as is the case in Sitka, no single hierarchical organization can effectively control risk assessment and management, making it useful to consider governance as involving "the structures, processes, and actions through which government institutions, markets, businesses, civil society, tribes, and others interact to address societal goals" [3] (Sect 1.4.2.2).

A warning system, as defined by the UNISDR, represents the capabilities needed to generate and disseminate timely and meaningful information to enable people in Sitka to prepare and act appropriately to reduce the possibility of harm. UNISDR finds that an effective warning system has four components [21]: (1) risk knowledge, which in this case includes scientific understanding of Sitka's landslide risk and community understanding of that risk, the warning system, and response options; (2) monitoring, which includes collecting, processing, and disseminating timely and actionable information on landslides; (3) communications, which includes sharing throughout the community of risk knowledge and the warning information from monitoring; and (4) response capability, which is the ability to act on warnings in such a way as to reduce risk.

It also proved useful to consider Sitka's landslide warning as a *decision support system* [7] (p. 36) consisting of information products, services, and systems. Decision support products include the dashboard which displays levels of risk and the data used to predict those risk levels. Services include training workshops to help residents use the dashboard. Decision support systems are defined as "the individuals, organizations, communication networks, and supporting institutional structures that provide and use decision support products and services" [7] (p. 37).

The concept of *co-design* has many manifestations. As noted above, here we use the term to mean a creative, participatory process of fashioning solutions to public policy challenges in which community members and scientists collaborate as equals, consistent with other definitions in the literature [5,6]. We chose co-design as the appropriate form for the engagement because we aimed to help Sitkans reduce their landslide risk through the provision of timely warning. Creating and linking the four elements of effective warning—knowledge, monitoring, communication, and response—in decision support products, services, and systems to best meet community needs presented a non-trivial design challenge.

In addition to the co-design of the landslide warning system, this project also included knowledge co-production, in which researchers and non-academic partners jointly develop a research agenda that serves their interests and needs [6]. In this paper, we focus on the co-design of the system and only discuss co-production of the research as it directly relates or results from the co-design.

The co-design workshops were implemented as a process of *deliberation with analysis*, an iterative learning process in which stakeholders deliberate on their objectives, options, and problem framings; researchers then provide decision-relevant information; and then the parties to the decision revisit their objectives, options, and problem framing influenced by this new information [7]. Deliberation with analysis represents an "iterative interaction" form of engagement [25] intended for situations in which the problem formulations, understanding of system functioning, and the set of promising solutions emerge gradually through interactions among the involved parties [26–28]. In addition to quantitative decision support products and tools, such deliberative engagements also can include visioning, storytelling, and serious games [29]. Such engagements have proven successful in addressing complex and controversial challenges, in particular in water and coastal management with workshop participants consisting of skilled professionals, often working in large government, business, or advocacy organizations [30–34]. In contrast, this Sitka engagement focused on lay participants from a small town.

To supplement the co-design process, we embedded the exercise in a broader process of community engagement. The US EPA guide to public participation identifies five levels of engagement: inform, consult, involve, collaborate, and decide [35]. In this effort, the co-design process operated at the collaboration level. A wide range of other engagement activities, including town halls, small group meetings, citizen science, and media interviews, operated at the inform, consult, and involve levels. Actual decisions were taken by city government agencies, the Science Center, federal agencies, Tribal government, and individual citizens as informed by the engagement processes.

Communication is a key step in risk governance and a key component of a warning system. The risk communication literature highlights that effective risk communication should clearly convey to individuals how risks are proximate for them in time and space, identify actions people can take with the information, inform mental models of the processes creating and modulating risks, and recognize that different people understand risk in different ways [36]. The risk communication literature also emphasizes that individuals are more likely to act on risk information if they receive similar information repeatedly through different channels [37] and that effective risk communication builds on existing communication pathways and social relationships to disseminate information. These understandings informed many aspects of this landslide warning effort, including the design of the dashboard, the content of the co-design workshops, the multi-layered public engagement activities, our dissemination and outreach activities, and plans for how actual warnings are disseminated.

4. Warning System Architecture

As of August 2022, Sitka now has a decentralized LWS, organized around the four elements of knowledge, warning, communications, and response, as shown in Figure 3.

A risk dashboard lies at the heart of the system and provides the warning component of the LWS. The dashboard is accessible online via mobile device or a computer (https://sitkalandslide.org, accessed on 16 February 2023) and at any moment in time provides hourly projections of landslide risk in Sitka over the next 24 h and for each of the next three days, as shown in Figure 4. The main page displays risk in one of three categories—low, medium, and high—and provides a brief description of each. These descriptions employ both probabilistic language (e.g., likely, unlikely) as well as historical context (e.g., over the last twenty years, no rainfall-induced landslides have occurred in Sitka with rain similar to the current intensity) as also shown in Figure 4.

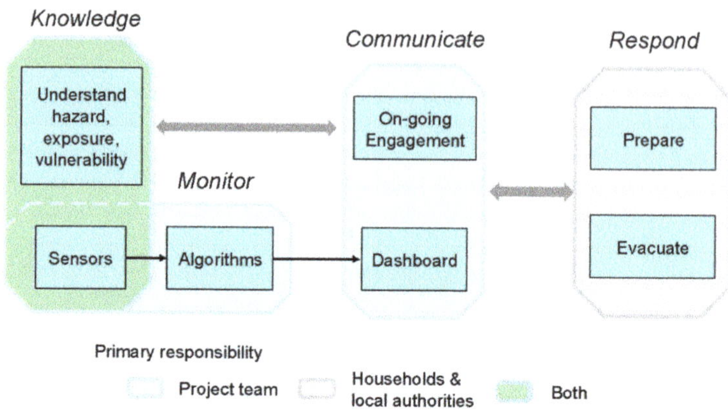

Figure 3. Elements of Sitka landslide warning system. Source: authors.

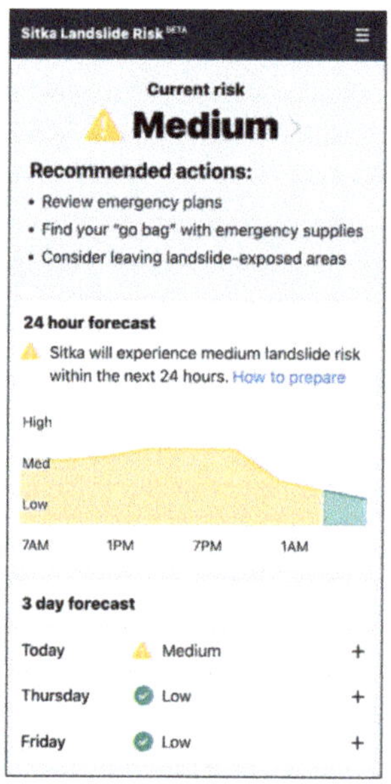

Risk Level	Description
High	Multiple landslides are very likely to occur in the Sitka area. There have been three storms in the last 20 years with similarly intense rainfall, and all three of them initiated multiple landslides
Medium	Landslides are possible. Isolated landslides have occurred in Sitka with similarly intense rainfall in the past.
Low	Landslides are unlikely based on historic observations in the last 20 years. Rainfall-induced landslides have not been documented for these rainfall conditions

Other resources available on Sitka Landslide site

Oral histories: Explore stories of landslides
Landslide 101: Find out more about landslide science
Areas at risk: Identify where you live relative to landslide exposure
How to prepare for a landslide: Learn how to prepare
Published work: Access articles by our research team
Report a landslide: Share new observations of landslides

Figure 4. Dashboard snapshot (left panel), text on risk levels (right panel), and other resources available on the Sitka landslide website. Source: authors, from https://sitkalandslide.org (accessed on 8 February 2023).

The intention is that community members will access the dashboard during periods of heavy rainfall or when such rainfall is expected in order to make informed decisions regarding whether they wish to alter their plans to reduce their landslide exposure. For instance, based on their risk aversion and the risk level reported by the dashboard, a family whose home is an area exposed to landslides might decide to sleep that night at a friend's house which has less exposure. NWS warnings, other public service announcements, or merely experiencing inclement weather might lead Sitkans to view the dashboard. The SSSC, supported by the project team, conducted a dissemination and outreach program (described below) to make the community aware of the landslide warning system and how to use it.

To enhance the community's risk knowledge, the dashboard also contains drill down menus which provide more detailed and contextual information. As listed in Figure 4, these include descriptions of how Indigenous peoples have thrived in this region amidst tumultuous geological forces including landslides, a primer on the science of landslides, maps showing which areas of Sitka are more or less exposed to landslides, information on how to prepare for landslides and respond to landslide warnings, a bibliography of relevant scientific literature, and links to an inventory of past landslides in and around Sitka, along with instructions for reporting new ones. Not shown in Figure 4, the drill down menus also provide more detail on the sensor data used by the dashboard for landslide predictions. In particular, the dashboard provides maps of landslide exposure from a debris flow model developed for the USFS by TerrainWorks [11] and links to other published works that document spatial patterns of landslide hazards [9].

Note that, formally, the dashboard reports a hazard level, not risk. Users can consult the exposure information provided by the dashboard's "Area at risk" drill down menu to begin to form a judgment about their level of risk. However, the project team decided to use the term risk on the dashboard as shown in Figure 4 for simplicity of communications with users.

The monitoring elements of the Sitka LWS rely primarily on sensors and forecasts from the local offices of the National Weather Service. The dashboard's risk estimates are calculated from the NWS's regional projections of three-hour rainfall intensity. The project's geoscientists developed the algorithm that converts estimates of rainfall intensity into the dashboard's landslide risk scale [38]. A network of sensors provides the data for the NWS rainfall predictions. These sensors include a rain gauge at the airport (sea level); a weather station atop Harbor Mountain (~2000 ft), one of the several peaks above the town; and a river gauge on the Kaasda Heen (also called Indian) River that flows through town.

The project team did not directly engage with identifying and planning response options that residents could implement based on the warning. Such responses might include relocating to a friend's house in a low landslide exposure area when landslide risks are high or not sending one's children to school on such high-risk days. The project's geoscience research concludes that medium and high-risk days are not likely to occur more than several times a year [38], suggesting that most residents may find that disruptions to daily life related to landslide warnings are relatively small. The project team worked with the Local Emergency Planning Committee to encourage them to include landslide preparedness in their portfolio of responsibilities and in their educational outreach to the community.

The LWS's development was supported by the project team's geoscience, social science, and information science research. The geoscience team informed the LWS's monitoring and risk knowledge elements. The team conducted a geomorphology survey to inform understanding of Sitka's landslide hazard, conducted an intercomparison of the four landslide runout models available for the Sitka area to inform exposure mapping [9], and developed landslide prediction algorithms [38]. In addition, the geoscience team deployed three moisture sensors in the hills above Sitka and ten tipping buckets as a citizen-science effort [39]. We had expected that these sensors would prove valuable to improving landslide prediction, but the predictions from rainfall intensity data alone appear to be surprisingly good. The moisture and tipping bucket sensors thus contributed primarily to improving risk knowledge. For instance, data from the tipping bucket rain gauges emphasize the spatial variability of rainfall intensity across Sitka's relatively small spatial area, thus supporting the decision to issue landslide risk warnings for the entire area and not specific hillsides. Data from the moisture sensors help confirm that drainage on the hillsides is sufficiently fast that instantaneous rainfall intensity alone, rather than instantaneous and cumulative rainfall intensity, is the best predictor of local landslide risk [38].

The social science and information science teams informed the LWS risk communications element. The social science team mapped Sitka's social networks using a survey in which respondents provided information on whom in the community they exchanged information on natural hazard risk [40]. Leveraging these data, the information science team developed an influence maximization algorithm that identified key influencers by weighting community members by geographical exposure to landslide hazards and considered demographic characteristics to ensure an equitable distribution of information [41]. This allowed us to identify key community connectors to receive training on the dashboard. The social network analysis also provided an empirical understanding of how information about risk flowed through the community and which groups might be most difficult to reach. In particular, the social network analysis revealed a pattern of landslide information sharing in which a highly connected community "core" was surrounded by several smaller groups apparently disconnected from this core.

The social science team also informed the risk knowledge associated with the LWS by gathering oral histories from the local indigenous community and contributed to the LWS response element by conducting, as described in Section 5, research on insurance response options.

The dashboard is operated and maintained by the SSSC. The software is designed to automatically update the risk projections twice per hour to reflect the most current rainfall forecasts. An important topic, unsettled at the project's end, involved funding for on-going maintenance, operations, and upgrades to the system. Initially the hope had been that the system would have been owned or adopted by either the city or by one of the federal agencies involved in the project. A new NSF-funded project awarded to the team (described below) provides some opportunities to maintain and continue operations of the Sitka dashboard. The SSSC is also writing grants for additional and longer-term funds to maintain the system.

5. Warning System Co-Design Process

Sitka's landslide warning system was developed through an intensive co-design process, embedded in a wide-ranging process of community engagement. The co-design focused on three in-person workshops, as shown in Figure 5, representing a collaborate level of engagement. Other engagement activities included on-going community outreach activities by SSSC staff, exploratory interviews with key stakeholders and representatives of different groups in Sitka conducted by the external social science team, surveys to support the social network analysis, as well as activities organized around three, week-long visits by the external project team to Sitka; and several extended stays in Sitka by individual external team members. Appendix A provides more comprehensive listing of workshops, engagements, and related activities.

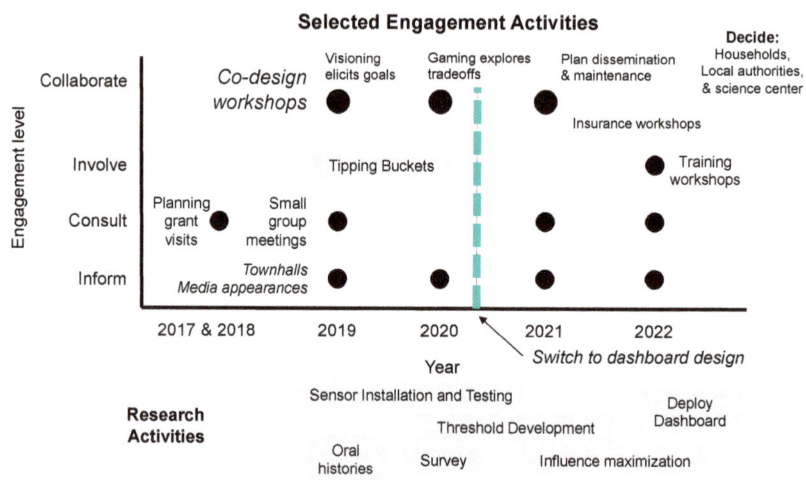

Figure 5. Selected engagement and research activities. Source: authors.

Landslides have been common in the Sitka area since the last ice age [9]. However, landslide risks were not on most Sitkans' minds until the 2015 slide killed three people and left the town confused and anxious. In response, the Sitka Sound Science Center convened a group of geoscience experts from academia and from federal and state agencies in what came to be called the Geotask force. The task force recommended that the town create a landslide warning system [42].

RAND had become involved in the later stages of the Geotask force and subsequently joined with the Science Center to write first a pilot grant and then a full grant to NSF to develop an LWS for Sitka. Recent USGS work in Washington State had demonstrated the utility of moisture sensors for improving landslide warning [43]. However, the USGS sensors were very expensive, on the order of $10,000 each. Our proposed project aimed to demonstrate the value of networks of low-cost, internet-enabled moisture sensors in the hills above Sitka. The project also aimed to improve risk communications with a novel social network analysis and influence maximization algorithms to better understand and to improve the flows of risk information through the community.

The full project began in the fall of 2018. A series of structured, day-long workshops in Sitka provided the core of the warning system co-design process. The external project team traveled to Sitka for a decision scoping workshop in May 2019, a design workshop in February 2020, and an update workshop in October 2021, as detailed in Appendix A. During that same October week in 2021, the project also held a separate workshop on landslide insurance options.

The May 2019 decision scoping workshop introduced the community to the project, and then used a backcasting exercise [44] focused on visioning and storytelling on the theme of a future Sitka with effective and well-regarded landslide warning. This exercise helped community members to articulate and research team members to understand goals for the warning system as well as potential actions to pursue those goals. Participants in the day-long workshop included about a dozen Sitkans, including community leaders and city officials responsible for emergency response. The workshop confirmed the project team's understanding that community goals for the warning system included reducing loss of life and the anxiety felt during intense rainfall events. However, the workshop also raised additional goals including offering opportunities for participation and access for everyone in the community, building and maintaining trust, clear and consistent communications, an orderly process of evacuation if and when needed, low cost, and minimal impacts on property markets.

Informed by the workshop and ongoing geoscience research, the project team developed an initial design for landslide warning based on a centralized, siren-based system similar to Sitka's existing Tsunami warning system. In the February 2020 design workshop, participants reviewed and suggested revisions to this initial plan. The workshop used serious games to help participants appreciate the challenges in balancing between failed and false warnings and to understand the potential role social network analysis could play in efficiently and effectively disseminating landslide risk information throughout the community.

The results of this workshop contributed to the warning system redesign shown in Figure 3, featuring the online, landslide risk dashboard enabling individuals to make their own evacuation decisions. The design workshop also informed changes to our geo- and social science research, as discussed in Section 7. The pandemic curtailed travel shortly after this design workshop, so for the next year and a half, subsequent discussions with the community were conducted in smaller, online groups. At the final design review and launch plans workshop in October 2021, the project team presented the emerging dashboard-based warning system and discussed with the community plans for disseminating and maintaining the system.

Concurrently to the co-design workshops, the project team engaged in a wide range of community engagement activities designed to increase knowledge and awareness in the community about the project, landslides, and the risks they generate. The project team participated in town halls organized by the Science Center, which were held parallel to each of the design workshops and provided a public forum for citizens to learn about the project and ask questions of the team. As one important purpose, these events aimed to engage a much larger number of Sitkans than could participate in the co-design workshops. The project team also provided regular updates to the city assembly and the tribal council through presentations at the regular meetings of those legislative bodies. The team also met several times a year with other parts of the city government including the LEPC, the Planning and Zoning Commission, and the School Board. The team also promoted a presence in local media. Team scientists were interviewed on the public radio station and sat in on the AM radio station's "problem corner" to answer audience questions. The project was profiled multiple times in the daily newspaper.

In addition, the team worked to forge more individual connections with community members. The Science Center's Scientists in the Schools program organized opportunities for the team scientists from external institutions to visit Sitka elementary and high school classrooms. Over the course of the project, external team members also spent one-month mini sabbaticals in Sitka, engaging in a breadth of community activities meant to help scientists learn about the community and help community members learn about the science. On their visits to Sitka, project team members participated in science cafes, met people at the local brewery and introduced themselves at tribal and city government meetings. They participated in trivia nights and shared in potlucks, community hikes, and boat rides, all with the intention of setting up a two way "getting to know each other" relationship so that scientists did not seem separate from the community but rather integrated into it.

To help implement the dashboard design, the project team conducted a range of dissemination activities and hired Azavea, a professional software developer, to build the dashboard. In the fall of 2021 Azavea met with the project team to understand our design concepts and then conducted interviews with eight individuals chosen to be broadly representative of the community to better understand their view of important design features. Based on this input, Azavea generated wireframe examples of dashboard designs and contents to facilitate discussions with the project team. These discussions helped finalize the dashboard content, including the names of the risk levels and the language connecting them to historical events. After several iterations, Azavea programmed the current dashboard including the data pipelines that feed the prediction algorithms and the information available in the drill down menus.

During the design process and as the dashboard neared completion, the project team launched its dissemination activities, intended to provide all members of the Sitka community with the knowledge and capacity to use the dashboard. The team employed multiple channels of communications including formal dissemination workshops informed by the social network and influence maximization analysis. The team also made presentations to various community groups, including boards and commissions for the city and tribal governments, as well as the Chamber of Commerce and the Rotary Club.

The SSSC and STA hosted the structured rollout workshops, which aimed to teach a select group of residents about the dashboard. The social network analysis and influence maximization research described in Section 4 helped to inform the invitation lists. We could only provide intensive training on the use of the dashboard to a small number of individuals (roughly forty total over four workshops), so we aimed to focus on inviting those individuals who could most efficiently and equitably disseminate information to the community. Most of the invitees (roughly 80 percent) were chosen by our community partners, based on their personal knowledge of their fellow residents. However, we supplemented the invitation lists for each workshop with two or three high-influence community connectors identified by the social network analysis. Using information on all the attendees at previous workshops and the invitees for the planned workshop, the influence maximization algorithms would suggest additional invitees who would provide the most effective and equitable dissemination of information, weighting community members by geographical exposure to landslide hazards and their demographic groups.

During the course of the co-design process and community engagement, it became clear that many Sitkans were concerned about the availability of landslide insurance and the effects of landslide risk information on local property values. In the aftermath of the Kramer slide, and the hazard mapping efforts it spawned, some homeowners had trouble insuring their homes and obtaining mortgages due to a lack of landslide coverage. The topic of landslide exposure maps had also become controversial, complicating the project team's decisions regarding what exposure information to provide on the dashboard. It became clear that the project team needed to address the landslide insurance issue to maintain the trust of the community.

The project team thus conducted an additional engagement to share information and identify potential options for landslide insurance. We conducted a nine-month effort from April 2021 to November 2021 consisting of twenty-four key informant interviews, a facilitated half-day workshop in Sitka on October. 4 described in Appendix A, and a post-workshop survey for participants [45,46]. The insurance workshop included a facilitated discussion regarding exposure maps, which led to the decision to include on the dashboard a map with sufficient resolution so that users could judge risk in their immediate neighborhood, but without sufficient resolution to identify specific properties.

6. What Changed during the Co-Design Process

Co-design aims to collaboratively fashion solutions to policy challenges. So, one measure of the process is the extent to which it shifts the design of Sitka's LWS to better align with scientific understanding and community goals. Here we document how the LWS design changed significantly during the project; how the team's geoscience, social science, and information science research evolved; and how the co-design process itself changed. In the next section we present evidence that the co-design process aligned with community goals was responsible for these changes.

The Sitka Fire Department has operated a tsunami warning system for over forty years. The system employs sirens to disseminate a city-wide alert when the fire chief determines, based on information warnings from the National Tsunami Warning Center, that an evacuation of low-lying areas in the city is warranted. At the start of the project, we envisioned that landslide warning would operate similarly [4]. The fire chief would make evacuation decisions, a siren would communicate the warning to residents, and a neighborhood buddy system would ensure everyone had heard and responded to the sirens. The co-design process shifted the LWS to one that empowers residents to make their own evacuation decisions using risk information provided by the online dashboard. In addition, the team's initial plans for the buddy system were supplanted by a more voluntary model, partially because community members during the co-design workshops indicated that they would not feel comfortable holding formal official responsibility for notifying others in their communities of imminent landslide risk.

Consistent with a co-production process, interactions with the community also shifted the research agendas of the project's research teams. The team's geoscientists adjusted their approach to what constituted a successful landslide prediction based on the community's criteria. At the start of the project, the geoscientists would have regarded a prediction as successful if a landslide subsequently occurred anywhere near Sitka, even if it didn't affect property within the city. For community members, however, successful prediction required much more spatial specificity, differentiating landslides that endangered people and property from those that did not [4].

Spatial hazards analysis in Sitka can identify susceptible hillslopes within Sitka and on nearby slopes [10,11,47], but current prediction tools do not allow practitioners to identify which of several susceptible hillslopes will experience failure during a particular storm. In response to community definitions of successful landslide warning, the project team's geoscientists fundamentally shifted the structure of the statistical models they used to predict landslides. Most LWSs around the world make predictions based on large landslide inventories (tens to thousands of landslide-inducing storms) across larger areas—mountain ranges, states, or even countries [48,49]. Instead, the geoscience team implemented novel statistical applications to train and evaluate probabilistic models with an extremely limited landslide inventory (5 storms) from only those hillslopes adjacent to (<2 km) the Sitka road network [38].

Community input also informed warning threshold selection. The team originally envisioned a single alert level for a siren-based system, but varying levels of risk tolerance documented in the workshop series demonstrated the need for two warning levels, one which minimized the probability of missed alarms and one which minimized the probability of false alarms. For example, the team chose a lower threshold for the "moderate" risk level at an estimated landslide probability of 1%, using a combination of traditional statistical metrics (Precision Recall) and a heuristic approach to select a conservative threshold with this very low probability of missed alarms [38].

The teams' social scientists also adjusted their research designs in response to community feedback. Initially, the social science team had intended to map existing patterns of communication and social relationships in the community and then use this information to help create the buddy system. However, Sitkans were skeptical of an outside research group mapping and analyzing their social networks. The project team thus adjusted to use the social network analysis primarily to inform our local partners, the SSSC and STA, as they created invitation lists for our dissemination workshops. In addition, the social science team had initially intended to gather oral histories from indigenous community members to improve our understanding of how members of the Sitka Tribe viewed landslide risk and to inform our risk communications. However, the Tribe was uncomfortable with the research team sharing the oral histories, so we agreed to just summarize their themes on a page on the dashboard.

Finally, the co-design process itself changed during the project. Initially, we had expected community members to deliberate on an appropriate threshold that the fire chief could use to issue a centralized warning. We expected these deliberations would be informed by a decision support tool the project team would build showing the tradeoff between lives at risk and disruptions from false warnings as a function of the warning threshold. However, as our understanding of the underlying geoscience and the LWS design evolved, we switched to a gaming format to sensitize community members to the false vs. failed warning tradeoffs inherent in landslide warning. We also added the insurance research stream in response to community feedback, as described in Section 5.

7. Did the Co-Design Process Succeed?

To what extent did the participatory co-design process accomplish its intended purposes? The project aimed to provide Sitka with an LWS that aligns with community goals while conducting research to improve understanding of new sensor technologies and of social networks and influence maximization.

For reasons described below, the project team could not formally evaluate Sitka's new LWS and the associated co-design process. We were, however, able to conduct a more formal evaluation of the insurance research and the extent to which the co-design process contributed to beneficial changes in LWS design and associated research activities. Here, we first report on our informal evaluation of goal alignment and then report on the more formal evaluation of the co-design process.

To consider goal alignment of the system and process, we documented in detail the co-design process, the goals articulated during that process, the changes that occurred during the process, and the resulting warning system. We then gathered available evidence on the extent to which the system and process align with the goals.

In the first co-design workshop, participants articulated the following goals for Sitka's LWS: (1) the system should help reduce loss of life from landslides, (2) reduce the anxiety felt during intense rainfall events, (3) offer opportunities for participation and access for everyone in the community, (4) build and maintain trust, (5) provide clear and consistent communications, (6) have low cost, (7) support an orderly process of evacuation, and

(8) have minimal impacts on property markets. Appendix A provides more detail on each goal.

The project team was unable to formally evaluate the extent to which the co-design process and resulting LWS aligns with these goals. In part, the warning system became operational in August 2022 only a month before the end of the project, making it difficult to measure community attitudes towards the deployed system. We have not evaluated the community response to the LWS in periods of heavy rain because, as of this writing, there has not yet been any landslide warnings, heavy rain events, or landslides since the dashboard became operational. In addition, other research groups as well as city, state, and federal agencies were reporting and acting on landslide risk concurrently with our project, making it difficult to isolate the impact of our efforts on residents' attitudes. Finally, the pandemic intervened, so that the community we surveyed at the start of the project was different from that at project's end. For instance, the co-design effort settled on a dashboard-based system in mid-February 2020 at a time when many community members were likely unfamiliar with the dashboard concept. When the landslide dashboard launched in August 2022, many community members had two years of experience gleaning information from ubiquitous COVID-related dashboards.

We did, however, gather evidence to suggest the co-design process and the warning system it produced were successful. First, the landslide prediction algorithms appear accurate and thus able to provide useful warning. During the winter of 2021, our geoscience team was able to use the then-current algorithms to disseminate to local officials' accurate warnings of intense landslide-producing storms.

The project team, in particular our community-based members, solicited extensive feedback during the final year of the project and its aftermath. This feedback suggests that the community is pleased with the results of this effort. Local officials and residents spoke favorably of the system. Since its launch, the dashboard has had about 6500 unique users, with about 25% of them from Alaska. The City Assembly and the Tribal Council both informally stated that they were extremely pleased and grateful for the work of the team and that they perceived the outputs of the research to be useful and valuable. This included positive feedback from the city's Fire Department, who had initially been skeptical of the practical value of the project but became impressed by how well the system seems to work and became comfortable with their role in the dashboard-centered system design.

Additional evidence is provided by the reactions of neighboring communities. Towards the end of the Sitka project grant, the project team received another NSF award to support co-design efforts for multi-hazard warning systems in six other communities in Southeast Alaska. This new project is called K̲UTI, both an acronym for the English phrase *Knowledge and Understanding, Technology and Institutions* and the Tlingit word for weather. In recruiting new partners for K̲UTI it became clear that many communities in the region were familiar with the work in Sitka, were eager to engage with our project team, and cited the reputation of the Sitka effort as one of their reasons.

A formal evaluation of the project's insurance-related activities proved possible. These activities occurred over a short time span, and no other group conducted similar work. After the insurance workshop, the project team fielded a web-based survey for workshop participants. The results suggested that almost all the participants found that the workshop improved their understanding of the challenges of securing landslide insurance, intended to share what they learned with others, and wanted to see landslide insurance become more available. However, less than a third of participants reported that the workshop increased their optimism that such insurance would become available [45,46].

Overall, this mostly informal evidence all suggests that the LWS and co-design process does align with the first five community goals. Regarding the other three goals: landslide risk has affected property values, but the project did help the community grapple with that challenge; orderly evacuation remains untested; and costs are discussed in the next section.

We did formally evaluate the extent to which the design process conformed to theories and best practices for stakeholder engagement in early warning system development [50].

As one component, this evaluation explored the ways in which the team's perceptions of the problem evolved over time and aimed to identify key moments that led to the changes in system design and research plans described in Section 6. These shifts were evaluated through analysis of nine interviews conducted with core project team members; notes taken during monthly team meetings, the three design workshops, as well as during public presentations to the city's political and administrative leaders and smaller discussions with representatives of Sitka's Police and Fire departments, Coast Guard, school district, mayor's office, city engineer and planning offices. The team also examined 41 local news articles and stories about the project, minutes from city council meetings where landslide risk was discussed before and during the project, and the Sitka Geoscience Taskforce report, which preceded the current co-design effort. By coding for key themes in these documents, we could track how the team's and community perceptions changed over time [50].

This evaluation suggested that research team members perceived a moderately clear breakpoint in the first year of the project, when it became clear that a top-down centralized system would not meet the needs of community stakeholders, specifically that warnings would not be accurate enough in the short term to enable emergency responders to feel comfortable issuing evacuation warnings. The interviews conducted in 2019 and early 2020 suggest that this realization was an important factor in shifting from a centralized, siren-based system to one that was decentralized and dashboard-based. Interviews with the team's geoscientists also confirmed that community interactions had been important in expanding their concept of what constituted a successful landslide prediction to include the community's criteria.

8. Lessons Learned and Implications for Climate Services

This paper describes a community-level, participatory process of co-design and community engagement that resulted in a landslide warning system for the small town of Sitka, Alaska. As a decision support system, the Sitka LWS includes information products and services and engages supporting networks and institutions. It includes the four warning system components: knowledge, monitoring, communications, and response. Intended as a risk governance exemplar, the co-design and community engagement aimed to embrace communication and inclusion, integrate multiple sources of knowledge, and promote reflection which revisits and reframes initial assumptions. Processes of deliberation with analysis, previously used in larger jurisdictions with expert participants, worked well in a small town. The co-design process helped to redirect the project's research; shift the design from a centralized, siren-based system to a decentralized, dashboard-based one; align the institutions and organizations needed to implement and operate the system going forward; and result in an LWS that the community recognizes as serving their needs.

To what extent does this Sitka LWS and its co-design process represent a model for U.S. climate services that could be scaled nationwide? This question has at least two parts. The first asks the extent to which its decentralized, dashboard-centered design might offer a model for U.S. disaster response and management. The second asks what it would entail for the participatory risk governance employed in Sitka to become the norm for U.S. climate services.

To the first question, the Sitka experience suggests that the appropriateness of individualized evacuation decision- making is highly context-dependent, and that US disaster response and management would require an extensive community co-design process to make location-specific judgments regarding its use. The appropriateness of individualized evacuation depends strongly, in Sitka at least, on the accuracy of warnings generated, potential spillover effects of evacuation decisions on first responders and others, the community's views of risk, and how best to ensure equity. In Sitka, each of these criteria favor the decentralized system.

The algorithms in the Sitka warning system seem unexpectedly accurate in predicting when landslides will occur in and around the town. However, the predictions lack any geographic specificity, due in part to the significant variability in rainfall intensity across the town, significant variation in soil composition even among nearby hillsides [9], and the current reliance on measuring rainfall intensity at a single location. The inability to predict whether landslides would occur within as opposed to nearby Sitka was interpreted by the community as a high false alarm rate and thus a reason for a decentralized system [4].

In Sitka, the spillover effects from evacuation lean in the direction of individual decisions. The town lacks sufficient space for everyone who would need shelter if evacuations were mandatory. Not evacuating, or choosing to evacuate at the last minute, does not put first responders at risk since last-minute evacuees generally do not need to travel far to safety. If community dissemination of the individualized warning system works as intended, there is little reason for first responders to enter an exposure zone until a landslide has occurred. This situation can be contrasted to that of a flood or wildfire in many communities, in which first responders might take on risks to save people still in the flood or fire zone and the geography is such that people fleeing at the last minute can block roads needed by the first responders for their own effectiveness and safety.

The workshop discussions and community engagement emphasized that many Alaskans have a strong sense of independence, so they prefer to make their own decisions about risk. Equitable access to warnings and responses in Sitka seems best handled through the social networks highlighted in this project, which appear at least as effective as would be a siren or other centralized warning.

Overall, while a decentralized landslide warning system seems appropriate for Sitka, any judgments about warning accuracy, potential spillover effects, community views of risk, and equity are context dependent and most usefully explored in a co-design process.

The Sitka experience with community level, participatory co-design also suggests what would be required to conduct nationwide such participatory governance at a local scale. These requirements include: (1) a strong local partner, (2) active engagement by state and federal agencies, (3) support for and skill with multidisciplinary research, (4) resources for ongoing operations and maintenance, and (5) trained personnel to conduct co-design processes and the supporting research.

This project's local partner, the Sitka Sound Science Center, anchored the warning system and its development with the community. While much of the research and project management was conducted by the team's external partners—RAND, University of Oregon, and University of Southern California—in the eyes of the community, the main focus of any credit and blame for the project's perceived successes and failures resided with SSSC. The Science Center enabled the co-design process and its embedding in widespread community engagement; hosted all the project's community meetings, provided voice for all the project's communications with the community; organized the project's citizen science, served as home base for external team members visiting the community; arranged the project's town halls, science in the schools, and other outreach activities; and brokered most of the project team's interactions with community groups. The STA also provided a vital connection to the community's Indigenous population, including representatives for the project's deliberative processes and liaising with the local tribal government.

Replicating this Sitka model nationwide would require a national effort to create and support more local organizations with the capabilities of SSSC.

This project was also fortunate to have specific individuals from the regional offices of the USFS and NWS who were personally committed and excited by community engagement. Some of these individuals lived in and were widely regarded as members of the Sitka community. These individuals acted as trusted members of the community, as active members of the research project team, and as liaisons to the Federal agencies which provide crucial information and expertise to the local Sitka warning system. Individual members of the Alaska state government played similar roles.

Replicating this Sitka model nationwide would require moving beyond these committed individuals to a strong institutional commitment to community engagement from the relevant federal and state agencies.

The project team supporting the Sitka warning system effort was broadly multidisciplinary, involving geoscience, social sciences, information sciences, and risk management research. The project's geo-science research was central, but the project also gathered local and Indigenous knowledge and conducted research on social science, risk governance, and policy analysis regarding options for landslide insurance. This information all also proved critical to the design, deployment, and dissemination of all four components of a decentralized warning system and to effectively engaging the community in its design.

However, the community and project team both presented barriers to this multidisciplinary research. Coordinating the multiple strands of research was an ongoing project management challenge, in particular as the research directions evolved through the course of the co-design process. Community members also lacked an initial appreciation for multidisciplinarity. They had a strong initial appreciation of the value of geoscience for reducing the risk from landslides and were often eager to engage with the team's geoscientists. Most community members, however, initially had less appreciation and some suspicion of the role of social sciences and policy analysis. The project's workshops and other community interactions thus needed to build such appreciation and trust.

Replicating the Sitka model would require ongoing and enhanced support of such multidisciplinary research and training in how to conduct it most effectively.

While ample federal funding was available to design and deploy the Sitka warning system, funds for maintaining it have been more difficult to come by. Our project team has estimated the staff time and financial resources that the SSSC would require to maintain Sitka's warning system on an ongoing basis. As detailed in Appendix B, such operations would cost about $65,000 per year. This includes a junior SSSC member working about one-third of their time organizing maintenance of the sensors and their data feeds, upgrading algorithms, community education and outreach programs, and maintaining contact with relevant city, tribal, state, and federal agencies. This also includes annual equipment costs of about $25,000 for website maintenance, telecommunications contracts for the sensors, and materials and depreciation on the sensors.

Finally, for this Sitka model to be scaled up nationwide, many of the project team's tasks, including some of the research and co-design facilitation, would need to be operationalized by trained staff working in government, NGOs, or related operational organizations, rather than by PhD's working in research institutions.

Replicating the Sitka model nationwide would require stable funding for maintaining such systems and a nationwide effort to train and support individuals and their institutions who could specialize in the work of helping communities to co-design them.

Author Contributions: Conceptualization, R.J.L. and L.B.; Formal analysis, R.B., A.P. and S.T.; Writing—original draft, R.J.L.; Writing—review & editing, S.T., J.S., A.P. and T.Y. All authors have read and agreed to the published version of the manuscript.

Funding: This research was funded by the Smart & Connected Communities program of the National Science Foundation under NSF award No. 1831770.

Institutional Review Board Statement: The study was conducted in accordance with the Declaration of Helsinki, and approved by the Institutional Review Board (or Ethics Committee) of the RAND Corporation Human Subjects Protection Committee, Approval #2018-0578 on 12 July 2018.

Informed Consent Statement: Informed consent was obtained from all subjects involved in the study.

Data Availability Statement: The data presented in this study draws primarily from notes, records, and other documentation of project activities, some of which can be shared in verbatim or summary form, pending agreement from participants and RAND's Institutional Review Board.

Acknowledgments: Numerous people, too many to name individually, gave generously of their time and wisdom to this project. We are particularly indebted to Katie Whipkey, Tory O'Connell Curran, and Jonathan Stock for helping to get the project started; our NSF program officer Sylvia Spengler for advice and encouragement; our federal partners Katherine Prussian, Dennis Landwehr, and Jacqui Foss at the US Forest Service, Aaron Jacobs at the National Weather Service; Ben Mirius at the US Geological Survey; our state partners Ronnie Daanen and Gabriel Wolken at the Alaska Division of Geological & Geophysical Surveys; and Jeff Feldpausch at the Sitka Tribe of Alaska. In addition, we are grateful to Sitka Tribe of Alaska Tribal Council, City and Borough of Sitka Assembly, the Sitka Fire Department, and the Sitka Planning Department for their time, interest, and advice.

Conflicts of Interest: The authors declare no conflict of interest.

Appendix A. Co-Design Workshops and Community Engagement Activities

The Sitka landslide co-design process was one component of a larger process of community engagement. The co-design process was organized around three in-person workshops in Sitka, Alaska, a decision scoping workshop in May 2019, a design workshop in February 2020, and a review and launch plans workshop in October 2021. Table A1 shows these workshops in bold type with other key events and engagement activities shown in regular typeface.

Table A1. Community engagement and co-design activities.

Date	Activity
2015	• Kramer slide
2016	• Geoscience task force
2017	• Planning grant visit (Dec)
2018	• Geoscience task force report • Submit proposal to NSF (Feb) • Project start (Sept)
2019	• **Decision scoping workshop (May)** • Scientist in residency (Fall)
2020	• **Design workshop (Feb)** • Landslide runout model workshop (Aug–virtual)
2021	• Scientist in residency (Spring) • Scientists in the schools (Aug-Sept) • Dashboard developer design workshops (Sept–Nov) • **Design review and launch plans workshop (Oct)** • Insurance workshop (Oct) • Scientists in the schools (Dec)
2022	• Town hall (Jan–virtual) • Scientists in the schools (Feb) • Presentations and work sessions with Sitka tribal council and Sitka city assembly (Spring–summer) • Dashboard dissemination workshops (May–July) • Dashboard launch (August)

Bold indicates a co-design workshop.

The May 2019 decision framing workshop agenda was organized around a backcasting exercise, a planning method that starts by defining a desirable future and then works backwards to identify policies and programs that will connect that future to the present situation. Workshop participants were asked to imagine that it is the year 2025 and Sitka has the nation's best landslide warning system. Participants crafted stories describing how Sitka had arrived at this happy situation, with a focus on describing key actions taken in the years 2019–2021.

This exercise generated a common understanding of goals for the warning system, including reducing potential loss of life, reducing anxiety about landslides, inclusion of all community members in the benefits of the warning system and opportunities to participate in its design and operations, an ability to inspire trust in the system, effective monitoring and communications, effective response to warnings, sustainable operations and costs, and an ability to protect property. The exercise also generated an understanding of actions the community would consider pursuing to achieve these goals, including developing appropriate knowledge, setting appropriate warning thresholds, effective communication, effective response actions, operations, and insurance pools and structural measures to protect property. The project team crafted initial versions of these lists based on the workshop discussions, and then shared them back with participants for revision and refinement. The final versions are shown in Tables A2 and A3.

Informed by the decision scoping workshop results and their ongoing geoscience research, the project team subsequently generated an initial design of the landslide warning system, as shown in Figure A1. This design envisioned a system in which the Sitka Fire Department would activate a siren when local conditions indicated landslides were sufficiently likely. The February 2020 design workshop presented participants with this initial design, engaged participants on scientific concepts and issues central to design choices, and provided an opportunity for review and comment on the initial LWS design.

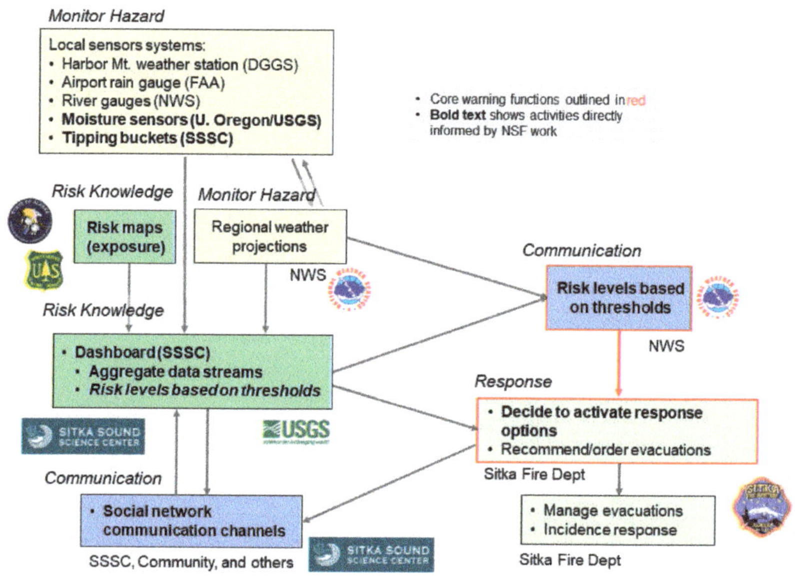

Figure A1. Warning system design used at start of second co-design workshop. Source: authors.

The design workshop was organized around two key project research areas: geoscience aimed at improved landslide predictions and social network analysis aimed at improving the efficiency and equity of landslide communications. The geoscience discussions provided an overview of progress to date with developing and testing the new moisture sensors. The social network discussions provided an overview of how better understanding such networks could improve risk communications in Sitka. Each segment included a "serious game" designed to illuminate and seek feedback on important design choices and tradeoffs for the LWS.

The geoscience research suggested the extent to which any improved landslide prediction would remain imperfect, thus highlighting a design tradeoff between failed and false warnings. The latter could increase community anxiety and disrupt daily life. The former could increase loss of life. In addition, there exist decisions as to the spatial resolution of the warning, that is, the extent to which warnings are generated for particular areas or provided more generally. For instance, workshop discussions revealed that the geoscientist and Sitka residents viewed differently the spatial scale of successful warning. The geoscientists would regard a warning to be accurate if a landslide occurred within many miles of Sitka. Many community members would regard a warning as failed if a landslide occurred outside the city bounds.

To address these issues with a serious game, participants were each given a plot with data showing maximum precipitation intensity for storms in Southeast Alaska that did and did not initiate landslides. Participants were asked to draw lines through the data representing a precipitation-duration threshold that might be used to generate warning. The answers were diverse. Some participants chose low thresholds that would generate no failed warnings, some chose high thresholds that would generate no false warnings, and some those thresholds in between.

Participants were then divided into groups based on where they drew their threshold and asked to choose a card which would reveal either a failed, false, or accurate warning. Each group chose from a card deck with probabilities corresponding to their chosen thresholds. Participants revealed their choices and were asked to describe how they felt about the results and how they would explain them to others. For instance, participants in the group with low thresholds would need to role play an official explaining a false alarm that had needlessly disrupted life in the community. Participants in the group with high thresholds would need to explain a failed warning that resulted in injuries in the community.

Social network analysis can improve the flow of appropriate warning and risk knowledge to all who need it; help ensure that all Sitkans receive useful, understandable, and actionable information; and improve message redundancy in which the same information reaches an individual through multiple channels. However, social network data collection and analysis raises issues of privacy, equity, and trust. To engage with these tradeoffs, we demonstrated a social network analysis with participants. We used a survey to collect data on who in the room participants would most trust for advice when clothes shopping. We used the resulting network map to animate a discussion of how social network analysis might improve the efficiency and equity of communications regarding Sitka's landslide warning system.

Table A2. Measures of success for landslide warning system discussed in visioning workshop.

Measures (M)	
Overall	Monitoring and Communications
· Limit loss of life · Low Stress	· Consistent with/doesn't conflict with other warning systems · Provides situation awareness in a timely and trusted manner · Gives measured levels of risk · Few false alarms/avoid alarm fatigue · Timely
Inclusion	Respond
· Clear lines of participation · Opportunities for volunteering · Accessible to entire population	· Orderly evacuation process · Timely
Trust	Operations and cost
· Familiar · Easily understood/explainable/simple · Compassionate/kind to people at most risk · Inspires public confidence · Sets reasonable expectations, but not create fear · Community buy-in · Good science · Trusted science, even when it changes · Acknowledge its own limitations	· Not cost much/cost effective/affordable · Maintainable · Reliable/System redundancy · Clear lines of responsibility · Sustainability · Capable of continual improvement Protect property · No excessive liability

On 4 October 2022, during the same visit as the final LWS co-design workshop, the project team also convened a landslide insurance workshop. This workshop aimed to

improve community understanding of landslide insurance issues and to gauge interest in various alternatives to the status quo. The eighteen workshop participants included insurance professionals, homeowners, real estate professionals, and banking professionals as well as State of Alaska officials responsible for natural hazards insurance. Prior to the workshop, the project team identified and estimated the potential economic implications of four alternative options for providing landslide insurance in Sitka.: (1) the State of Alaska requires private insurers to provide landslide insurance, (2) the State of Alaska or the Federal government share risk with private insurers by reimbursing private insurers if claims exceed a specified level, (3) the State or Federal governments directly offer landslide insurance, and (4) a landslide insurance pool in which homeowners from multiple SE Alaska communities join together to pool risks. At the workshop, the project team presented these options, described analogues elsewhere in the United States, and facilitated a discussion among participants of pros and cons. The team also facilitated a discussion which led to the decision to provide exposure maps on the dashboard, with sufficient resolution so that users could judge risk in their immediate neighborhood but without sufficient resolution to identify specific properties.

Table A3. Levers discussed in visioning workshop with which one or more actors in Sitka could engage to pursue goals for landslide warning system.

Policy Levers (L)	
Knowledge	Response
· Lots of education · Hazard maps · Media Management · What *citizen* science	· Evacuations plans/places to go · Training: community, first responders, mental health (even at command centers) · Table-top scenarios/drills · Post-event care/trauma · Are evacuations mandatory? · Ways to deal with volunteers, especially when they aren't helpful · Contingency plans for events outside expectations
Monitoring	Operations
· Where the thresholds are set	· Trust fund · Source and amount of funding · Clear lines of responsibility for action/communications · Who runs app/website · Who is in charge of training
Communication	Protect property

· Dissemination channels for warning
  - Use paper system
  - Flashing lights (to avoid areas)
  - Street signs/warning markers
· Attributes of warning
  - Focused warning
  - Timed warnings
  - How far our warning occurs? 48 h? 6 h, etc.
· App or website with real time data
· Good ways to get accurate information out to people that don't impede incident command
· Ways to recover from alarm fatigue
	· Insurance pools
	· Physical mitigation/diversion of flows

Appendix B. Cost Estimates for Maintaining the Sitka System

What would it take to operate and maintain the Sitka landslide warning system as an ongoing operation? Based on our experience to date, our project team made the following estimates of the resources required.

- STAFF TIME: The operations and maintenance would require Sitka Sound Science Center staff person ~1/3 time to:

- Organize maintenance of local environmental monitoring instruments (about a month of time per year)
  - Harbor Mountain Weather System
  - 10 Tipping Buckets (community hosted)
  - 3 Soil Moisture Sensors (ONSET)
  - FAA/NWS Airport Weather Station—This facility is run by the Federal Aviation Administration (FAA) and National Weather Service (NWS), so SSSC isn't responsible for its maintenance, but close contact should be maintained with the Sitka Flight Service operators as it feeds the webpage
- Implement any upgrades to data feeds from NWS and other sensors and implement any upgrades to dashboard prediction algorithms as more data becomes available.
- Produce annual summaries of instrumental data and dashboard predictions
- Annual lesson in high school class—they host one of the tipping buckets
- Maintain community contacts
  - USFS, Fire Chief, STA Safety Officer, etc.
  - Prepare for and attend:
    - Local Emergency Planning Commission (monthly, summer off)
    - Sitka Tribe of Alaska—Natural Resource Committee (attend annually)
    - City and Borough of Sitka Assembly (attend annually with update)
    - Planning Commission, if commission becomes interested in applying landslide hazard to city planning, zoning etc.
    - Southeast Environmental Conference (annually)
- Respond to questions
- Be point of contact for geoscience-related issues in Sitka (e.g., work with Volcano Center)
- Create physical materials like rack cards, maps, posters
- Storm Reports—during and after extreme rainfall events, compile and send to community contact list
- Annually send updates to USFS with updates for the Landslide Inventory based on storm, community reports
- Upgrade and update site; sitkalandslide.org and SSSC Landslide Research page
- Oversee contracts associated with telecoms and dashboard

- Materials for maintaining the system would require:
  - Website: $20,000 per year for website mainatinance service contract
  - Tipping buckets: $100/year for batteries for tipping buckets
  - $3500 telecommunication contracts for tipping buckets
  - Moisture sensors:
    - $100/year data connection fee
    - $300/year repair and replacement ($3000 sensor amortized over 10 years)
  - $1000/year IT support
  - $350/year office consumables

TOTAL COSTS

- Labor ~ $25,000/year
- Materials: $25,250
- Indirect: $13,100
- Total: $63,350

References

1. IPCC. Summary for Policymakers. In *Climate Change 2022: Impacts, Adaptation, and Vulnerability. Contribution of Working Group II to the Sixth Assessment Report of the Intergovernmental Panel on Climate Change*; Pörtner, H.-O., Roberts, D.C., Poloczanska, E.S., Mintenbeck, K., Tignor, M., Alegría, A., Craig, M.T., Langsdorf, S., Löschke, S., Möller, V., et al., Eds.; Cambridge University Press: Cambridge, UK, 2022.
2. Reed, M.S. Stakeholder participation for environmental management: A literature review. *Biol. Conserv.* **2008**, *141*, 2417–2431. [CrossRef]
3. Ara Begum, R.; Lempert, R.; Ali, E.; Benjaminsen, T.A.; Bernauer, T.; Cramer, W.; Cui, X.; Mach, K.J.; Nagy, G.; Stenseth, N.C.; et al. Point of Departure and Key Concepts. In *Climate Change 2022: Impacts, Adaptation, and Vulnerability, Contribution of Working Group II to the Sixth Assessment Report of the Intergovernmental Panel on Climate Change*; Pörtner, H.-O., Roberts, D.C., Poloczanska, E.S., Mintenbeck, K., Tignor, M., Alegría, A., Craig, M., Langsdorf, S., Löschke, S., Möller, V., et al., Eds.; Cambridge University Press: Cambridge, UK; New York, NY, USA, 2022. [CrossRef]
4. Busch, L.; Lempert, R.; Izenberg, M.; Patton, A. Run Uphill for a Tsunami, Downhill for a Landslide. *Issues Sci. Technol.* **2021**, *38*, 40–47.
5. Blomkamp, E. The Promise of Co-Design for Public Policy. In *Routledge Handbook of Policy Design*; Howlett, M., Mukherjee, I., Eds.; Routledge: New York, NY, USA, 2018; pp. 59–73.
6. Moser, S.C. Can science on transformation transform science? *Lessons from co-design. Curr. Opin. Environ. Sustain.* **2016**, *20*, 106–115. [CrossRef]
7. National Research Council (NRC). *Informing Decisions in a Changing Climate*; Panel on Strategies and Methods for Climate-Related Decision Support, Committee on the Human Dimensions of Climate Change, Division of Behavioral and Social Sciences and Education: Washington, DC, USA, 2009.
8. Schoen, D.; Rein, M. *Reframing: Towards the Resolution of Intractable Policy Controversies*; Basic Books: New York, NY, USA, 1994.
9. Patton, A.I.; Roering, J.J.; Orland, E. Debris flow initiation in postglacial terrain: Insights from shallow landslide initiation models and geomorphic mapping in Southeast Alaska. *Earth Surf. Process. Landf.* **2022**, *47*, 1583–1598. [CrossRef]
10. Booth, A.M.; Sifford, C.; Vascik, B.; Siebert, C.; Buma, B. Large wood inhibits debris flow runout in forested southeast Alaska. *Earth Surf. Process. Landf.* **2020**, *45*, 1555–1568. [CrossRef]
11. Miller, D. Modeling Susceptibility to Landslides and Debris Flows in Southeast: TerrainWorks Report to the U.S. Forest Service, Tongass National Forest. 2019. Available online: https://terrainworks.sharefile.com/share/view/s8532031dfcf44ab0a87bc8a83b2b3b1c (accessed on 23 February 2023).
12. City and Borough of Sitka (CBS). About Sitka. Available online: https://www.cityofsitka.com/about-sitka (accessed on 8 February 2023).
13. City and Borough of Sitka (CBS). City and Borough of Sitka (CBS) Charter. Sitka, AK. Available online: https://www.cityofsitka.com/general-code-charter (accessed on 6 February 2023).
14. Sitka Tribe of Alaska. Sitka Tribe of Alaska: About Us. Available online: https://www.sitkatribe.org/pages/about-us-about-us (accessed on 6 February 2023).
15. Local Emergency Planning Committee. Available online: https://www.cityofsitka.com/LocalEmergencyPlanningCommittee (accessed on 16 December 2022).
16. Dasgupta, P.; Edenhofer, O.; Amezquita, A.M.A.; Bento, A.; Caney, S.; Croix, D.D.l.; Fosu, A.; Jakob, M.; Saam, M.; Shrader-Frechette, K.; et al. Economic Growth, Human Development, and Welfare. IPSP, Ed.; In *Rethinking Society for the 21st Century: Report of the International Panel on Social Progress*; Cambridge University Press: Cambridge, UK, 2018; pp. 141–186. [CrossRef]
17. Guston, D.H. Boundary Organizations in Environmental Policy and Science: An Introduction. *Sci. Technol. Hum. Values* **2001**, *26*, 399–408. [CrossRef]
18. Fischer, F. *Reframing Public Policy: Discursive Politics and Deliberative Practices*; Oxford University Press: Oxford, UK, 2003. [CrossRef]
19. Fung, A. Varieties of participation in complex governance. *Public Adm. Rev.* **2006**, *66*, 66–75. [CrossRef]
20. Fung, A. Putting the public back into governance: The challenges of citizen participation and its future. *Public Adm. Rev.* **2015**, *75*, 513–522. [CrossRef]
21. United Nations International Strategy for Disaster Reduction [UNISDR]. *Terminology: Basic Terms of Disaster Risk Reduction;* United Nations: Geneva, Switzerland, May 2009.
22. Renn, O. *Risk Governance: Coping with Uncertainty in a Complex World*; Earthscan: London, UK, 2008.
23. Van Asselt, M.B.; Renn, O. Risk governance. *J. Risk Res.* **2011**, *14*, 431–449. [CrossRef]
24. International Organization for Standardization (ISO). *Risk Management—Principles and Guidelines*; International Organization for Standardization: Geneva, Switzerland, 2009; p. 24.
25. Bremer, S.; Meisch, S. Co-production in climate change research: Reviewing different perspectives. *WIREs Clim. Chang.* **2017**, *8*, e482. [CrossRef]
26. Dewulf, A.; Craps, M.; Bouwen, R.; Taillieu, T.; Pahl-Wostl, C. Integrated management of natural resources: Dealing with ambiguous issues, multiple actors and diverging frames. *Water Sci. Technol. J. Int. Assoc. Water Pollut. Res.* **2005**, *52*, 115–124. [CrossRef]

27. Edenhofer, O.; Kowarsch, M. Cartography of pathways: A new model for environmental policy assessments. *Environ. Sci. Policy* **2015**, *51*, 56–64. [CrossRef]
28. Kwakkel, J.H.; Walker, W.E.; Haasnoot, M. Coping with the Wickedness of Public Policy Problems: Approaches for Decision Making under Deep Uncertainty. *J. Water Resour. Plan. Manag.* **2016**, *142*, 01816001. [CrossRef]
29. Lempert, R.J.; McDonald, T.; Popper, S.W.; Prosdocimi, D.; Small, T.A. *Implementing a New Mobility Vision for Rancho Higuera in a Deeply Uncertain, Fast-Changing World: Results from RAND Corporation Participatory Local Planning Workshops*; RAND Corporation: Santa Monica, CA, USA, 2020.
30. Groves, D.G.; Fischbach, J.R.; Bloom, E.; Knopman, D.; Keefe, R. *Adapting to a Changing Colorado River: Making Future Water Deliveries More Reliable Through Robust Management Strategies*; RAND Corporation: Santa Monica, CA, USA, 2013.
31. Fischbach, J.R.; Siler-Evans, K.; Wilson, M.; Tierney, D.; Cook, L.; May, L. *Resilient Stormwater Management in Allegheny County: A Pilot Study*; RAND Corporation: Santa Monica, CA, USA, 2017.
32. Bouwer, L.M.; Haasnoot, M.; Wagenaar, D.; Roscoe, K. *Assessment of Alternative Flood Mitigation Strategies for the C-7 Basin in Miami, Florida*; the South Florida Water Management District: West Palm Beach, FL, USA, 2017.
33. Groves, D.G.; Molina-Perez, E. *Developing a Robust Water Strategy for Monterrey, Mexico: Diversification and Adaptation for Coping with Climate, Eco-nomic, and Technological Uncertainties—Planning Support Tool*; RAND Corporation: Santa Monica, CA, USA, 2019.
34. Wong-Parodi, G.; Mach, K.J.; Jagannathan, K.; Sjostrom, K.D. Insights for developing effective decision support tools for environmental sustainability. *Curr. Opin. Environ. Sustain.* **2020**, *42*, 52–59. [CrossRef]
35. Public Participation Guide: Selecting the Right Level of Public Participation. Available online: https://www.epa.gov/international-cooperation/public-participation-guide-selecting-right-level-public-participation#goal (accessed on 1 February 2023).
36. Pidgeon, N.; Rischhoff, B. The role of social and decision sciences in communicating uncertain climate risks. *Nat. Clim. Chang.* **2011**, *1*, 35–44. [CrossRef]
37. Janoske, M.; Liu, B.F.; Sheppard, B. *Understanding Risk Communication Best Practices: A Guide for Emergency Managers and Communicators*; U.S. Department of Homeland Security: College Park, MD, USA, 2012.
38. Patton, A.I.; Luna, L.V.; Roering, J.J.; Jacobs, A.; Korup, O.; Mirun, B.B. Landslide initiation thresholds in data sparse regions: Application to landslide early warning criteria in Sitka, Alaska. *Nat. Hazards Earth Syst. Sci.* **2019**, *in preparation*.
39. Chu, M.; Patton, A.; Roering, J.; Siebert, C.; Selker, J.; Walter, C.; Udell, C. SitkaNet: A low-cost, distributed sensor network for landslide monitoring and study. *HardwareX* **2021**, *9*, e00191. [CrossRef]
40. Izenberg, M.; Brown, R.; Siebert, C.; Heinz, R.; Rahmattalabi, A.; Vayanos, P. A community-partnered approach to social network data collection for a large and partial network. *Field Methods* **2022**. [CrossRef]
41. Rahmattalabi, A.; Jabbari, S.; Lakkaraju, H.; Vayanos, P.; Rice, E.; Tambe, M. Fair Influence Maximization: A Welfare Optimization Approach. *arXiv* **2020**, arXiv:2006.07906. [CrossRef]
42. Sitka Geotask Force. Sitka Geotask Force Summaries: August 2015 Sitka Landslides. Available online: https://sitkascience.org/sssc/wp-content/uploads/2019/02/Sitka-Geotask-Force-Summary-Final-2016.pdf (accessed on 6 February 2023).
43. Mirus, B.B.; Morphew, M.D.; Smith, J.B. Developing Hydro-Meteorological Thresholds for Shallow Landslide Initiation and Early Warning. *Water* **2018**, *10*, 1274. [CrossRef]
44. Robinson, J.B. Unlearning and backcasting: Rethinking some of the questions we ask about the future. *Technol. Forecast. Soc. Change* **1988**, *33*, 325–338. [CrossRef]
45. Izenberg, M.; Clark-Ginsberg, A.; Clancy, N.; Busch, L.; Schmidt, J.; Dixon, L. Efforts to end a stalemate in landslide insurance availability through inclusive policymaking: A case study in Sitka, Alaska. *Int. J. Disaster Risk Reduct.* **2022**, *80*, 103202. [CrossRef]
46. Dixon, L.; Izenberg, M.; Lempert, R.; Patton, A.; Roering, J. *Residential Landslide Insurance in Sitka, Alaska: Market Conditions and Options for Improving Availability*; RAND Corporation: Santa Monica, CA, USA, 2023. (In review)
47. Shannon; Wilson, I. South Kramer Avenue Landslide: Jacobs Circle to Emmons Street. 2016, p. 22. Available online: https://www.cityofsitka.com/media/Public%20Works/SouthKramerLandslideReport-S.W2016.pdf (accessed on 23 February 2022).
48. Guzzetti, F.; Gariano, S.L.; Peruccacci, S.; Brunetti, M.T.; Marchesini, I.; Rossi, M.; Melillo, M. Geographical landslide early warning systems. *Earth-Sci. Rev.* **2020**, *200*, 102973. [CrossRef]
49. Segoni, S.; Piciullo, L.; Gariano, S.L. A review of the recent literature on rainfall thresholds for landslide occurrence. *Landslides* **2018**, *15*, 1483–1501. [CrossRef]
50. Turner, S. *Beliefs, Needs, and Participation: Aligning Risk Governance and Decision Making under Deep Uncertainty*; Pardee RAND Graduate School: Santa Monia, CA, USA, 2023.

sustainability

MDPI

Article

The Role of Non-Climate Data in Equitable Climate Adaptation Planning: Lessons from Small French and American Cities

Elena Lioubimtseva [1,*] and Charlotte da Cunha [2]

[1] Department of Geography and Sustainable Planning, Grand Valley State University, Allendale, MI 49401, USA
[2] CEARC Laboratory, University of Versailles Saint-Quentin-en-Yvelines/Université Paris Saclay, 11 Boulevard d'Alembert, 78280 Guyancourt, France
* Correspondence: lioubime@gvsu.edu

Abstract: There is a growing consensus that to effectively adapt to climate change, cities need user-friendly tools and reliable high-resolution biophysical and socio-economic data for analysis, mapping, modeling, and visualization. This study examines the availability of various types of information used in climate adaptation plans of 40 municipalities with a population of less than 300,000 people in the United States and France, probing into the choice and usage of relevant information by small municipalities. We argue that non-climatic spatial data, such as population demographic and socio-economic patterns, urban infrastructure, and environmental data must be integrated with climate tools and datasets to inform effective vulnerability assessment and equitable adaptation planning goals. Most climate adaptation plans examined in this study fail to address the existing structural inequalities and environmental injustices in urban infrastructure and land use. Their challenges include methodological and ideological barriers, data quality issues, and a lack of meaningful community connections. Adaptation methodological approaches should be reassessed in the context of much-needed societal transformation. Lessons learned from our studies offer valuable insights for the potential development of national and state-level climate adaptation information services for cities.

Keywords: climate change adaptation; adaptation plan; small municipality; France; United States; climate services; information

Citation: Lioubimtseva, E.; da Cunha, C. The Role of Non-Climate Data in Equitable Climate Adaptation Planning: Lessons from Small French and American Cities. *Sustainability* **2023**, *15*, 1556. https://doi.org/10.3390/su15021556

Academic Editors: Charles Herrick, Jason Vogel and Glen Anderson

Received: 9 November 2022
Revised: 31 December 2022
Accepted: 4 January 2023
Published: 13 January 2023

1. Introduction and Background

This study contributes to the growing international body of knowledge on climate services and data intended for climate change adaptation planning at a local scale. Climate services for adaptation have been defined as all public and private sector services supporting adaptation to climate change [1,2]. Based on [3] "the aim of climate services is to provide people and organizations with timely, tailored climate-related knowledge and information that they can use to reduce climate-related losses and enhance benefits, including the protection of lives, livelihoods, and property" (p. 588). The European Union further defines climate services as a process of "transforming climate-related data and other information into customized products such as projections, trends, economic analyses, advice on best practices, the development and evaluation of solutions, and any other climate-related services that may be of use for society" [4]. There has been significant progress toward improved climate change scenarios, downscaling, theoretical and methodological development, and production of applied tool-kits, and online clearinghouses intended to support climate adaptation planning at a city scale, produced collaboratively by national and international governmental entities and research institutions [5–8]. There is also a growing recognition that effective climate adaptation planning requires the analysis of multidisciplinary data, which is not limited to climate change trends and scenarios alone. The integration of climate and weather data with social, economic, cultural, and environmental data is paramount to evaluate the present and future human vulnerability to climate change, addressing disproportionate socioeconomic risk to climate impacts and engaging overburdened communities in the planning process [9–11]. A growing number of organizations have developed various services to assist local governments and communities with climate adaptation planning. Examples of international platforms include the Global Framework for Climate Services (GFCS) of the World Meteorological Organization [12], Copernicus Climate Change Service (C3), the EUMETSAT (European Organization for the Exploitation of Meteorological Satellites), and Climate-ADAPT [6] of the European Union [13]. National and regional instruments, such as the U.S. Climate Resilience Toolkit [14], Climate Adaptation Knowledge Exchange (CAKEX) by Eco-Adapt [15], Adapt West [16], the Great Lakes Integrated Science and Assessment (GLISA) [17], the French National Observatory on the Effects of Global Warming (ONERC), the National Ecological Transition Agency (ADEME), and the platform ClimatHD by Meteo France offer more specific country-wide or region-wide data coverage. Some U.S. states, such as California, provide state-level open access peer-reviewed cross-disciplinary data and collaboration opportunities to stakeholders, including infrastructure managers, municipal planners, community-based organizations, state agencies, scientists and climate experts, educators, and the public via Cal Adapt [7]. These databases provide state-wide data on temperature, rainfall, wind, soil moisture, and ocean conditions, as well as maps, risk and vulnerability analyses, assessments, and long-term projections and scenarios. They can be combined with socio-economic variables and non-meteorological data such as agricultural production, health trends, human settlement in high-risk areas, and road and infrastructure maps for the delivery of goods, depending on user needs and other relevant information.

In France, since 2011, ten regional working groups of independent experts have been created to support regional climate change monitoring efforts. These include five existing groups—AcclimaTerra in Nouvelle-Aquitaine, GREC-SUD in Provence-Alpes-Côte d'Azur, Ouranos-AuRA in Auvergne-Rhône-Alpes, RECO in Occitanie, GREC Guadeloupe and five more groups still being formed in Brittany, Normandy, Pays de La Loire, Ile-de-France, and Hauts-de-France. These multidisciplinary committees are modeled after IPCC Working Groups and are positioned at the interface of academic and non-academic spheres, constituting a catalyst for action in response to the impacts of climate change. The Nouvelle-Aquitaine region, which currently holds the most climate adaptation plans in the country, is home to AcclimaTerra [18], the precursor group that pioneered this initiative in 2018 [19]. Similar regionalization of metropolitan climate adaptation planning has been observed in the United States for many years with Regional Adaptation Planning (RAP) initiatives evolving around major cities and involving municipalities of various sizes. Metropolitan RAPs assume diverse organizational arrangements and operate in a variety of political and geographical contexts, including the development of their own climate services. Their spatial scale varies from parts of urban agglomerations to bioregional watersheds, but in the United States they most commonly reflect the boundaries of existing "metropolitan regional" entities, such as counties and regional planning organizations [20].

Since the purpose of climate adaptation is the reduction in vulnerability to adverse climate impacts, any climate adaptation plan should be based on a thorough assessment of human vulnerability and principles of climate justice. Therefore, climate services for climate adaptation planning are inherently multidisciplinary and must include demographic, social, economic, and environmental justice data and tools as well. Justice is a legal term closely related to the social concept of equity, offering a human rights perspective on the climate crisis, acknowledging that climate change has differing social, economic, public health, and other adverse impacts on underprivileged populations [21]. Developing transparent planning strategies that eliminate disparities would be impossible without reliable social and economic data about race, class, gender, and other dimensions of diversity. To address this need, several U.S. states, e.g., Michigan [22], are currently developing Environmental Justice Screens—online platforms providing environmental justice spatial data at a much higher resolution than the already existing U.S. EPA EJ Screen [23]. In France, the discourse on environmental justice remains mostly confined to academia. While a focus on racial discrimination has been at the heart of the U.S. environmental justice movement [24,25], the concept of "race" as a major factor of environmental injustice is still barely acknowledged in France, mostly due to the effort of its republican ideology to erase any recognition of racial inequalities [26].

Despite progress in the development of climate services for adaptation planning, including some non-climatologic data, there is still a significant gap between the actual data needs and existing products and services offered by various organizations. This gap is particularly problematic for small municipalities, which have limited capacity to locate, access, and interpret adequate information, compared to large high-capacity cities. Cross-national peer-learning experience is rarely available to smaller communities [27] and very few scholarly studies have compared the provision of climate services for local climate adaptation planning between different countries [28,29]. One significant challenge is that climate services largely develop through interaction between the scientific and non-scientific communities, whereas scientific literature is built mostly through exchange within the academic sphere. As Vaughan et al. point out "While several outlets allow members of specific research communities to communicate with each other, there are far fewer mechanisms that allow operational climate service providers and consumers to engage in two-way dialog on the questions they would like addressed by the research community. This two-way communication is essential given the overwhelming evidence that climate services are most useful when they are developed as part of an iterative process of "co-discovery", "co-development," and "co-evaluation" involving the producers and users of climate information" [30]. Thus, an analysis driven by the users' perspective is necessary to go beyond the academic discussions and incorporate knowledge and data generated by communities themselves.

The primary objective of our study is to examine climate adaptation data needs from the perspective of small municipalities (defined here as urban areas with populations less than 300,000). The U.S. and France provide an especially interesting case due to fundamental differences in their approaches to local climate adaptation planning and provision of climate services, with the French system being highly centralized and a variety of community-driven approaches across the United States. Our secondary objectives are to investigate what information, methods, and tools have been used in local vulnerability assessments and climate adaptation plans in both countries, to identify major gaps, and to synthesize insights from these two different national models of local climate adaptation planning. We do not aim here to compare different national approaches. Instead, our goal is to use this cross-national case study to provide some insights into common challenges faced by small municipalities and emerging solutions in both countries.

2. Methodology

2.1. Climate Adaptation Plans

This inquiry on the role of multidisciplinary data and tools available for municipal climate adaptation planning is informed by the analysis of climate adaptation plans and vulnerability assessment reports developed by urban and rural municipalities with populations less than 300,000 people in the U.S. and France. To investigate the content, sources, and scale of climatic and non-climatic tools, services, and data used by the local communities we examined 40 published climate adaptation plans (23 in the U.S. and 17 in France) of small cities, towns, and counties. The selection of planning documents for the U.S. part of the dataset is described in detail in [31], while the selection of both U.S. adaptation plans and French PCAETs (Plan Climat Air Energie Territorial) is the most recent update of our earlier dataset published in [27] and [32]. Climate adaptation planning in France has been fully integrated into local territorial climate-air-energy plans, which are now mandatory for all communities with more than 20,000 inhabitants [33]. On the contrary, climate adaptation efforts in the United States have been voluntary and mostly driven by state, local, and tribal initiatives [34]. The sample of municipalities is not meant to be exhaustive and aims to reflect the geographic diversity of both countries. These cities, towns, and counties are listed in Table 1, with a summary of climate change impacts addressed in their climate adaptation plans.

Table 1. U.S. and French climate adaptation plans.

Municipality	Source	Impacts of Climate Change Addressed							
		Coastal Changes	**Severe Storms**	**Extreme Heat**	**Extreme Cold**	**Flooding**	**Drought**	**Wildfires**	**Seasonal Shifts**
U.S. climate adaptation plans									
Albany, NY	[35]		X	X	X	X			X
Alger County, MI	[36]	X	X	X	X	X	X	X	X
Boulder County, CO	[37]			X	X	X	X	X	
Chula Vista, CA	[38]	X	X	X		X	X	X	
Corte Madera, CA	[39]	X	X	X		X	X	X	
Flagstaff, AZ	[40,41]		X	X		X	X	X	X
Georgetown, ME	[42]	X	X			X			
Groton, CT	[43]	X	X	X		X	X		X
Iowa City, IA	[44,45]		X	X		X	X		X
Keene, NH	[46,47]		X	X	X	X	X		X
Laguna Woods, CA	[48]		X	X		X	X	X	
Marquette, MI	[49]	X	X	X		X	X		X
Marquette County, MI	[50]	X	X	X	X	X	X	X	X
Marshfield, MA	[51]	X	X	X	X	X			X
North Kingston, RI	[52]	X	X	X	X	X			X
Punta Gorda, FL	[53,54]	X	X	X		X			
Salem, MA	[55]	X	X	X		X			
Santa Cruz, CA	[56,57]	X	X	X		X	X	X	
Sarasota, FL	[58,59]	X	X	X		X			
Taos County, NM	[60]		X	X		X	X	X	X
Tybee Island, GA	[61]	X	X			X			
Tompkins County, NY	[62]		X	X		X	X		
Watsonville, CA	[63]	X	X	X		X	X	X	
French climate adaptation plans									
Brest métropole, Bretagne	[64,65]	X	X	X		X	X		
Clermont Auvergne Métropole, Auvergne-Rhône-Alpes	[66–69]			X		X	X	X	
Cordais et Causse (4 C), Occitanie	[70–72]					X		X	
Golfe du Morbihan, Bretagne	[73]	X	X			X			
La ivière du Levant, Guadeloupe	[74]	X		X					X
Le Grand Chalon, Bourgogne-Franche-Comté	[75,76]			X		X			X
Niortais, Nouvelle-Aquitaine	[77,78]			X		X	X		
Pays de Barr, Grand Est	[79–81]			X		X	X		X

Table 1. *Cont.*

Municipality	Source	Impacts of Climate Change Addressed							
		Coastal Changes	**Severe Storms**	**Extreme Heat**	**Extreme Cold**	**Flooding**	**Drought**	**Wildfires**	**Seasonal Shifts**
Pays Dieppois—Terroir de Caux (PDTC), Normandie	[82–85]	X		X		X			
Pays Voironnais, Auvergne-Rhône-Alpes	[86]			X		X		X	X
Perpignan Méditerranée Métropole, Occitanie	[87]	X		X		X	X		
Saint Omer—CAPSO, Hauts-de-France	[88–90]			X		X	X		
St-Quentin-en-Yvelines—CASQY, Ile de France	[91,92]			X					
Sud-Estuaire, Pays de la Loire	[93–96]					X			
Sundgau, Grand Est	[97–99]					X	X		X
Vallée de Chamonix-Mont-Blanc, Auvergne-Rhône-Alpes	[100,101]								X
Var Esterel Méditerranée, Provence-Alpes-Côte d'Azur	[102,103]	X	X	X		X			

2.2. Conceptualization of Vulnerability

An adaptation plan is a road map to reducing human vulnerability to the current and future impacts of climate change. Adaptations seek to adjust human–environmental systems in response to actual or expected climatic stimuli to minimize their harm or exploit beneficial opportunities [104]. Therefore, adaptation planning always starts with an assessment of existing and projected vulnerabilities to climate impacts. The need to assess vulnerability to climate change is based on the acknowledgment that actual losses caused by hazard events such as storms, floods, or droughts are not solely a result of climate change but also determined by societal and economic preconditions that shape the way in which people are prepared for or respond to such events [105]. How vulnerability is defined and assessed largely shapes the agenda and priorities of adaptation planning and provides an essential baseline for measurable goals. It also determines the content of information and tools necessary to establish present and future climate adaptation goals. The discourse about vulnerability within climate change adaptation and climate risk scholarly literature encompasses various interpretations of the concept of vulnerability. Since the 2012 IPCC SREX report [106] and within the newer conceptualization of climate risks in the IPCC Assessment Reports Five [104] and Six [107], there is an emerging consensus that vulnerability is better framed as a starting point rather than an outcome. Approaches that conceptualize vulnerability as an outcome often include hazard information and therefore do not sufficiently differentiate between vulnerability and risk [105]. In the pre-SREX conceptual framework, vulnerability was considered as "a function of the character, magnitude, and rate of climate change and variation to which a system is exposed, its sensitivity, and its adaptive capacity" [108]; while under the newer framework, sensitivity and adaptive capacity are considered internal aspects of vulnerability, as opposed to exposure, conceptualized as an external factor [104,109]. This shift reflects the reconceptualization of vulnerability as a socioeconomic variable. In practice, both frameworks have been operationalized by scholars and agencies have continued to follow the IPCC AR4 definition of vulnerability [31,110]. In place-based community-scale assessments [111–113] vulnerability is most commonly conceptualized as a composite variable defined by both biophysical and socioeconomic factors of exposure, sensitivity, and adaptive capacity as a combination of geographical, demographic, and socioeconomic indicators [114].

In the absence of national or international climate adaptation and vulnerability assessment standards, it is inevitable for different municipalities to adopt diverse approaches to their vulnerability assessments. These different conceptual frameworks are summarized in Table 2. In theory, vulnerability assessment is meant to be objective to provide a reliable baseline for adaptation planning. In practice, however, vulnerability assessments are highly subjective because they depend on the philosophies and value orientations of organizations and stakeholders who conduct them [31]. Therefore, we use here the term "perceived vulnerability", commonly used in social and clinical psychology [115,116] as a measure of subjective perception of vulnerability by groups of the population. In health behavior theories, perceived vulnerability reflects a belief about the likelihood of a health threat's occurrence or the likelihood of developing a health problem or being exposed to infections or natural disasters [116,117]. We find this concept highly relevant for describing the collective beliefs of communities about the likelihood of being vulnerable to climate change.

Table 2. Information used in climate adaptation plans.

Area of Interest	Information Used in Climate Adaptation Plans
(a) **Conceptualization and assessment of vulnerability**	
As a synonym of exposure (omit sensitivity and adaptive capacity)	Climate change trends, climate change scenarios, risk analysis
As a combination of exposure, sensitivity, and adaptative capacity (pre-IPCC-SREX)	Climate change trends, climate change scenarios, risk analysis, demographic, health, and socio-economic data
As a combination of sensitivity and adaptive capacity to projected climate risks (post-IPCC-SREX)	Demographic, socio-economic, health statistics and risk analysis based on climate change trends and scenarios
As a combination of exposure and sensitivity (omit adaptive capacity)	Climate change trends, climate change scenarios, risk analysis, demographic data
As a combination of exposure and adaptive capacity (omit sensitivity)	Climate change trends, climate change scenarios, risk analysis, socio-economic data
(b) **Consideration of climate justice in adaptation goals related to:**	
Green and blue infrastructure	Climate, ecological and environmental data
Housing	Housing inventory and plans
Energy security	Energy access, cost, and future projections
Public transportation	Transportation networks and plans
Utilities	Utilities infrastructure and plans
Emergency services	Emergency infrastructure and plans
Food security	Food access, safety, and security data and projections
Water quality	Water quality data and scenarios
Air quality	Air quality data and scenarios
Community education	Information about education attainment and community education resources
Insurance	Insurance access data and scenarios
Community health	Health statistics trends, data about access to health care, and projections
(c) **Groups of stakeholders involved in data co-production and planning**	
Local citizens	Stories, survey, and focus group input, art, traditional knowledge, citizen science
Environmental and climate advocacy groups	Environmental and climate data, case studies, stories, non-scientific articles, blogs
Social justice advocacy groups	Environmental and climate data, case studies, stories, non-scientific articles, blogs
Local government officials	Policy connection, litigation, public mobilization, public funding
City planners	Urban, land-use, environmental spatial data, case studies, ordinances, litigation
Members of state or federal/national agencies	Guidelines, toolkits, case studies, science/policy connection, public funding, training materials
Academic institutions	Guidelines, toolkits, scholarly literature, spatial data, scenarios, public lectures
Local businesses	Surveys and focus group input, private funding
External consulting firms	Climate, geoscience, and environmental data, risk analysis, impact scenarios

2.3. *Conceptualization of Equity and Inclusion*

Consideration of climate justice is fundamental to reducing human vulnerability and providing adaptation benefits for all residents and neighborhoods. Climate justice can have distributive and procedural forms [118], where the former relates to the distribution of adverse impacts of climate change and the latter to how and by whom adaptation planning decisions should be made [21]. In climate adaptation planning, equity and justice imply planning strategies to eliminate disparities and create physical and social environments that aim to ensure a fairer distribution of community resources along race, class, gender, and other dimensions of diversity [119]. Municipalities that examine their vulnerability beyond biophysical climate impacts and consider the demographic, social, and economic characteristics of their populations appear to be more likely to develop specific measures focusing on vulnerable groups [31,120]. To identify cities' information needs for equitable planning, we consider twelve (12) climate adaptation domains frequently addressed in climate adaptation plans: *green and blue infrastructure, housing, energy security, public transportation, utilities, emergency services, food security, water quality, air quality, community education, insurance,* and *community health* (Table 2).

There is also growing consensus that transparent, actionable, and equitable adaptation planning requires inclusivity [121], engagement of diverse stakeholders, especially vulnerable groups, and integration of scientific and community knowledge [122], including traditional and indigenous knowledge in the process of climate service co-production [123,124]. In this study, we examine the participation of nine (9) types of stakeholders, directly and indirectly, in the co-production of information used in vulnerability assessment and the co-development of local climate adaptation plans. These are *local citizens, environmental and climate advocacy groups, social justice advocacy groups, elected officials, planners, members of state, federal/national agencies, academic institutions, local businesses,* and *external consulting firms* (Table 2).

2.4. Data Analysis

Each climate adaptation plan, including its bibliographic sources and metadata, was screened for information about the content and sources of methodologies and data used in vulnerability assessment and formulation of adaptation goals. The qualitative assessment includes three components driven by the following questions:

- How is the concept of human vulnerability defined and what information is used to assess it?
- How climate justice is addressed in climate adaptation goals across various sectors, and what information is used to formulate the goals?
- What groups of stakeholders are involved in the co-production of information used in vulnerability assessment and the co-development of local climate adaptation plans?

3. Results and Discussion

Table 3 provides a summary of our findings about vulnerability assessment, used as a basis for adaptation planning, consideration of justice, and the participation of stakeholders and co-production of information and climate adaptation plans.

Table 3. Areas of interest and information addressed in climate adaptation plans.

Area of Interest	U.S. Plans	French Plans
1. **Conceptualization and assessment of vulnerability**		
As a synonym of exposure (omit sensitivity and adaptive capacity)	5 (22%)	6 (35%)
As a combination of exposure, sensitivity, and adaptive capacity(pre-IPCC-SREX)	9 (38%)	3 (18%)
As a combination of sensitivity and adaptive capacity to projected climate risks (post-IPCC-SREX)	2 (9%)	0 (0%)
As a combination of exposure and sensitivity (omit adaptive capacity)	2 (9%)	0 (0%)
As a combination of exposure and adaptive capacity (omit sensitivity)	5 (22%)	8 (47%)
2. **Consideration of justice in climate adaptation goals related to:**		
Green and blue infrastructure	11 (48%)	4 (24%)
Housing	8 (35%)	14 (82%)
Energy security	6 (20%)	10 (59%)
Public transportation	8 (35%)	6 (35%)
Utilities	5 (22%)	2 (12%)
Emergency services	12 (52%)	7 (41%)
Food security	4 (17%)	13 (76%)

Table 3. *Cont.*

Area of Interest	U.S. Plans	French Plans
Water quality	4 (17%)	5 (29%)
Air quality	3 (13%)	1 (6%)
Community education	11 (48%)	9 (53%)
Insurance	2 (9%)	0 (0%)
Community health	7 (30%)	7 (41%)
3. **Groups of stakeholders involved in data co-production and planning**		
Local citizens	18 (78%)	9 (53%)
Environmental and climate advocacy groups	16 (70%)	8 (47%)
Social justice advocacy groups	6 (26%)	2 (12%)
Local government officials	22 (96%)	16 (94%)
City planners	21 (91%)	17(100%)
Members of state or federal/national agencies	12 (52%)	8 (47%)
Academic institutions	15 (65%)	3 (18%)
Local businesses	14 (61%)	13 (76%)
External consulting firms	13 (57%)	5 (29%)

3.1. Assessment of Human Vulnerability

All municipalities examined in this study conducted their vulnerability assessments, either prior to or as a part of their climate adaptation process. However, using different guidelines from various sources based on different schools of thought, they define and interpret vulnerability in a variety of ways.

Figure 1a,b illustrate how the definition of vulnerability chosen by municipalities can pre-determine their focus on different dimensions of vulnerability and, consequently, different types of information used as a basis for their adaptation strategies. Out of 23 U.S. municipalities (Figure 1a), only two recently revised plans (9% of the sample) followed the post-SREX IPCC framework differentiating between social vulnerability (sensitivity and adaptive capacity) and external hazard exposure. Nine plans (38%) adopted the pre-SREX IPCC definition combining metrics of exposure, sensitivity, and adaptive capacity. The older conceptual framework appears to be by far the most popular in climate adaptation guidelines and municipal plans. Likewise, in the scholarly literature on adaptation planning, the newer IPCC framework was not quite as well accepted, and a vast majority of research articles published after SREX and the IPCC Fifth Assessment Report adopted the earlier conceptualization [110,125].

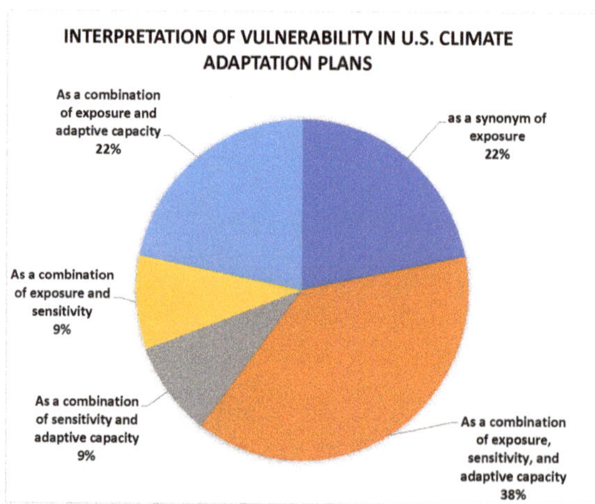

(a)

Figure 1. *Cont.*

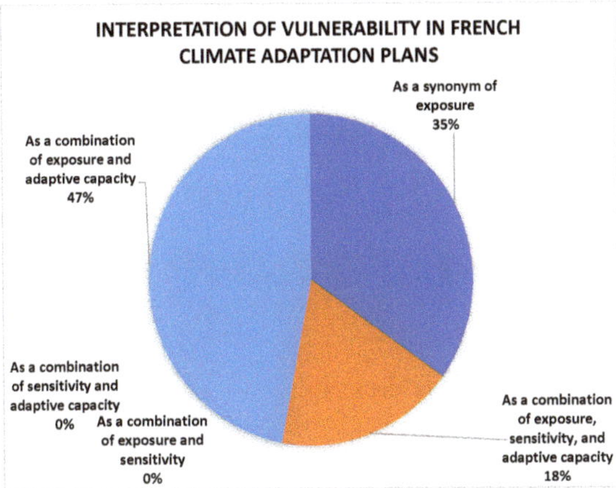

(b)

Figure 1. Interpretation of vulnerability in (**a**) U.S. climate adaptation plans, and (**b**) French climate adaptation plans.

Interestingly, five more U.S. plans (22%) refer to the older IPCC framework in their methodologies, but in practice, address only exposure and adaptive capacity metrics and entirely omit sensitivity variables (such as age, gender, race, disability status, and wellness). In addition, two U.S. plans (9%), which refer to the same definition, address only exposure and sensitivity metrics and omit adaptive capacity. Finally, five remaining U.S. plans (22%) omit social and economic factors altogether, assessing vulnerability as *exposure* to various biophysical climate-change-related hazards. Four of these five were among the very first climate adaptation plans in the country, developed in the 2000s, reflecting the interpretation of this concept in the scholarly literature prior to the Third Assessment Report of the IPCC [126], but one of these plans was published in 2017.

French climate adaptation plans follow more uniform national guidelines and adopt only three versions of vulnerability assessment frameworks (Figure 1b). Almost half of them (47%) interpret vulnerability as a combination of biophysical factors of exposure to climate impacts and economic factors of adaptive capacity. While IPCC reports are routinely cited in plans' introductions, none of them follow the post-SREX IPCC framework and only three French plans (18%) adopted the pre-SREX IPCC framework. Six French plans (37%) equate vulnerability with exposure. Although the term *sensibilité* is frequently used in all plans, which can be literally translated into English as *sensitivity*, it is understood and assessed solely as biophysical exposure. For example, the "sensitivity" of a city's population to flooding risk is discussed and assessed based on precipitation scenarios rather than a differentiated analysis of population demographics as might be expected in the English-language climate adaptation literature.

3.2. *Consideration of Justice in Climate Adaptation Goals*

Adaptation plans must be equitable and fairly protect all residents, especially the most vulnerable groups. However, it is apparent that many adaptation plans do not set justice-centered priorities (Figure 2). Equitable access to emergency services and community climate education come up as the top activities addressed in both countries, yet only about half of all plans set such goals. French plans are more frequently concerned with equity in housing (82%), food security (76%), and energy security (59%); whereas 48% of U.S. plans set objectives for more equitable access to green infrastructure and ecosystem services.

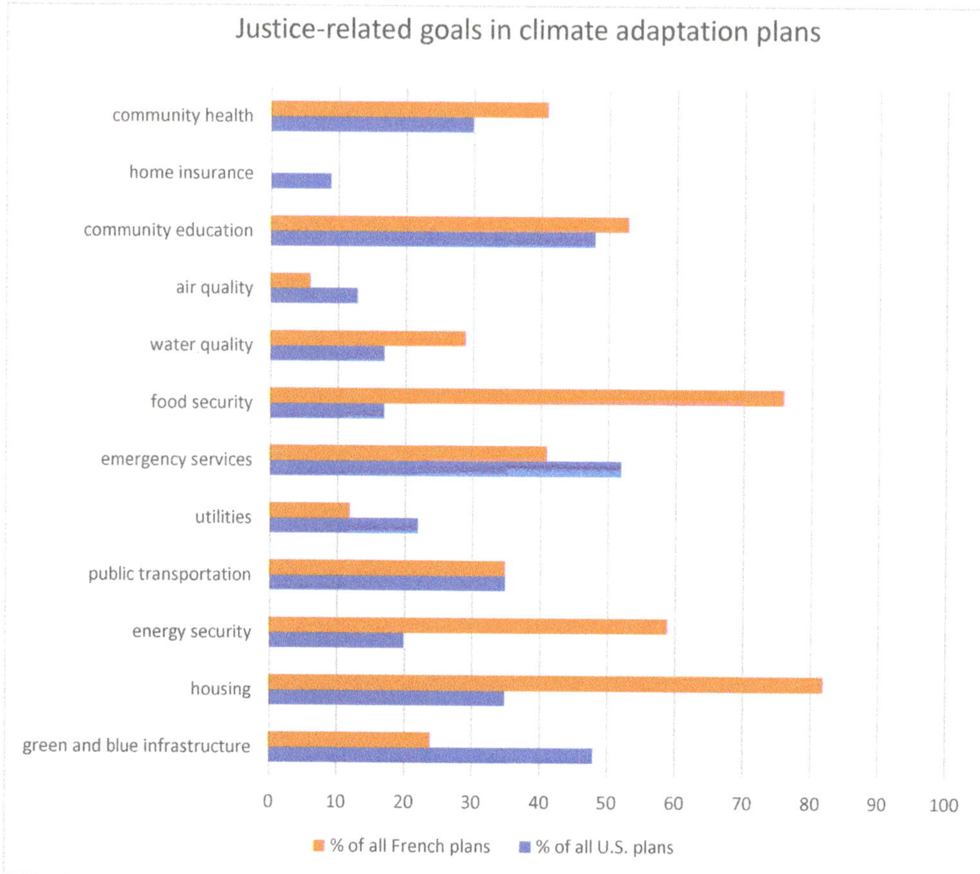

Figure 2. Justice-related goals in various domains of climate adaptation plans in the U.S. and French plans.

To monitor the implementation of these and other climate adaptation objectives, planners need accurate local data and tools to assess patterns of existing vulnerabilities, develop plausible scenarios, and formulate equitable adaptation strategies. For example, goals for equitable access to emergency response services related to weather extremes and hazards appear in 52% of U.S. and 41% of French plans. Yet, in both countries, it is apparent that factors such as race and ethnicity matter when it comes to the provision of governmental assistance [25,127]. "Years after Katrina, it is clear that the slow and incompetent emergency response was a disaster that overshadowed the deadly storm itself, and while Katrina brought governmental racial injustice to the forefront, this disparity has been affecting African American communities long before the storm. For decades, African Americans and other people of color have borne disproportionate environmental burdens—from pollution and poorly maintained neighborhoods to unsafe drug testing and lead poisoning—and for decades government regulators have largely ignored these injustices" [24]. To address these injustices block-by-block and neighborhood-by-neighborhood, climate adaptation strategies must rely on accurate information and be driven by fair planning policies [128,129].

Although 52% and 48% of French and U.S. plans, respectively, formulate adaptation goals related to inclusive community education and access to climate change information, they rarely contain specific metrics which could help track their implementation. For example, such goals may include communication of climate data in more diverse and accessible formats, such as community workshops, flyers, and brochures translated into

Spanish and other languages of predominant immigrant communities (in the U.S.), climate festivals, informal education and citizens science projects, and other community education programs and events.

The majority (76%) of French plans address justice in their food security adaptation goals, focusing on support of local agriculture, especially sustainably grown and organic, and local food sourcing for school cafeterias and pre-schools. Such an approach is multifaceted and pragmatic—to reduce carbon emissions from agriculture and food transportation, to support local agricultural markets, and provide children with nutritious, sustainably grown food.

Justice-centered adaptation strategies related to housing focus on energy efficiency and affordability of residential heating and cooling (in the U.S.) for low-income households and energy conservation with more efficient building materials, insulation, and sustainable design. Bridging climate change adaptation, community resilience, and GHG mitigation goals, 82% of French plans and 35% of American plans set specific goals for the housing sector aiming to reduce the share of energy expenditure in household budgets and improve energy conservation. Fifty-nine percent (59%) of French and 30% of U.S. plans also mention specific energy security measures, such as for example, the development of community solar projects and local microgrids. There is some inevitable overlap between equity-related objectives in utilities infrastructure and housing sectors, causing double counting of adaptation measures in these domains.

Institutional studies about social justice in the housing sector are also linked to the cost of public transportation, particularly in urban areas [130]. However, very few plans state objectives for free or otherwise subsidized transportation to improve mobility options for their less well-off populations. Adaptation objectives calling for equitable access to green and blue infrastructure and ecosystem benefits appear in 48% and 24% of U.S. and French plans, respectively. Examples of such strategies in the U.S. plans include urban afforestation and wildfire management measures, flood risk management through the river valley and coastal restoration; green infrastructure development, such as green roofs, green walls, rain gardens, and bioswales, collectively known as Nature-Based Solutions (NBS). In French climate plans, ecosystem-based adaptation strategies are mostly limited to the preservation of or creation of green spaces in urban areas. Nevertheless, the new National Strategy for Climate Adaptation Planning in France has prioritized nature-based climate solutions. As explored in detail by Pathak and others [131] (this issue), the implementation and monitoring of NBS require local-scale ecological data (such as soils, hydrology, microclimate, indigenous, endangered, and culturally significant species), integrated with climate services and tools.

Only 41% of French and 30% of American plans in our sample set goals related to climate adaptation measures supporting community health, such as extreme weather preparedness, extreme heat preparedness, and prevention of water-borne and vector-borne infections. Clearly, adequate planning tools, data integration, and collaboration between local health departments and planners are urgently needed to address the impacts of climate change on community health. Insufficient attention to public health in municipal climate adaptation planning has been reported in other studies. For example, the recent analysis of climate adaptation plans of 22 large cities in 14 countries, including 16 cities in high-income countries [132] indicated that even "highly health-adaptive large cities report fairly modest public health engagement in climate adaptation plans, and very few seem to have integrated a health perspective across thematic or sectoral climate adaptation priorities" (p.14).

Air quality is a key determinant of community health and is directly linked to temperature changes. Yet only 13% of U.S. and 6% of French plans set any justice-focused targets for air quality. Numerous studies indicate that racial and ethnic minorities and low-income people both in the United States [133] and France [127] are being disproportionally exposed to higher levels of air pollution. Ozone- and fine particle-related mortalities are expected to increase due to climate change, especially affecting vulnerable populations [134]. One of the key challenges for equitable planning is the lack of readily available large-scale monitoring data raising public awareness about glaring spatial correlations between environmental pollution, health, income, and race. Climate services need to be designed to uncover these existing spatial relationships between climate vulnerability and institutional racism, which continue to be rooted in unfair practices in urban planning. However, government regulatory agencies, such as the United States Environmental Protection Agency (EPA), the European Environment Agency (EEA), and the French Central Laboratory for Air Quality Monitoring (LCSQA), operate air quality monitoring networks of fixed monitoring stations that focus on assessing background levels in relatively large regions, grossly neglecting variabilities at a higher spatial resolution. Air pollution can be as much as eight times higher at one end of a city block than the other, according to the Environmental Defense Fund [135]. Local action requires local-scale data, integrating micro-level community-operated air monitoring networks, such as, for instance, Just Air Solutions, who, in partnership with the University of Michigan, is working directly with low-income communities in Detroit and Grand Rapids, MI on neighborhood-scale mapping, monitoring, and data visualization using ground sensors and GIS [136]. Another example of monitoring spatial inequalities in air quality at a high spatial resolution includes mapping projects by Institut Ecocitoyen Pour La Connaissance de Pollution [137] based in Fos-sur-Mer in France, monitoring communities exposed to air, water, and ecosystem pollution associated with industrial zones [138].

Similarly, only a handful of plans in our sample adopt a justice-related lens in addressing the vulnerability of their water resources. Adaptation goals targeting water shortage and water quality are typically generalized for the entire municipality. Although water supply in both countries is generally considered well-managed and safe, it presents problems associated with inequality in the distribution of water resources across different regions and unhealthy drinking water quality, which are likely to be exacerbated by climate change. Water quality problems are more likely in smaller, minority, and low-income communities that are socially, economically, and politically disempowered [139]. The recent drought episodes in France have prompted the government to develop guidelines for water prioritization [140], such as irrigation, swimming pools, and others, which raise many questions about equity, for example, irrigation of private golf courses at the expense of public green spaces in underprivileged communities [141]. Planning decisions based on transparent data would also require improved mapping and monitoring systems integrating water quality and allocation data.

As risks to hazards caused by the effects of climate change continue to increase, the current approaches to spreading financial responsibility need to be re-evaluated. Equitable access to home insurance appears to be the least represented sector in our sample of plans. Public–private insurance programs, however, could play an important role in managing the cost of adaptation and hazard mitigation measures. This would also require more sophisticated climate services for insurance companies to anticipate how their market will evolve in response to climate change, and specifically to provide risk modeling expertise, capital market solutions, actuarial services, and reinsurance design [142]. The U.S. National Flood Insurance Program (NFIP) managed by FEMA and delivered to the public by a network of more than 50 insurance companies and the NFIP Direct [143] plays an important role in reducing climate-related losses. Some increasingly important strategies used by the NFIP include mandatory flood insurance, insurance rate subsidization, and public–private cooperation to prevent the withdrawal of private insurers from high-risk areas.

3.3. *Stakeholders' Role in Knowledge Development*

While professional city planners and government officials lead local climate adaptation planning in both countries, many other groups participate in various stages of climate adaptation planning and co-creation of relevant information, methodologies, and tools (Figure 3). Municipalities in the U.S. appear to involve broader coalitions of stakeholders with local citizens (78%), environmental and climate advocacy groups (70%), academic institutions (65%), local businesses (61%), and private consulting firms (57%) being the most prominent participants. French plans more frequently involve local businesses (76%), followed by local citizens (53%), and environmental and climate advocacy groups (47%). The degree of stakeholder participation varies from attending community workshops and responding to local surveys to active engagement in data collection, community-based research, and other forms of direct and indirect contribution to adaptation plans and, increasingly, co-production and dissemination of information.

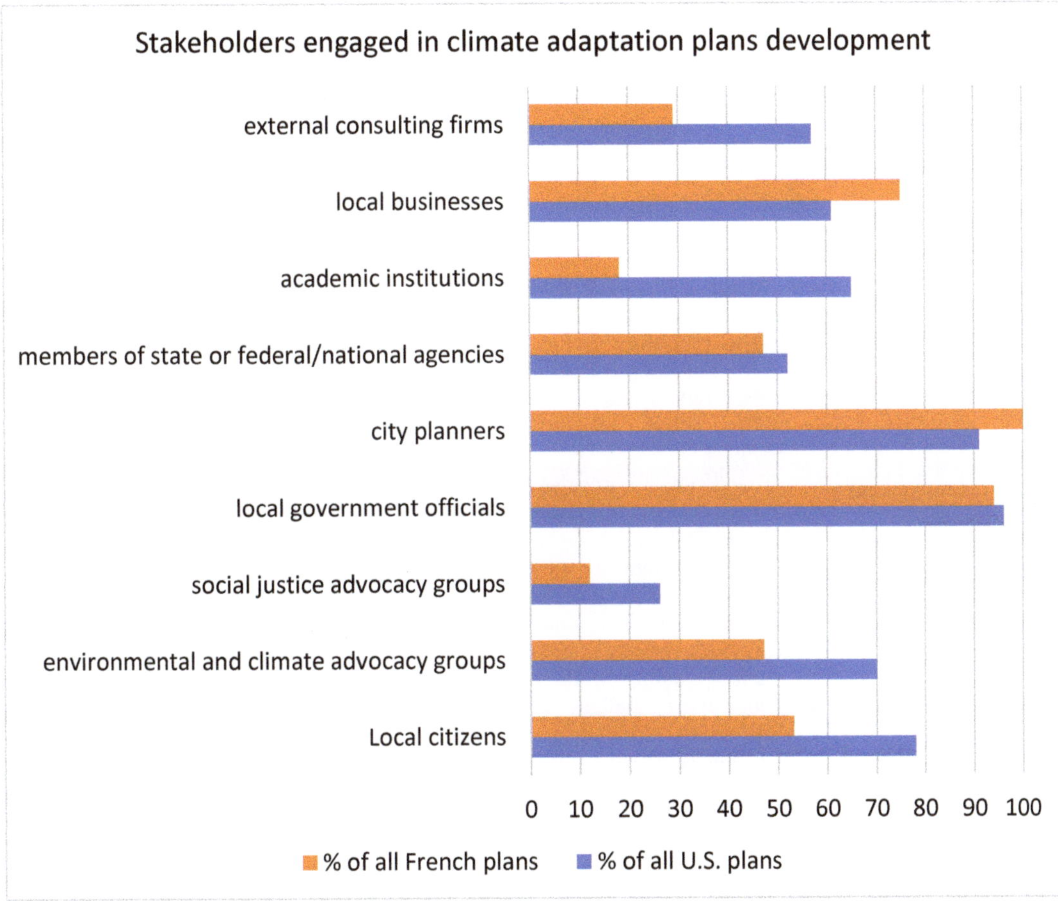

Figure 3. Categories of stakeholders involved in development of climate adaptation plans in the U.S. and France.

Although most climate adaptation planning methodologies recommend that municipalities engage community members in their vulnerability assessments and climate adaptation planning, opportunities for meaningful engagement of local citizens and especially vulnerable groups are quite low. One possible reason for this may be the lack of inclusive user-friendly collaborative engines tailored to non-expert participants, connecting local communities with relevant climate services and tools. A promising example of such a platform outside of our study areas is the Climate Just platform in the UK [144], connecting users and producers through high-resolution mapping of community vulnerability to climate change. Involvement of broad coalitions of various groups of stakeholders including citizens, schools, universities, environmental organizations, and private firms in the co-creation and analysis of knowledge is possibly the only realistic way to bridge the gap between the national and state-scale data providers and the city-scale and neighborhood-scale data needs. In this context, citizen science and collaborative crowdsourcing platforms have great potential for data collection, dissemination, and social participation [11,145]. By being involved in local citizen science projects, people value their role as being a part of the solution and become active contributors to climate services. Studies in the United States, France, and other countries suggest that data co-production not only can provide local-scale information for early warning and climate adaptation planning but also build community trust and support for climate policies [122,146].

Our analysis has several limitations. Although rigorous and formal, it is nevertheless based on a small sample of 40 cities and is meant to provide examples of information produced and used in climate adaptation planning. The results should not be extrapolated to generalize patterns and trends of climate adaptation planning. The list of municipalities used as case studies in our study is not exhaustive and is meant to provide insights from the two different national models. It should be noted that a priori French plans in our sample are more representative than their U.S. counterparts. Climate plans, including both climate change mitigation and adaptation components, are now required in France for all municipalities with populations of more than 20,000 people. There are now 307 urban and rural municipalities of various sizes in France that are following this national requirement, representing about one-quarter of French municipalities, with most of them being relatively large cities [33]. Even though many of the U.S. plans have been enabled and supported by federal policies [147], they are voluntary, driven by local and state circumstances and initiatives, and are less typical for the entire nation.

4. Conclusions

The planning, implementation, and monitoring of climate adaptation strategies rely on a broad range of constantly evolving multidisciplinary spatial data, generated at various scales. While traditional climate services provide useful background information for generalized long-term climate preparedness, they still offer minimal, if any, social, economic, and environmental data, typically being limited to climate data trends and scenarios.

Municipalities face numerous challenges in developing relevant methodologies, keeping up with scholarly literature, and obtaining adequate information for their climate adaptation planning efforts, which may result in the low quality of plans and mediocre implementation. Small municipalities have especially limited technological, human, and financial capacity. In France, municipalities receive significant support from the national agency overseeing local climate adaptation planning—ADEME, while in the U.S., many climate adaptation plans of small cities have been developed in partnership with local university partners through various grants. Despite these major differences, we have identified several major challenges hindering effective local climate adaptation planning in both countries and possibly worldwide.

Methodological challenges. Although numerous methodological resources for local governments have evolved during the past ten years, including brochures, toolkits, and clearinghouses featuring examples of existing adaptation plans, sorting through them in search of clear guidelines could be an insurmountable task of its own. In the absence of national and international standards for vulnerability assessment, municipalities adopt diverse methodological frameworks, definitions, and protocols, or skip the assessment altogether. Such conceptual fragmentation presents a major challenge for long-term monitoring, comparison, and data sharing among the cities. In many ways, such methodological ambiguity mirrors the continuous rift between adaptation planning and risk assessment communities in the scholarly literature [105,109,125]. The re-conceptualization of "vulnerability", introduced in the IPCC SREX and the Working Group Two Fifth Assessment Report has not been well received and provoked a split in the scientific community [109]. The most recent IPCC Sixth Assessment Report [107] further uses the concept of risk of the potential adverse impacts of, and response options to, climate change, treating exposure as a precondition rather than a dimension of vulnerability. Many vulnerability researchers, however, argue that treating exposure as a precondition of vulnerability or completely disassociating biophysical contexts from vulnerability limits the analysis of differential vulnerability caused by differences in biophysical components associated with geographic location, which can influence both the sensitivity and adaptive capacity of a system [125]. Most institutional guidelines including the ADEME methodologies used in France are based on the over twenty-year-old framework of the IPCC Third Assessment Report [126]. The simplicity and applicability of this framework made it popular with climate adaptation practitioners.

Ideological challenges. In the absence of methodological requirements to formulate adaptation objectives targeting climate justice, municipalities rarely do so. Many sectors of adaptation planning, such as community health, transportation, air quality, water quality, and many others are systematically overlooked in both countries. Even disaster emergency planning, where a focus on equity comes most frequently as a top priority, is absent in 48% of the U.S. and 69% of French plans. Equity in housing and food security adaptation is grossly overlooked in U.S. plans. Equitable access to green infrastructure and ecosystem services is mentioned in less than one-quarter of French plans and only 48% of U.S. plans. Even when such objectives are formulated, implementation strategies are often vague, lacking quantitative metrics for monitoring and evaluation. Further research is necessary to understand if these shortcomings are caused by local political ideologies, outdated methodologies, lack of adequate data, lack of involvement of vulnerable stakeholders, or all the above.

Data quality challenges. Adaptation planning and implementation monitoring require acquisition, analyses, and timely interpretation of high-quality multi-disciplinary data of relevant spatial and temporal resolutions, integrated into user-friendly formats, understandable for planners and the public. This includes not only macro-, meso-, and micro-climatological data but also agroecological, hydrological, demographic, cultural, economic, community health, zoning, land use, and other information. While many interesting high-quality products have been developed by academic and private data providers, they are not typically integrated with each other, are often hard to locate, and are rarely directly accessible to local planning departments, especially in small municipalities. Local air pollution, water quality, soil contamination, food security, community health, socio-economic, and demographic data, necessary to reveal their spatial correlations, are rarely available at the neighborhood and census-block scale.

Community connection challenges. Collection, analysis, and timely interpretation of relevant information require active community participation, especially at the scale of municipalities. National agencies and large for-profit data providers are unlikely to be able to fulfill these needs. Adaptation planning requires information, which is constantly evolving, relevant, local, transparent, open-access, and collected at the block or even household scale. We need active, truly diverse, and inclusive networks of local stakeholders engaging schools, universities, private and public organizations, community groups, and volunteers in the co-production of data, including local stories and indigenous knowledge, to inform collective co-construction of climate adaptation strategies.

We draw several recommendations for climate adaptation researchers and decision makers:

(a) Municipalities need flexible, user-friendly, and reliable tools for comprehensive vulnerability assessment, mapping, and monitoring, informed by the up-to-date body of knowledge and best practices around the world, and relevant to their geographical context. Many currently existing products are based on outdated literature and offer rigid step-by-step guidelines, rather than interactive analytical tools. Cities need the best common standards, which are currently lacking, but not necessarily common data sources or guidelines.

(b) Centralized approaches to data monitoring for climate adaptation planning often fail to provide information at relevant temporal and spatial scales. Produced by different agencies and groups of experts, these databases are often hard to integrate and downscale. Decentralized interdisciplinary monitoring networks equipped with digital applications allowing local citizens to engage in knowledge production may offer promising alternatives.

(c) Climate-adaptation design tools for local governments should prioritize climate justice in all adaptation contexts and sectors. GIS-based online mapping tools and mobile applications are very helpful in visualization and analysis of spatial correlations between income, race, environmental justice issues, and various dimensions of vulnerability to climate impacts, helping to inform difficult conversations about resource allocations in climate adaptation planning.

(d) Long-term funding programs are necessary to provide financial and other resources and incentives for stakeholders' collaboration and community engagement in local knowledge co-production. Funding agencies should prioritize active local and regional partnerships involving academic institutions, schools, advocacy groups, local businesses, and especially citizens and organizations representing the most vulnerable communities. Funding programs that prioritize mainstreaming climate adaptations into neighborhood revitalization, food-security, community wellness, environmental education, and citizen-science projects should be designed to support long-term partnerships among all local actors.

(e) Climate education networks, local working groups, and other boundary organizations connecting experts and non-experts would play an increasingly important role in merging community-based education, scientific research, climate action, and co-design of digital technologies, tools, and data for local climate adaptation planning.

Author Contributions: Conceptualization of this study has been developed by E.L.; methodology, formal analysis, investigation, resources, data curation, and draft writing by E.L. and C.d.C.; final review and editing and data visualization has been completed by E.L. All authors have read and agreed to the published version of the manuscript.

Funding: This research received no external funding.

Institutional Review Board Statement: Not applicable.

Informed Consent Statement: Not applicable.

Data Availability Statement: Not applicable.

Acknowledgments: We would like to thank the Guest Editors of this special issue and three anonymous reviewers for their thoughtful comments, thorough proofreading, and numerous edits that have allowed us to improve the quality of this final manuscript. We are also grateful to Grand Valley State University Open Access Publication Fund and the CEARC Center (Cultures, Environment, Arctic, Representations, Climate) for covering the open access publication fee for this article.

Conflicts of Interest: The authors declare no conflict of interest.

References

1. Cavelier, R.; Borel, C.; Charreyron, V.; Chaussade, M.; Le Cozannet, G.; Morin, D.; Ritti, D. Conditions for a market uptake of climate services for adaptation in France. *Clim. Serv.* **2017**, *363*, 4–40. [CrossRef]
2. Hewitt, C.; Mason, S.; Walland, D. The Global Framework for Climate Services. *Nat. Clim. Chang.* **2012**, *2*, 831–832. [CrossRef]
3. Vaughan, C.; Dessai, S. Climate services for society: Origins, institutional arrangements, and design elements for an evaluation framework. *WIREs Clim. Chang.* **2014**, *5*, 587–603. [CrossRef] [PubMed]
4. Panenko, A.; George, E.; Lutoff, C. Towards the development of climate adaptation knowledge-action systems in the European Union: An institutional approach to climate service analysis. *Clim. Serv.* **2021**, *24*, 100265. [CrossRef]
5. United States Environmental Protection Agency. Climate Adaptation. Available online: https://www.epa.gov/climate-adaptation (accessed on 19 June 2022).
6. European Commission and the European Environment Agency. Climate-ADAPT. Available online: https://climate-adapt.eea.europa.eu/ (accessed on 19 June 2022).
7. California Energy Commission, State of California. Cal-Adapt. 2021. Available online: https://cal-adapt.org/ (accessed on 5 July 2022).
8. NOAA Climate Program Office. The Adaptation Science Program. 2020. Available online: https://cpo.noaa.gov/Divisions-Programs/Climate-and-Societal-Interactions/The-Adaptation-Sciences-Program (accessed on 19 June 2022).
9. Webber, S. Putting climate services in contexts: Advancing multi-disciplinary understandings: Introduction to the special issue. *Clim. Chang.* **2019**, *157*, 1–8. [CrossRef]

10. Steynor, A.; Lee, J.; Davison, A. Transdisciplinary co-production of climate services: A focus on process. *Soc. Dyn.* **2020**, *46*, 414–433. [CrossRef]
11. Neset, T.-S.; Wilk, J.; Cruz, S.; Graca, M.; Rod, J.K.; Maarse, M.J.; Wallin, M.J.; Andersson, L. Co-designing a citizen science climate service. *Clim. Serv.* **2021**, *24*, 100273. [CrossRef]
12. WMO. What Are Climate Services. 2022. Available online: https://gfcs.wmo.int/what-are-climate-services (accessed on 5 July 2022).
13. Copernicus Earth Observation Programme. Copernicus Climate Change Services. 2022. Available online: https://climate.copernicus.eu/ (accessed on 15 December 2022).
14. U.S. Climate Resilience Toolkit. Meet the Challenges of a Changing Climate. 2021. Available online: https://toolkit.climate.gov/ (accessed on 5 July 2022).
15. Eco-Adapt. About the Climate Adaptation Knowledge Exchange. 2022. Available online: https://www.cakex.org/ (accessed on 13 July 2022).
16. Conservation Biology Institute. Adapt West—The Product: A Bi-National Database and Synthesis. 2022. Available online: https://adaptwest.databasin.org/ (accessed on 5 July 2022).
17. Great Lakes Integrated Sciences and Assessments. The Great Lakes Adaptation Data Suite (GLADS). Available online: https://glisa.umich.edu/ (accessed on 5 July 2022).
18. AcclimaTerra. AcclimaTerra. 2022. Available online: https://www.acclimaterra.fr/en/ (accessed on 9 September 2022).
19. AcclimaTerra. *Regional Scientific Committee, Anticipating Climate Change in Nouvelle-Aquitaine to Guide Policy at Locla Level. Edition AcclimaTerra*; Impremerie LaPlante: Bordeaux, France, 2020.
20. Shi, L. Promise and paradox of metropolitan regional climate adaptation. *Environ. Sci. Policy* **2019**, *92*, 262–274. [CrossRef]
21. Thomas, D.S.; Twyman, C. Equity and justice in climate change adaptation amongst natural-resource-dependent societies. *Glob. Environ. Chang.* **2005**, *15*, 115–124. [CrossRef]
22. Michigan Department of Environment, Great Lakes, and Energy. MiEJscreen: Environmental Justice Screening Tool (DRAFT). Available online: https://www.michigan.gov/egle/maps-data/miejscreen (accessed on 5 July 2022).
23. United States Environmental Protection Agency. EJScreen: Environmental Justice Screening and Mapping Tool. 2022. Available online: https://www.epa.gov/ejscreen. (accessed on 5 July 2022).
24. Bullard, R.D.; Wright, B. *The Wrong Complexion for Protection. How the Government Response to Disaster Endangers African American Communities*; New York University Press: New York, NY, USA, 2012; p. 304.
25. Bullard, R.D.; Johnson, G.S.; Torres, A.O. *Environmental Health and Racial Equity in the United States: Building Environmentally Just, Sustainable, and Livable Communities*; American Public Health Association: Washington, DC, USA, 2011.
26. Deldrève, V.; Lewis, N.; Moreau, S.; Reynolds, K. Les nouveaux chantiers de la justice environnementale. *Vertigo* **2019**, *19*, 1–13. [CrossRef]
27. Lioubimtseva, E.; Da Cunha, C. Local climate change adaptation plans in the US and France: Comparison and lessons learned in 2007–2017. *Urban Clim.* **2020**, *31*, 100577. [CrossRef]
28. Kundzewicz, Z.W.; Førland, E.J.; Piniewski, M. Challenges for developing national climate services—Poland and Norway. *Clim. Serv.* **2017**, *8*, 17–25. [CrossRef]
29. Baklanov, A.; Cárdenas, B.; Lee, T.-C.; Leroyer, S.; Masson, V.; Molina, L.T.; Müller, T.; Ren, C.; Vogel, F.R.; Voogt, J.A. Integrated urban services: Experience from four cities on different continents. *Urban Clim.* **2020**, *32*, 100610. [CrossRef] [PubMed]
30. Vaughan, C.; Buja, L.; Kruczkiewicz, A.; Goddard, L. Identifying research priorities to advance climate services. *Clim. Serv.* **2016**, *4*, 65–74. [CrossRef]
31. Lioubimtseva, E. The role of inclusion in climate vulnerability assessment and equitable adaptation goals in small American municipalities. *Discov. Sustain.* **2022**, *3*, 3. [CrossRef]
32. da Cunha, C.; Lioubimtseva, E. Metadata for climate change adaptation plans of small and mid-size French and American cities. *Data Brief* **2021**, *36*, 106981. [CrossRef] [PubMed]
33. ADEME—Agence de la Transition Ecologique. Observatoire Territoires et Climat. 2022. Available online: https://www.territoires-climat.ademe.fr/observatoire (accessed on 1 December 2022).
34. Bierbaum, R.; Smith, J.; Lee, A.; Blair, M.; Carter, L.; Chapin, F.; Fleming, P.; Ruffo, S.; Stults, M.; McNeeley, S.; et al. A comprehensive review of climate adaptation in the United States: More than before, but less than needed. *Mitig. Adapt. Strateg. Glob. Chang.* **2013**, *18*, 361–406. [CrossRef]
35. New York State Department of State. *Albany Climate Change Vulnerability Assessment and Adaptation Plan*; Mayor's Office of Energy and Sustainability: Albany, NY, USA, 2013.
36. Klipp, J.; Thaler, T.; Griffith, G.; Crossett, T. *Climate Adaptation and Mitigation Plan for Alger County, Michigan. Preparing Communities, Forests and Water Resources for a Changing Climate*; Model Forest Policy Program: Sagle, ID, USA, 2011.
37. Stratus Consulting. *Boulder County Climate Change*; Boulder County: Boulder, CO, USA, 2012.
38. City of Chula Vista. *Climate Adaptation Strategies Final Implementation Plans*; City of Chula Vista: Chula Vista, CA, USA, 2011.
39. Town of Corte Madera. *Town of Corte Madera Climate Adaptation Assessment: A Roadmap to Resilience*; Town of Corte Madera: Corte Madera, CA, USA, 2021.
40. City of Flagstaff. *City of Flagstaff Resiliency and Preparedness Study*; City of Flagstaff: Flafstaff, AZ, USA, 2012.
41. City of Flagstaff. *Flagstaff Climate Action and Adaptation Plan*; City of Flagstaff: Flagstaff, AZ, USA, 2018.

42. Georgetown Conservation Commission. *Climate Change Adaptation Report: Georgetown, Maine. A Special Publication by the Georgetown Conservation Commission*; Georgetown Conservation Commission: Georgetown, ME, USA, 2015.
43. Stults, M.; Pagach, J. *Preparing for Climate Change in Groton, Connecticut: A Model Process for Communities in the Northeast*; ICLEI-Local Governments for Sustainability and Connecticut Department of Environmental Protection: Groton, CT, USA, 2011.
44. City of Iowa City. *Iowa City Climate Action and Adaptation Plan*; City of Iowa City: Iowa City, IA, USA, 2018.
45. Chen, Z.; Delp, M.; Felber, O.; Ferrier, U.; Freund, H.; Grant, A.; Harris, J.; Loring, N.; Neel, H.; Ribikawskis, M.; et al. *Iowa City Climate Action and Adaptation Implementation Strategies*; The University of Iowa Office of Outreach and Engagement: Iowa, IA, USA, 2019.
46. City of Keene. *Adapting to Climate Change: Planning a Climate Resilient Community*; City of Keene & ICLEI Local Governments for Sustainability: Keene, NH, USA, 2007.
47. City of Keene. *Keene, NH Climate Adaptation Action Plan Final Report*; City of Keene & ICLEI Local Governments for Sustainability: Keene, NH, USA, 2010.
48. City of Laguna Woods City Council. *City of Laguna Woods Climate Adaptation Plan*; City of Laguna Woods: Laguna Woods, CA, USA, 2014.
49. Michigan State University. *Adapting to Climate Change and Variability. Marquette, Michigan*; Michigan State University, City of Marquette, Superior Watershed Partnership and Land Trust: Marquette, MI, USA, 2013.
50. King, H.; Thaler, T.; Griffith, G.; Crossett, T.; Rasker, R. *Forest and Water Climate Adaptation: A Plan for Marquette County, Michigan*; Model Forest Policy Program: Sagle, ID, USA, 2013.
51. Chase, J.H.; Cooper, J.G.; Fitzgerald, R.E.; Lima, F.A.; Miller, S.R.; Pignatelli, T.M. *Climate Change Adaptation Chapter: Marshfield, Massachusetts*; ScholarWorks@UMass Amherst: Amherst, MA, USA, 2012.
52. Rhode Island Sea Grant & University of Rhode Island Coastal Resources Center. *Adaptation to Natural Hazards and Climate Change in North Kingstown, Rhode Island*; Rhode Island Statewide Planning Program: Providence, RI, USA, 2015.
53. Beever, J., III; Gray, W.; Trescott, D.; Cobb, D.; Utley, J.; Hutchinson, D. *City of Punta Gorda Adaptation Plan*; Southwest Florida Regional Planning Council, Charlotte Harbor National Estuary Program: Fort Myers, FL, USA, 2009.
54. Taylor Engineering, I. *City of Punta Gorda Adaptation Plan Update*; City of Punta Gorda: Punta Gorda, FL, USA, 2019.
55. City of Salem Department of Planning & Community Development. *Ready for Tomorrow: The City of Salem Climate Change Vulnerability Assessment and Adaptation Plan*; City of Salem: Salem, MA, USA, 2014.
56. City of Santa Cruz. *City of Santa Cruz Climate Adaptation Plan*; City of Santa Cruz: Santa Cruz, CA, USA, 2011.
57. City of Santa Cruz Climate Action Program. *City of Santa Cruz 2018 Climate Adaptation Plan Update*; City of Santa Cruz: Santa Cruz, CA, USA, 2018.
58. City of Sarasota. *City of Sarasota Climate Change Vulnerability and Adaptation Plan*; Technical Memo; City of Sarasota: Sarasota, FL, USA, 2016.
59. City of Sarasota. *City of Sarasota Climate Adaptation Plan*; Final Report; City of Sarasota: Sarasota, FL, USA, 2017.
60. Foster, M.; Williams, R.; Thaler, T.; Griffith, G. *Forest and Water Climate Adaptation: A Plan for Taos County, New Mexico*; Model Forest Policy Program: Sagle, ID, USA, 2010.
61. Evans, J.; Gambill, J.; McDowell, R. *Tybee Island Sea Level Rise Adaptation Plan. Final Report*; National Sea Grant College Program: Tybee Island, GA, USA, 2013.
62. Tompkins County Planning Department. *Tompkins County Comprehensive Plan. Planning for Our Future: Adaptation Chapter*; Tompkins County: Ithaca, NY, USA, 2015.
63. Harris & Associates. *City of Watsonville Climate Action and Adaptation Plan*; City of Watsonville Public Works & Utilities: Watsonville, CA, USA, 2021.
64. Brest Métropole. *Diagnostic de Vulnérabilité au Changement Climatique de Brest Métropole*; BMO: Brest, France, 2019.
65. Brest Métropole. *Brest Métropole—Plan climat 2019–2025—Programme D'actions*; BMO: Brest, France, 2020.
66. Clermont Auvergne Métropole, & Ville de Clermont-Ferrand. *Schéma de Transition Energétique et Ecologique de Clermont Auvergne Métropole et de la Ville de Clermont-Ferrand. Diagnostic Air, Énergie, Climat, Écologie*; Ville de Clermont-Ferrand: Clermont-Ferrand, France, 2017.
67. Clermont Auvergne Métropole, & Ville de Clermont-Ferrand. *Schéma de Transition Energétique et Ecologique de Clermont Auvergne Métropole et de la Ville de Clermont-Ferrand. Bilan de la Concertation*; Ville de Clermont-Ferrand: Clermont-Ferrand, France, 2018.
68. Clermont Auvergne Métropole; Ville de Clermont-Ferrand. *Schéma de Transition Energétique et Ecologique de Clermont Auvergne Métropole et de la Ville de Clermont-Ferrand. Fiche Actions*; Ville de Clermont-Ferrand: Clermont-Ferrand, France, 2019.
69. Clermont Auvergne Métropole, & Ville de Clermont-Ferrand. *Schéma de Transition Energétique et Ecologique de Clermont Auvergne Métropole et de la Ville de Clermont-Ferrand. Plan d'Actions*; Ville de Clermont-Ferrand: Clermont-Ferrand, France, 2019.
70. Pôle Territorial de l'Albigeois et des Bastides. *Concertation—Stratégie—Plans d'Actions*; Pôle Territorial de l'Albigeois et des Bastides: Albi, France, 2019.
71. Pôle Territorial de l'Albigeois et des Bastides. *Diagnostics Climat Air Energie*; Pôle Territorial de l'Albigeois et des Bastide: Albi, France, 2019.
72. Pôle Territorial de l'Albigeois et des Bastides. *Tableau de Bord des Actions PCAET de la Communauté de Communes du Cordais et du Causse*; Pôle Territorial de l'Albigeois et des Bastides: Albi, France, 2019; Available online: http://4c81.fr/wp-content/uploads/2019/07/TDB_ACTIONS_PCAET_4C_vf.pdf (accessed on 22 July 2022).

73. Golfe du Morbihan Vannes Agglomération. *Plan Climat Air Energie Territorial—Diagnostic, Stratégie et Plan D'actions*; Golfe du Morbihan Vannes Agglomération: Vannes, France, 2020.
74. Communaute d'Agglomeration la Riviera du Levant. *Plan Climat Air Energie Territorial (Pcaet) de la Communaute d'Agglomeration la Riviera du Levant*; Communaute d'Agglomeration la Riviera du Levant: Le Gosier, France, 2019.
75. Le Grand Chalon. *Diagnostic/Stratégie*; Le Grand Chalon: Chalon-sur-Saone, France, 2018.
76. Le Grand Chalon. *Programme d'Actions*; Le Grand Chalon: Chalon-sur-Saone, France, 2018.
77. Agglomération du Niortais. *Diagnostic Territorial Climat Air Energie*; Agglomération du Niortais: Niort, France, 2018.
78. Agglomération du Niortais. *Plan Climat Air Énergie Territorial (PCAET) 2018–2024*; Agglomération du Niortais: Niort, France, 2019.
79. Communauté de Communes du Pays de Barr. *Plan Climat Air Energie Territorial. Diagnostic Territorial—Cahier 1: Analyse de la Vulnérabilité Climatique*; Communauté de Communes du Pays de Barr: Barr, France, 2019.
80. Communauté de Communes du Pays de Barr. *Plan Climat Air Energie Territorial. Plan d'Actions*; Communauté de Communes du Pays de Barr: Barr, France, 2019.
81. Communauté de Communes du Pays de Barr. *Plan Climat Air Energie Territorial. Stratégie—Cahier 3: Articulation des Objectifs*; Communauté de Communes du Pays de Barr: Barr, France, 2019.
82. Dieppe Pays Normand. *Plan Climat Air Energie Territorial*; Dieppe Pays Normand: Dieppe, France, 2019.
83. Dieppe Pays Normand. *Plan Climat Air Energie Territorial. Plans d'Actions*; Dieppe Pays Normand: Dieppe, France, 2019.
84. Dieppe Pays Normand. *Plan Climat Air Energie Territorial. Stratégie Territorial*; Dieppe Pays Normand: Dieppe, France, 2019.
85. Dieppe Pays Normand. *Une Démarche Participative*; Dieppe Pays Normand: Dieppe, France, 2019.
86. Communauté du Pays Voironnais. *Plan Climat Air Energie Territorial 2019–2025*; Communauté du Pays Voironnais: Voiron, France, 2019.
87. Perpignan Méditerranée Métropole. *Programme ODD—Plan Climat-Air-Energie Territorial—2018–2024*; Perpignan Méditerranée Métropole: Perpignan, France, 2019.
88. CAPSO. *Plan Climat Air Energie Territorial (PCAET) de la Communauté d'Agglomération du Pays de Saint-Omer (CAPSO)—2020–2026*; Communauté d'Agglomération du Pays de Saint-Omer: Saint-Omer, France, 2020.
89. CAPSO. *Plan Climat Air Energie Territorial (PCAET)—2020–2026. Diagnostic Territorial*; Communauté d'Agglomération du Pays de Saint-Omer: Saint-Omer, France, 2020.
90. CAPSO. *Plan Climat Air Energie Territorial (PCAET)—2020–2026. Plan d'Actions. 98*; Communauté d'Agglomération du Pays de Saint-Omer: Saint-Omer, France, 2020.
91. CASQY. *Diagnostic du Plan Climat Air Energie de Saint-Quentin-en-Yvelines*; Communauté d'Aglomération de Saint-Quentin-en-Yvelines: Saint-Quentin-en-Yvelines, France, 2018.
92. CASQY. *Elaboration du Plan Climat Air Energie Territorial de Saint-Quentin-en-Yvelines*; Communauté d'Aglomération de Saint-Quentin-en-Yvelines: Saint-Quentin-en-Yvelines, France, 2020.
93. CCSE. *Plan Climat Air Energie Territorial. Diagnostic, Communauté de Communes Sud Estuaire*; Communauté de Communes Sud Estuaire: Paimbœuf, France, 2019.
94. CCSE. *Plan Climat Air Energie Territorial. Programme d'actions adoptée par Sud Estuaire*; Communauté de Communes Sud Estuaire: Paimbœuf, France, 2019.
95. CCSE. *Plan Climat Air Energie Territorial. Stratégie adoptée par Sud Estuaire*; Communauté de Communes Sud Estuaire: Paimbœuf, France, 2019.
96. CCSE. *Plan Climat Air Energie Territorial. Fiches Actions*; Communauté de Communes Sud Estuaire: Paimbœuf, France, 2020.
97. CCS. *Plan Climat Air Energie Territorial. PARTIE II—Synthèse de la Démarche*; Communauté de Communes Sundgau: Altkirch, France, 2017.
98. CCS. *Plan Climat Air Energie Territorial. PARTIE IV—Diagnostic Climat-Air-Energie*; Communauté de Communes Sundgau: Altkirch, France, 2017.
99. CCS. *Plan Climat Air Energie Territorial. PARTIE V—Stratégie et Programme d'Actions*; Communauté de Communes Sundgau: Altkirch, France, 2017.
100. Vallée de Chamonix-Mont-Blanc. *Plan Climat Energie Territorial de la Vallée de Chamonix-Mont-Blanc. Plan d'Action*; Vallée de Chamonix-Mont-Blanc: Serris, France, 2012.
101. Vallée de Chamonix-Mont-Blanc. *Plan Climat Haute Montagne*; Vallée de Chamonix-Mont-Blanc: Serris, France, 2019.
102. CAVEM. *Plan Climat Air Energie Territorial (PCAET)—Document 1. Diagnostic Territorial*; Communauté d'Agglomération Var Estérel Méditerranée: Saint-Raphaël, France, 2019.
103. CAVEM. *Plan Climat Air Energie Territorial (PCAET)—Document 2: Stratégie Territoriale*; Communauté d'Agglomération Var Estérel Méditerranée: Saint-Raphaël, France, 2019.
104. IPCC WGII. *Climate Change 2014: Impacts, Adaptation, and Vulnerability, Contribution of Working Group II to the Fifth Assessment Report of the Intergovernmental Panel on Climate Change*; IPCC: Geneva, Switzerland, 2014.
105. Birkmann, J.; Jamshed, A.; McMillan, J.M.; Feldmeyer, D.; Totin, E.; Solecki, W.; Zaiton Ibrahim, Z.; Roberts, D.; Bezner Kerr, R.; Poertner, H.-O.; et al. Understanding human vulnerability to climate change: A global perspective on index validation for adaptation planning. *Sci. Total Environ.* **2022**, *803*, 150065. [CrossRef]
106. IPCC. *Managing the Risks of Extreme Events and Disasters to Advance Climate Change Adaptation. A Special Report of Working Groups I and II of the Intergovernmental Panel on Climate Change*; Cambridge University Press: Cambridge, UK, 2012.

107. IPCC. *Climate Change 2022: Impacts, Adaptation, and Vulnerability. Contribution of Working Group II to the Sixth Assessment Report of the Intergovernmental Panel on Climate Change*; Cambridge University Press: Cambridge, UK, 2022.
108. IPCC. *Climate Change 2007: Impacts, Adaptation and Vulnerability. Contribution of Working Group II to the Fourth Assessment Report of the Intergovernmental Panel on Climate Change*; Cambridge University Press: Cambridge, UK, 2007.
109. Sharma, J.; Ravindranath, N. Applying IPCC 2014 framework for hazard-specific vulnerability assessment under climate change. *Environ. Res. Commun.* **2019**, *1*, 051004. [CrossRef]
110. Zuniga-Teran, A.A.; Mussetta, P.C.; Lutz Ley, A.N.; Díaz-Caravantes, R.E.; Gerlak, A.K. Analyzing water policy impacts on vulnerability: Cases across the rural-urban continuum in the arid Americas. *Environ. Dev.* **2021**, *38*, 100552. [CrossRef]
111. Inostroza, L.; Palme, M.; de la Barrera, F. A Heat Vulnerability Index: Spatial Patterns of Exposure, Sensitivity, and Adaptive Capacity of Santiago de Chile. *PLoS ONE* **2016**, *11*, e0162464. [CrossRef]
112. Weis, S.W.M.; Agostini, V.N.; Roth, L.M. Assessing vulnerability: An integrated approach for mapping adaptive capacity, sensitivity, and exposure. *Clim. Chang.* **2016**, *136*, 615–629. [CrossRef]
113. Swami, D.; Parthasarathy, D. Dynamics of exposure, sensitivity, adaptive capacity and agricultural vulnerability at district scale for Maharashtra, India. *Ecol. Indic.* **2021**, *121*, 107206. [CrossRef]
114. Howe, P.D.; Yarnal, B.; Coletti, A.; Wood, N.J. The participatory vulnerability scoping diagram—Deliberative risk ranking for community water systems. *Ann. Assoc. Am. Geogr.* **2013**, *2*, 343–352. [CrossRef]
115. Duncan, L.A.; Schaller, M.; Park, J.H. Perceived vulnerability to disease: Development and validation. *Personal. Individ. Differ.* **2009**, *47*, 541–546. [CrossRef]
116. Weinstein, N.D.; Lyon, J.E.; Rothman, A.J.; Cuite, C.L. Changes in pereceived vulnerability following natural disaster. *J. Soc. Clin. Psychol.* **2000**, *19*, 372–395. [CrossRef]
117. Gerard, M.; Gibbons, F.X.; Bushman, B.J. Relation Between Perceived Vulnerability to HIV and Precautionary Sexual Behavior. *Psychol. Bull.* **1996**, *119*, 390–409. [CrossRef] [PubMed]
118. Bulkeley, H.; Carmin, J.; Castan Broto, V.; Edwards, G.A.; Fuller, S. Climate justice and global cities: Mapping the emerging discourses. *Glob. Environ. Chang.* **2013**, *23*, 914–925. [CrossRef]
119. Northridge, M.; Freeman, L. Urban planning and health equity. *J. Urban Health* **2011**, *88*, 582–597. [CrossRef]
120. Lioubimtseva, E.; da Cunha, C. Community Engagement and Equity in Climate Adaptation Planning: Experience of Small-and Mid-Sized Cities in the United States and in France. In *Justice in Climate Action Planning. Strategies for Sustainability*; Petersen, D.H.B., Ed.; Springer: Berlin/Heidelberg, Germany, 2022; pp. 257–276.
121. Uittenbroek, C.; Mees, H.; Hegger, D.; Driesses, P. The design of public participation: Who participates, when and how? Insights in climate adaptation planning from the Netherlands. *J. Environ. Plan. Manag.* **2019**, *62*, 2529–2547. [CrossRef]
122. Baztan, J.; Vanderlinden, J.-P.; Jaffrès, L.; Jorgensen, B.; Zhu, Z. Facing climate injustices: Community trust-building for climate services through arts and sciences narrative co-production. *Clim. Risk Manag.* **2020**, *30*, 100253. [CrossRef] [PubMed]
123. Petzold, J.; Andrews, N.; Ford, J.D.; Hedemann, C.; Postigo, J.C. Indigenous knowledge on climate change adaptation: A global evidence map of academic literature. *Environ. Res. Lett.* **2020**, *15*, 113007. [CrossRef]
124. da Cunha, C.; Nikulkina, I.; Vanderlinden, J.-P.; Shadrin, V.; Doloisio, N.; Salakhova, D. Adaptive capacity for climate change: Local initiatives and federal planning. The case of Tiksi, Sakha Republic, Russia. *Polar Sci.* **2022**, *31*, 100761. [CrossRef]
125. Ishtique, A.; Estoque, R.; Eakin, H.; Parajuli, J.; Rabby, Y.W. IPCC's current conceptualization of 'vulnerability' needs more clarification for climate change vulnerability assessments. *J. Environ. Manag.* **2022**, *303*, 114246. [CrossRef] [PubMed]
126. IPCC. *Climate Change 2001: Impacts, Adaptation, and Vulnerability. Contribution of Working Group Two to the Third Assessment Report of the IPCC*; Cambridge University Press: Cambridge, UK, 2001.
127. Charles, L.; Emelianoff, C.; Ghorra, C. Les multiples facettes des inégalités écologique. *Développement Durable Territ.* **2020**, *11*, 1–16. [CrossRef]
128. Lambert, M.-L.; Arnaud, A.; Claeys, C. Justice climatique et démocratie environnementale—Les inégalités d'accès au droit des populations vulnérables aux risques littoraux—Quelques éléments de comparaison. *Ertigo—La Rev. Electron. En Sci. De L'environnement* **2022**, *19*, 1–31. [CrossRef]
129. Larson, P.S.; Gronlund, C.; Thompson, L.; Sampson, L.; Washington, R.; Steis Thorsby, J.; Lyon, N.; Miller, C. Recurrent Home Flooding in Detroit, MI 2012–2020: Results of a Household Survey. *Int. J. Environ. Res. Public Health* **2021**, *18*, 7659. [CrossRef]
130. Institut d'aménagement et d'urbanisme de la région d'Île-de-France. *Vulnérabilité Energétique: Les Pavillonnaires Modestes sous Tension*; Institut D'aménagement et D'urbanisme de la Région d'Île-de-France: Paris, France, 2015.
131. Pathak, A.; Hilberg, L.E.; Hansen, L.J.; Stein, B.A. Key Considerations for the Use of Nature-Based Solutions in Climate Services and Adaptation. *Sustainability* **2022**, *14*, 16817. [CrossRef]
132. Sheehan, M.C.; Khudairi, F.; Swaich, G.S.; Hines, W.; Mehta, S. Urban climate-health governance: Charting the role of public health in large global city adaptation plans. *PLoS Clim.* **2022**, *1*, e0000012. [CrossRef]
133. Jbaily, A.; Zhou, X.; Liu, J.; Lee, T.-H.; Kamareddine, L.; Verguet, S.; Dominici, F. Air pollution exposure disparities across US population and income groups. *Nature* **2022**, *601*, 228–233. [CrossRef]
134. Orru, H.; Ebi, K.; Forsberg, B. The Interplay of Climate Change and Air Pollution on Health. *Curr. Environ. Health Rep.* **2017**, *4*, 504–513. [CrossRef] [PubMed]
135. Environmental Defense Fund. *Impact Through Innovation. Annual Report 2018*; EDF: New York, NY, USA, 2018.
136. JustAir. The Data We Need to Craft the Future We Deserve. Available online: https://www.justair.co/ (accessed on 10 December 2022).

137. Institut Ecocitoyen. Institut Ecocitoyen pour la Connaissance des Pollutions. Available online: http://www.institut-ecocitoyen.fr/ (accessed on 3 November 2022).
138. Dron, J.; Chamaret, P.; Marchand, N.; Temime-Roussel, B.; Ravier, S.; Sylvestre, A.; Wortham, H. Variabilité physico-chimique des épisodes de pollution atmosphérique à proximité de la zone industrialo-portuaire de Fos-sur-Mer. *Pollut. Atmosphérique* **2017**, 2268–3798. [CrossRef]
139. Bae, J.; Lynch, M.J. Ethnicity, Poverty, Race, and the Unequal Distribution of US Safe Drinking Water Act Violations, 2016–2018. *Sociol. Quaterly* **2022**. [CrossRef]
140. Ministère de la Transition Ecologique et de la Cohésion des Territoires. Origine et Gestion de la Sécheresse. Available online: https://www.ecologie.gouv.fr/secheresse (accessed on 10 December 2022).
141. d'Allens, G. Arrosage des Golfs: Malgré la Sécheresse, Les Dérogations Pleuvent. *Reporterre*. 8 August 2022. Available online: https://reporterre.net/Arrosage-des-golfs-malgre-la-secheresse-les-derogations-pleuvent (accessed on 19 June 2022).
142. Brasseur, G.B.; Gallardo, L. Climate services: Lessons learned and future prospects. *Earth's Future* **2016**, *4*, 79–89. [CrossRef]
143. The Federal Emergency Management Agency. FEMA. 2022. Available online: https://www.fema.gov/flood-insurance (accessed on 1 December 2022).
144. The University of Machester. Map Tool. Available online: https://www.climatejust.org.uk/map (accessed on 3 November 2022).
145. Albagli, S.; Iwama, A.Y. Citizen science and the right to research: Building local knowledge of climate change impacts. *Humanit. Soc. Sci. Commun.* **2022**, *9*, 39. [CrossRef]
146. See, L. A Review of Citizen Science and Crowdsourcing in Applications of Pluvial Flooding. *Front. Earth Sci.* **2019**, *7*, 44. [CrossRef]
147. Herrick, C.; Vogel, J. Climate Adaptation at the Local Scale: Using Federal Climate Adaptation Policy Regimes to Enhance Climate Services. *Sustainability* **2022**, *14*, 8135. [CrossRef]

sustainability

Article

When Do Climate Services Achieve Societal Impact? Evaluations of Actionable Climate Adaptation Science

Aparna Bamzai-Dodson [1,2,*] and Renee A. McPherson [2,3]

1 U.S. Geological Survey North Central Climate Adaptation Science Center, Boulder, CO 80303, USA
2 Department of Geography and Environmental Sustainability, University of Oklahoma, Norman, OK 73071, USA
3 U.S. Geological Survey South Central Climate Adaptation Science Center, Norman, OK 73071, USA
* Correspondence: abamzai@usgs.gov

Abstract: To cope with complex environmental impacts in a changing climate, researchers are increasingly being asked to produce science that can directly support policy and decision making. To achieve such societal impact, scientists are using climate services to engage directly with stakeholders to better understand their needs and inform knowledge production. However, the wide variety of climate-services outcomes—ranging from establishing collegial relationships with stakeholders to obtaining specific information for inclusion into a pre-existing decision process—do not directly connect to traditional methods of measuring scientific impact (e.g., publication citations, journal impact factor). In this paper, we describe how concepts from the discipline of evaluation can be used to examine the societal impacts of climate services. We also present a case study from climate impacts and adaptation research to test a scalable evaluation approach. Those who conduct research for the purposes of climate services and those who fund applied climate research would benefit from evaluation from the beginning of project development. Doing so will help ensure that the approach, data collection, and data analysis are appropriately conceived and executed.

Keywords: climate change; climate services; adaptation; actionable science; stakeholder engagement; societal impact; evaluation

Citation: Bamzai-Dodson, A.; McPherson, R.A. When Do Climate Services Achieve Societal Impact? Evaluations of Actionable Climate Adaptation Science. *Sustainability* **2022**, *14*, 14026. https://doi.org/10.3390/su142114026

Academic Editors: Charles Herrick, Jason Vogel and Glen Anderson

Received: 26 August 2022
Accepted: 26 October 2022
Published: 28 October 2022

1. Introduction

1.1. Background

The defining characteristic of the past century is the impact of human activities on environmental systems, such as global climate change [1,2], that result in challenging and uncertain policy and decision contexts. To support policy and decision making, scientists are being asked to provide climate services—the provision of timely climate data and information created in a form that is useful, usable, and used (i.e., actionable) [3,4]. To generate such climate services, scientists are interacting "out in the world" with information end-users, known more broadly as stakeholder engagement [5,6]. Engagement of stakeholders in research projects has a demonstrated positive impact on subsequent information use for decision making [7]. However, traditional definitions of research success most often focus on agency or academic metrics, such as number of publications and citation metrics [8,9], and do not capture societal impacts well [10,11].

Defining success for societal impact can be challenging because the needs of stakeholders can vary from learning how to work collaboratively with researchers (collegial engagement) to being generally better informed (conceptual information use) to taking specific on-the-ground action (instrumental information use) [12,13]. Additionally, climate service providers have a wide range of engagement approaches available to them to meet these varying needs—spanning from informing stakeholders of results to empowering them as co-equal project investigators [14]. To accommodate this diverse range of engagement needs and approaches, evaluation processes need to be specifically tailored to examine the impact and actionability of such information to a community [15,16].

In this paper, we introduce concepts from the field of evaluation and describe how they may be used to help define indicators for and evaluate the societal impacts of climate services. We focus on climate impacts and adaptation research, as it is one area where the provision of climate services is growing at a rapid pace. We present results from a case study application of these concepts to research funded by the U.S. Geological Survey Climate Adaptation Science Center network and discuss how these findings can be further developed. The deliberate consideration of success and explicit attention to evaluation can improve the actionability of science.

1.2. Evaluation Theory and Practice

Evaluation helps individuals and organizations learn and improve program operations by testing the effectiveness of or changes in activities; it differs from assessment, which is intended to grade or score performance [17,18]. The field of evaluation uses several different theoretical approaches and methods for operationalization. It has a rich set of literature that differentiates between the advantages and limitations of these approaches and techniques and identifies the appropriate contexts for their use [19–21]. Therefore, no matter the context, it is incumbent upon the evaluator to initially determine the kind of evaluation required and ensure that they draw upon the appropriate best practices when designing the evaluation process.

Figure 1 summarizes a few key concepts from the discipline of evaluation for consideration when designing and conducting an evaluation of climate services. Evaluations benefit from beginning with an appraisal of the following: (1) what it is that specifically needs to be evaluated (the evaluand), (2) what aspects of the evaluand (process, outputs, or outcomes) are most appropriate for evaluation, and (3) when (summatively or formatively) and at what organizational scale (program or project) the evaluation will be conducted [22,23]. As part of this appraisal, the evaluator identifies the purpose of the evaluation, inputs to the activity, and other contextual factors, such as the level of analysis or precision [22,23]. Once the approach and method are identified, the evaluator selects suitable variables for measurement and analysis. These variables cover necessary aspects of the evaluand that are to be evaluated, while also being scientifically sound (e.g., measured reliably, scaled appropriately) [22,23]. If the approach is quantitative or mixed-methods, then the evaluator also ensures that statistical assumptions and analyses are logically sound and allow sufficient statistical power.

These approaches intersect in layered ways when operationalized, and an evaluator can make intentional selections among them to meet the goals of the evaluation. For example, to enhance the investment of public funding for climate services, an evaluator may elect to conduct a formative evaluation of processes at the program level. This approach would help the funding program iteratively improve funding opportunities, proposal reviews, and project management to increase alignment with the overall goal of use of information for policy and decision making. In contrast, to improve their understanding of the operational practices necessary for successful delivery of climate services and pitfalls to avoid, an evaluator may instead elect to conduct a summative evaluation of project outputs and outcomes. Traditionally, evaluation of climate impacts and adaptation research has occurred in an ad hoc summative manner that has not been robustly informed by evaluation theory and practice, although resources are emerging to help climate service providers bridge this gap [8].

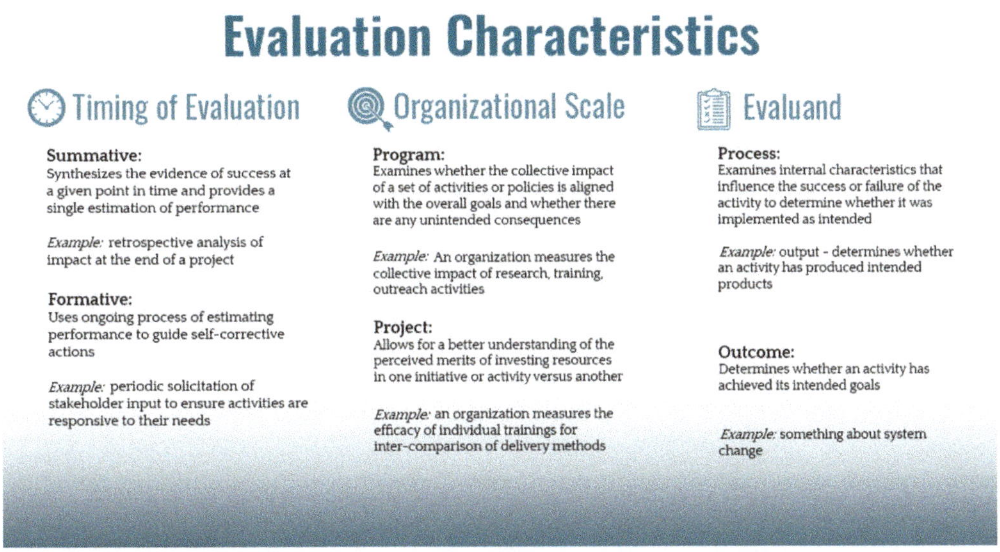

Figure 1. Evaluation key concepts relevant to what specific aspects of an activity require evaluation, when the evaluation will occur, and at what organizational scale it will take place [22,23].

1.3. Success and Evaluation for Climate Services

Here, we present some applications of evaluation to better understand the societal impact of climate services. Several traditional models and mechanisms are available for gathering quantitative measures of scientific impact, including research inputs such as the amount of funding obtained and research outputs such as the number of publications, their associated journal impact factor, and number of citations [24] or the number of downloads of products from websites [25]. However, none of these measures identifies whether or how the stakeholder used the information to make a decision because knowledge delivery does not equal knowledge use [11]. To evaluate climate services, an evaluator can focus on stakeholder perception of the process (e.g., workshop evaluations) or, more important to actionability, how well stakeholder input increases the usability. Even better, the evaluation can examine the actual use of the research outputs.

In practice, evaluating information usability and use by policy and decision makers is notoriously difficult. Wall et al. [26] provide an initial direction for evaluating the societal impacts of climate services, such as if agencies and managers find the science credible and if the findings are explicitly applied in agency planning, resource allocation, or a policy decision. McNie [27] suggests other options, such as evaluating whether "all relevant information was considered" or "whether the science was understood and interpreted correctly". Quantifying these impact metrics is difficult, but options include conducting follow-up interviews with decision makers engaged in projects [28] and analyzing the language in plans and decisions [29].

More nuanced approaches to incorporating perspectives from stakeholders require deeper engagement and focus primarily on understanding how the stakeholder experienced or perceived the engagement. These approaches can include examining factors such as the following: (1) the time required to build the relationship, (2) an understanding of how the project might influence the person or their community, and (3) the nature of the interactions between scientists and users, including building trust [30,31]. Data collection may include surveys (particularly those using open-ended questions that allow people to describe what they experienced or why they hold a certain view) or semi-structured interviews. The iterative nature of some stakeholder engagement in climate services means formative evaluation is possible through using longitudinal evaluation designs. For example, the same survey can be administered multiple times during the development of a decision support tool to ensure that updates to the tool enhance usability [32].

In situations where stakeholder engagement yields neither scientific nor societal impact, success may be defined in more intangible ways, including the depth of integration of stakeholders into the investigator team and their satisfaction with the process [26]. Here, methods for evaluation can focus on identifying and monitoring measurable outcomes on intermediary time scales. For example, a project team can design a conceptual logic model that captures stakeholder impact as a long-term outcome and identifies how to measure change at interim checkpoints [33]. Or the team can apply a theory of change-based framework where establishing and maintaining relationships are key social learning outcomes for an entire community of practice [34]. Regardless of the approach selected, thinking strategically about evaluation from the front-end of a project ensures that appropriate information is collected throughout to monitor whether goals are being achieved and take corrective actions as needed.

2. Case Study

In this paper, we share case study data and results to demonstrate how a climate services boundary organization with the goal of funding the production of actionable science examined its projects to understand their societal impact. This work is part of a broader evaluation of climate impacts and adaptation research projects funded by the U.S. Geological Survey (USGS) South Central and North Central Climate Adaptation Science Centers (CASCs), two regional centers within a nationwide network (Figure 2). Our thoughts on the strengths and weaknesses of the selected evaluation methods and results are included in the Discussion to aid other climate services organizations in their evaluation planning efforts.

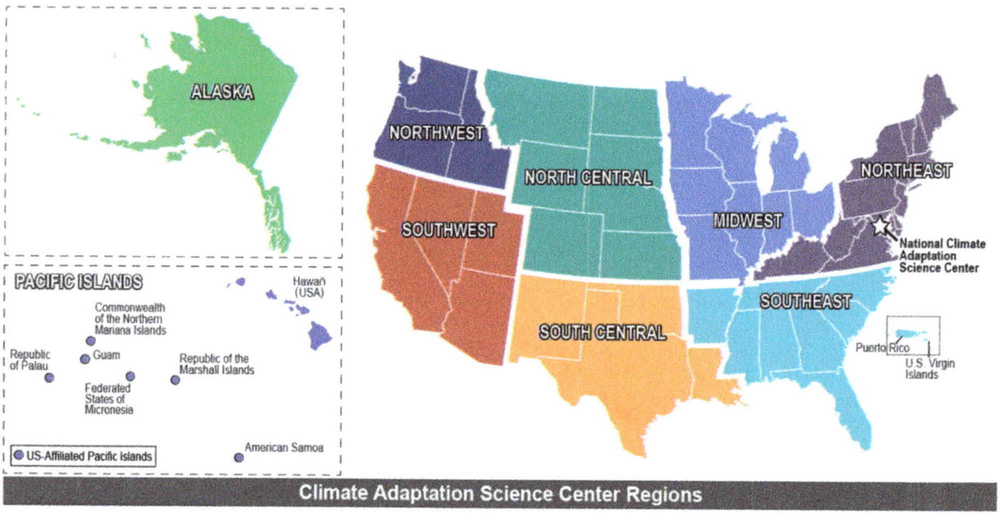

Figure 2. Map of the National and Regional Climate Adaptation Science Centers (CASCs). This analysis focuses on projects funded by the North Central and South Central CASCs.

The CASC network was established by the U.S. Department of the Interior to "provide climate change impact data and analysis geared to the needs of fish and wildlife managers as they develop adaptation strategies in response to climate change" [35]. To achieve this mission, CASC project solicitations are intended to fund research that creates products and tools that directly support resource managers in their development and implementation of climate adaptation plans and actions. Although funded projects are usually research activities of two to three years in length, this emphasis on research use means that they also result in the provision of climate services and can generate partnerships that last beyond the length of an individual project. Examples of climate services activities from prior funded projects include (1) researchers and Tribal water managers working together to better understand micro-drought onset conditions to inform drought adaptation planning, (2) scientific synthesis of information on future fire regimes delivered to managers via training, and (3) the implementation by researchers of small-scale adaptation demonstration projects to illustrate the retention of water on the landscape to resource managers.

From 2013 to 2016, the CASC network was guided by the Federal Advisory Committee on Climate Change and Natural Resource Science, which produced a report providing recommendations on how to improve operations [36]. A key recommendation in this report was for USGS to develop an evaluation process to ensure that programmatic activities and funded projects align with the mission [36]. Suggested evaluation categories include "relevance, quality, processes, accessibility, and impact of science products and services", although no framework or method for implementing this evaluation process was provided [36]. USGS headquarters conducts annual internal and five-year external program-level reviews of the regional centers to examine overall operations and impact [37] but does not pursue project-level evaluation. As a result, regional CASCs are developing and piloting their own supplemental project evaluation processes.

The broader evaluation of South Central and North Central CASC projects included an analysis of project documentation, a survey of stakeholders engaged in the projects, and a focused set of interviews with highly engaged stakeholders. This paper focuses on the survey, which was intended to provide a summative project-level evaluation of process, outputs and outcomes, and broader impacts based on the perspectives of stakeholders. This approach was chosen because formative evaluation was not a consideration in the development of the funding program. Furthermore, enough time had elapsed that multiple years of projects had reached completion. Our expectation was that evaluation of the entire suite of projects by the program office would provide us with sufficient data to compare characteristics between dissimilar types of projects (e.g., projects carried out at local scales in comparison to projects to create data at broad regional scales). We used an electronic survey of project stakeholders because it was a no-cost option; no resources other than limited staff capacity were dedicated to this evaluation effort. These limitations are commonplace in federal science programs, making this case a suitable proxy for conditions faced by other funders of climate services.

3. Methods

We contacted the primary investigators for 28 South Central CASC projects and 16 North Central CASC projects to identify the stakeholders whom they engaged during the project, resulting in a total of 186 unique contacts for the South Central CASC and 188 unique contacts for the North Central CASC. All contacts were invited via email from the research team to complete the survey, the protocol for which is publicly available from Bamzai-Dodson et al. [38] and the design for which is based on published indicators of usable science [26]. Institutional Review Board (IRB) approval was obtained via The University of Oklahoma (IRB number 7457). Paperwork Reduction Act approval was obtained from the U.S. Office of Management and Budget (control number 1090-0011).

The survey was divided into four sections: process, outputs and outcomes, impacts, and demographics. The survey protocol was pre-tested by 20 staff from across the nationwide CASC network, and their feedback was incorporated into the final form. Six questions asked respondents about the process of creating new knowledge together among investigators, resource managers, and decision makers, focusing on the nature and timing of interactions. Nine questions asked respondents about perceptions of the products developed through this project, including factors that promoted or limited their use by the individual or their agency. Six questions asked respondents about their partnership with the investigators, including what made it likely or unlikely for them to work together again. Four questions asked respondents for demographic information, such as the geography, sector, and professional role that they worked in. Questions were a mix of multiple choice, Likert scale, open-ended, and matrix table, based on accepted practices for effective survey design [39,40].

Survey dissemination and collection of responses was carried out electronically using Qualtrics [41], with a release date of 7 December 2018 and a 90-day dissemination window. Data collection was hampered due to the U.S. federal government shutdown from 22 December 2018 to 25 January 2019. Federal contacts were re-invited on 1 July 2019 to take the survey during a second 90-day dissemination window, but response rates remained low. Table 1 provides the response rate information per region, and Table 2 summarizes the demographics of respondents. All survey questions were optional to complete, so the total responses per question does not always equal the total number of complete responses (49). A public summary of the survey results is published in Bamzai-Dodson et al. [42].

Table 1. Survey response rates for each Climate Adaptation Science Center (CASC) region.

	South Central CASC	North Central CASC
Responses solicited	186	188
Completed responses	24 (12.9 percent)	25 (13.3 percent)

Table 2. Respondent demographics for both CASC regions by organization type and organizational role. "Other" self-identified as part of a "federally supported partnership".

	Local, State, Federal, or Tribal Agency	University or College	Non-Governmental Organization (NGO) or Private	Other
Resource manager/decision maker/planner	12	0	8	1
Scientist/technician/researcher	2	12	4	0
Equally both	5	0	2	0

4. Results

4.1. Process: Engagement in the Process of Knowledge Production

Questions in this section of the survey were designed to examine the nature and focus of interactions between stakeholders and investigators during the process of knowledge production. Research indicates that when, how, and how often scientists and stakeholders interact with each other during a project can be important factors to the perceived success of the project [26]. More than half of the respondents (57.1 percent) indicated their engagement began prior to proposal development, with an additional 12.2 percent engaged during proposal development. Engagement during a project ranged from never (zero times per year) to at least every week (52 or more times per year), although most respondents (67.4 percent) were engaged between one to eight times per year. No respondents said that the level of interaction was too much; however, 16 percent said that there was too little interaction. These results indicate that early and ongoing interactions were common factors in CASC projects and that even high frequency engagement was not perceived as too much interaction by stakeholders. One respondent described their experience being engaged in a project late and expressed appreciation for the investigators' responsiveness to their input: "The investigative team was slow to involve those of us who were able to provide more local expertise into the design process, however they did exhibit remarkable flexibility in inviting/allowing that input and then adapting their process to better include such material/knowledge".

The phases of a project during which the most stakeholders reported interaction were definition of the problem (87.5 percent), selection of products (85.7 percent), and dissemination of findings (87 percent). No interaction was most often reported by stakeholders during the design of research methods (27.1 percent), the collection of project data (27.1 percent), and the analysis of project data (32.6 percent). Only one respondent (3.85 percent) indicated that a formal needs assessment was done as part of the project, and 10 respondents (38.5 percent) indicated that needs were determined through informal conversation. Eleven respondents (23.4 percent) indicated that a formal risk or vulnerability assessment was conducted, and 18 respondents (38.3 percent) indicated that risk or vulnerability were assessed through informal conversation. These findings indicate that stakeholders are primarily engaging in CASC projects at key decision points related to the context, scoping, and products of a project and not when decisions such as method selection, data collection, and data analysis are made about research design. CASC project teams are also preferentially choosing to use informal approaches when determining the management context of a research project instead of following established formal strategies for assessing needs, risk, or vulnerability (e.g., scenario planning, structured decision making, systems engineering).

The responses to these questions were informative for describing the frequency, timing, and intent of engagement. However, we found that our evaluation and survey design missed identifying who had initiated each stage of engagement, evaluating the perceived quality of interactions at those points, understanding why engagement was lower during design decisions, and whether a lack of engagement at those points was detrimental to project outcomes. One possibility is for funding programs or climate service providers to identify key decision points regarding the formation of research goals and questions and the development and dissemination of products during which the quality and outcomes of the engagement process can be evaluated in an ongoing manner. Such an approach would strengthen the alignment between stakeholder aspirations, priorities, and needs and project goals, outputs, and outcomes.

4.2. Outputs and Outcomes: Production and Use of Outputs

Questions in this section of the survey were designed to determine the types of outputs and knowledge produced by projects and understand how they were used by stakeholders. Research indicates that the number, type, quality, usability, and use of outputs from projects can be important factors to the perceived success of the project [26]. The most common project output reported by respondents was data provision, ranging from disseminating observations (e.g., place-based phenological data) to projections (e.g., climate model data) (Figure 3). Respondents also reported receiving summarized information from investigators, such as two-page overviews of new findings and quarterly newsletters. Notably, some respondents remarked on more subtle relational outcomes such as "many relationships" and "a new world view". One respondent provided the following feedback on the networking opportunities that their project provided: "The most fruitful and beneficial outcomes from this project will be the connections established between collaborators. It is difficult to quantify [the potential outcomes of new relationships] but I think bringing people to the table is, nonetheless, extremely valuable and worth supporting".

All respondents indicated that projects helped them both be better informed broadly about an issue and be better informed specifically about a particular problem. However, stakeholders indicated that projects were not useful to gain a new technical skill (25.8 percent), formulate policy (23.8 percent), and implement adaptation plans (13 percent). Respondents indicated that projects helped them to understand changes in weather and climate observations and model projections and to link those changes to impacts on resources or places that they manage; however, no respondents indicated that projects helped them identify, evaluate, or select potential adaptation strategies to cope with such impacts. These results indicate that although knowledge and outputs produced by these projects were used by stakeholders to inform adaptation planning, they were not used to make specific climate adaptation decisions (although they may have been used in the implementation of other resource management decisions).

Twenty-four respondents indicated that there were specific factors that they felt contributed to their use of project outputs and provided descriptions of these factors in open-ended replies. The most common factor was a strong partnership between the investigator and stakeholder, illustrated as "trust, relationships, open-mindedness on all sides" and "an attention to the relationship, protocol, transparency, and communication". Some respondents described contexts with a very clear management challenge linked to a demonstrated information need, such as a "well defined management need to be explored" and "Federal mandated water settlement legislation". Respondents also mentioned several different ways in which investigators were able to make broad results relevant to their specific management challenge. Examples include the creation of "fine spatial resolution climate products" and the provision of "alternatives to traditional drought indices".

Figure 3. Word cloud generated from 40 open-ended responses to the question "What kinds of information, data, tools, or other products did this project provide you?".

Thirteen respondents indicated that specific factors limited their use of project results. The most common barriers were a need for additional time to use results (19.2 percent) and resource constraints (15.4 percent). One respondent described how late engagement in a project could act as a barrier to information use: "The one area that could have been improved would have been upfront discussion of delivery mechanisms to achieve broader impacts. The proposal included a component of incorporating results into specific agency products, without talking to the agency manager for all of those products before the proposal was submitted". Respondents also described a need for "continued data collection and processing," especially in places where extreme weather events disrupted data continuity. No respondents indicated an issue with the quality of the science provided by investigators.

These results indicate that while funded projects resulted in conceptual use of outputs (informing) by stakeholders and may have resulted in instrumental use (implementation) for general resource management [13], they fell short of their intended goal of instrumental use for climate adaptation. Stakeholders had confidence in the quality and integrity of scientific outputs and understood their broad relationship to management contexts but lacked time and resources to apply such information to specific climate adaptation decisions, plans, or actions. However, it has been noted that moving from conceptual to instrumental use of information can partly be a factor of the maturity of the project and the relationship between the investigator and stakeholder [43], and thus it is possible that revisiting respondents after additional time has passed may reveal stronger instrumental use of information. To capture long-term use of outputs by stakeholders, funding programs and climate service providers may need to implement evaluation processes that continue on for multiple years after the formal conclusion of a single activity.

4.3. Impacts: Building of Relationships and Trust

Questions in this section of the survey were designed to examine the impacts of participating in a project to the building of relationships and trust between stakeholders and investigators. Research indicates that trust between investigators and stakeholders is foundational to two-way communication and accountability during the project and to sustain further work after the project [26]. Respondents reported positive feelings overall about their engagement in South Central and North Central CASC projects. Respondents felt satisfied with their experiences with the investigator team (93.6 percent) and felt satisfied with their experiences with the project (87.2 percent).

All respondents agreed that investigators were honest, sincere, and trustworthy, and 91.5 percent of respondents agreed that investigators were committed to the engagement process. The same percentage of respondents (91.5 percent) agreed that investigators appreciated and respected what they brought to the project, while 89.4 percent of respondents agreed that the investigators took their opinion seriously during the discussions. Furthermore, all respondents said it was likely that they would use additional results generated by this investigator team. These results indicate that stakeholders still felt goodwill towards investigators as individuals, even when engagement processes and integration of their input into the project might have fallen short of expectations.

Respondents provided a range of reasons that would make it likely for them to work with the investigators again in the future (Figure 4). Many respondents mentioned the nature of their relationship as a team, citing a desire to work with "good people" where the "collaborative spirit and tone of mutual respect is great". In addition to a positive team atmosphere, respondents mentioned the level of expertise of investigators. One project investigator was identified as an "outstanding scientist and human being," with the respondent adding that "[their] humility despite [their] great knowledge and intellect is inspiring". Finally, respondents mentioned the importance of the relevance of findings, such as the "ability to provide useful products" and "good, practical, implementable results that were directly applicable to my agency's goals and strategies".

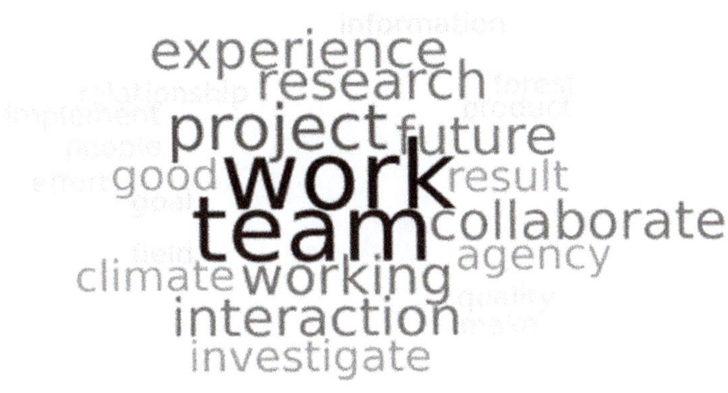

Figure 4. Word cloud generated from 37 open-ended responses to the question "From your perspective, what reason(s) would make it likely for you to work with this investigator team or the CASC again in the future?".

When provided the opportunity to give any other feedback on their experience, several respondents noted their appreciation for the integration into the project of informal knowledge or results. One respondent highlighted the investigators' "willingness to more readily recognize and respond to non-peer reviewed (nascent) local research," and another acknowledged that investigators were willing to implement "a demonstration project" for local stakeholders. A third respondent stated that they valued support for a project "that was not firmly deliverables based" because one of the main outputs was the creation of a collaborative network of individuals.

Results from this section demonstrate the perceived value to stakeholders of building trusted partnerships and communities of practice. In particular, funding programs and climate service providers would benefit from identifying empirical methods for measuring and monitoring trust between the producers and users of climate services, as trust plays a key role in the uptake of information for policy and decision making [30,31]. Beyond trust, formative evaluation during a project could help investigators identify instances where stakeholders may feel that their input is not being appreciated or their opinions are not being taken seriously. This would allow for the institution of corrective actions to improve the flow of communications and provide more responsive climate services.

5. Discussion and Recommendations

In this paper, we summarized a variety of approaches from the discipline of evaluation and described their relevance to defining success and evaluating the societal impacts of climate services within the context of climate impacts and adaptation research. We presented a case study to demonstrate how to operationalize selected approaches from this literature using a survey of stakeholders engaged in projects funded by the South Central and North Central CASCs. Funders of climate services, such as the CASCs, are positioned to influence the form and goals of research across many stages of the process, from setting the priorities that appear in a solicitation to identifying appropriate proposal review criteria to selecting which projects receive funding. Evaluation of and by funders of climate services is critical to understanding whether actions taken across each of these stages and by individual projects support the overall goal of societal impact [44,45].

Because virtually all respondents indicated satisfaction with projects and investigators, our ability to contrast projects and interpret differences among them was limited. Additionally, our case study was limited by the low survey response rate and relatively small sample size, possibly resulting from the immediate and lingering impact of the 2018–2019 U.S. federal government shutdown. As a result, although we were not able to use the collected data the way in which we originally intended, we still were able to examine the characteristics of investigators and projects that stakeholders found satisfactory. Describing these characteristics allowed us to meet our intended program objectives and provided lessons learned from completed projects that can be applied to subsequent similar projects.

Our results corroborated previous studies that have demonstrated that stakeholders prefer being engaged in projects early, often, and consistently [43,46]. Previous research has shown that stakeholders may become fatigued or stressed with interactions that do not result in perceptible changes to the research agenda to prioritize stakeholder benefits [47,48]. Our findings showed that even interacting with investigators more than once a week was perceived as satisfactory and not as too much interaction, opening the possibility that stakeholder fatigue may not be an issue when there is an obvious connection between the reason for the interaction and a benefit to the stakeholder. Almost all stakeholders left these interactions feeling better informed by the knowledge and outputs produced by projects and able to apply such knowledge and outputs to general resource management. However, very few of them were able to directly implement this information into climate adaptation planning or action, with several mentioning a need for additional time to use the results. Even so, stakeholders placed value on participation in these projects due to the relational benefits that they gained, such as growing their professional network and conversing with scientific experts in informal settings. Stakeholders also emphasized investigators' personal collaborative natures such as their ability to demonstrate mutual respect and humility, illustrating the importance of an investigator's willingness to take an "apprentice" role and learn from the decision makers [49].

Importantly, attempting to generate a summative "one size fits all" survey for such a broad set of objectives prevented us from examining the societal impact of individual projects, even if it helped identify characteristics of projects found satisfactory by stakeholders. Although we took care to design a single evaluation process that built on appropriate theory, methods, and survey design, we discovered that each project came with its own unique objective regarding societal impact, which ideally needed an individually tailored evaluand and measures. Surveys such as ours are an increasingly common way for programs to evaluate the societal impact of their activities, but the results can fall short of achieving that goal. Instead, we recommend that future initiatives to examine societal impact for the CASCs, and other climate services funding programs, consider that evaluation for each project be integrated up front into proposal development, such as asking investigators to create a logic model with measurable attributes. Such an approach would help ensure that subsequent project evaluations would then be designed with a specific purpose in mind and could ameliorate the issue of a low response rate.

This additional request for inclusion of evaluation design and implementation, however, can only be met with a matching provision of additional resources from funders. Doing so would allow climate service providers to work with relevant evaluation experts to conduct an initial appraisal and design and implement an evaluation process. Smart [50] suggests seven key questions to consider when planning for evaluation, which we map to concepts useful for answering these questions in Table 3. These questions range from the big picture (why is it needed?) to the practical (who will I collect data from?). When combined, answers to the questions aid in the selection of evaluation approaches that provide meaningful information and guide improvement. Investigators, funders, and evaluators can use these questions as a common starting point when discussing evaluation.

Table 3. Seven key questions to consider when planning an evaluation process, and relevant concepts useful to answering these questions.

Key Question	Relevant Concepts to Consider
1. Why?	Assessment: score performance (grade) Evaluation: test the effectiveness of activities (diagnostic)
2. What do I need to find out?	Program: collective impact of a set of activities Project: one initiative or activity
3. What will I measure?	Process: internal characteristics of activity Output: produced intended products Outcome: achieved intended goals
4. How will I measure it?	Metrics: specific data to be collected Methods: e.g., interviews, surveys, document analysis
5. Who will I collect data from?	Sample: program staff, investigators, stakeholders Ethics: adhere to basic principles that protect study participants
6. When will I collect data?	Summative: single estimation of performance Formative: ongoing process of estimating performance
7. What will I do with the data?	Change organizational activities going forward (relate back to Q. 1) Use to inform long-term goal setting: e.g., Theory of Change, logic model

One unintended benefit of this study was that it fed into the broader conversation across the regional CASCs about whether it was possible to quantitatively measure the societal impacts of research projects that they fund. Since development and dissemination of this survey protocol, the Southeast CASC has carried out additional quantitative and qualitative research from which findings are still emerging. To date, their evaluation initiative has described the differing ways in which individuals and organizations use climate adaptation science [51] and the distinct pathways which projects that aim for societal impact can follow in comparison to projects that aim for high scientific impact [52]. These network-wide conversations are a continued effort to apply concepts from evaluation theory and practice to the challenge of funding and providing climate services.

6. Conclusions

Evaluation is a critical component of understanding the societal impact of the provision of climate services, yet many existing approaches fall short of achieving this goal. We set out to do program-wide evaluation at the project-level by creating a single survey instrument. While analyzing our data, we noted that the diverse array of project objectives meant that the single overarching survey did not contain enough nuance to evaluate individual projects. Instead, each project needed a tailored measurement tool that was developed with its unique objectives in mind. For example, we found that our survey could not capture the differing definitions of success between place-based projects for targeted stakeholders and projects producing large, regional-scale products for many stakeholders. Nor could our survey capture the differences between projects designed to build relationships and trust between people and those designed to provide context for making a specific decision. This study demonstrates the limitations of a program summatively evaluating projects and that embedding evaluation in each project from the start would be beneficial. In particular, funders of science can encourage applicants to proactively consider evaluation during proposal development and provide the resources to bring in relevant and necessary evaluation expertise to an investigator team.

Author Contributions: Conceptualization, A.B.-D. and R.A.M.; Investigation, A.B.-D.; Supervision, R.A.M.; Writing—original draft, A.B.-D.; Writing—review & editing, A.B.-D. and R.A.M. All authors have read and agreed to the published version of the manuscript.

Funding: This work was funded by the U.S. Geological Survey South Central and North Central Climate Adaptation Science Centers. Any use of trade, firm, or product names is for descriptive purposes only and does not imply endorsement by the U.S. Government.

Institutional Review Board Statement: The study was conducted in accordance with U.S. Geological Survey Fundamental Science Practices, the Declaration of Helsinki, and approved by the Institutional Review Board of The University of Oklahoma (IRB number 7457). Paperwork Reduction Act approval was obtained from the U.S. Office of Management and Budget (control number 1090-0011).

Informed Consent Statement: Informed consent was obtained from all subjects involved in the study.

Data Availability Statement: Bamzai-Dodson, A.; Lackett, J.; McPherson, R.A. North Central and South Central Climate Adaptation Science Center Project Evaluation: Survey Data Public Summary; U.S. Geological Survey data release: Reston, VA, USA 2022. https://doi.org/10.5066/P93WPOS5.

Acknowledgments: We thank the members of A.B.-D.'s dissertation committee, the Climate Adaptation Science Center Evaluation Working Group, and the Science of Actionable Knowledge Working Group for insightful discussions around what constitutes successful actionable science for climate adaptation. Special thanks are given to Kate Malpeli and Louise Johansson for their design work on Figure 1.

Conflicts of Interest: The authors declare no conflict of interest.

References

1. Crutzen, P.J. The "Anthropocene". In *Earth System Science in the Anthropocene*; Ehlers, E., Krafft, T., Eds.; Springer: Berlin/Heidelberg, Germany, 2006; pp. 13–18. ISBN 978-3-540-26590-0.
2. National Research Council. *Meeting the Challenge of Climate*; The National Academies Press: Washington, DC, USA, 1982.
3. Dilling, L.; Lemos, M.C. Creating Usable Science: Opportunities and Constraints for Climate Knowledge Use and Their Implications for Science Policy. *Glob. Environ. Change* **2011**, *21*, 680–689. [CrossRef]
4. Brasseur, G.P.; Gallardo, L. Climate Services: Lessons Learned and Future Prospects. *Earth's Future* **2016**, *4*, 79–89. [CrossRef]
5. Reed, M.S.; Vella, S.; Challies, E.; de Vente, J.; Frewer, L.; Hohenwallner-Ries, D.; Huber, T.; Neumann, R.K.; Oughton, E.A.; del Ceno, J.S.; et al. A Theory of Participation: What Makes Stakeholder and Public Engagement in Environmental Management Work? *Restor. Ecol.* **2018**, *26*, S7–S17. [CrossRef]
6. National Research Council. *Informing Decisions in a Changing Climate*; The National Academies Press: Washington, DC, USA, 2009.
7. Nguyen, V.M.; Young, N.; Brownscombe, J.W.; Cooke, S.J. Collaboration and Engagement Produce More Actionable Science: Quantitatively Analyzing Uptake of Fish Tracking Studies. *Ecol. Appl.* **2019**, *29*, e01943. [CrossRef] [PubMed]
8. Meadow, A.M.; Owen, G. *Planning and Evaluating the Societal Impacts of Climate Change Research Project: A Guidebook for Natural and Physical Scientists Looking to Make a Difference*; UA Faculty Publications: Tucson, AZ, USA, 2021. [CrossRef]
9. Robinson, J.A.; Hawthorne, T.L. Making Space for Community-Engaged Scholarship in Geography. *Prof. Geogr.* **2018**, *70*, 277–283. [CrossRef]
10. Cozzens, S.E. The Knowledge Pool: Measurement Challenges in Evaluating Fundamental Research Programs. *Eval. Program Plan.* **1997**, *20*, 77–89. [CrossRef]
11. National Research Council. *Using Science as Evidence in Public Policy*; The National Academies Press: Washington, DC, USA, 2012.

12. Meadow, A.M.; Ferguson, D.B.; Guido, Z.; Horangic, A.; Owen, G.; Wall, T. Moving toward the Deliberate Coproduction of Climate Science Knowledge. *Weather Clim. Soc.* **2015**, *7*, 179–191. [CrossRef]
13. VanderMolen, K.; Meadow, A.M.; Horangic, A.; Wall, T.U. Typologizing Stakeholder Information Use to Better Understand the Impacts of Collaborative Climate Science. *Environ. Manag.* **2020**, *65*, 178–189. [CrossRef]
14. Bamzai-Dodson, A.; Cravens, A.E.; Wade, A.; McPherson, R.A. Engaging with Stakeholders to Produce Actionable Science: A Framework and Guidance. *Weather Clim. Soc.* **2021**, *13*, 1027–1041. [CrossRef]
15. Block, D.R.; Hague, E.; Curran, W.; Rosing, H. Measuring Community and University Impacts of Critical Civic Geography: Insights from Chicago. *Prof. Geogr.* **2018**, *70*, 284–290. [CrossRef]
16. Ford, J.D.; Knight, M.; Pearce, T. Assessing the 'Usability' of Climate Change Research for Decision-Making: A Case Study of the Canadian International Polar Year. *Glob. Environ. Chang.* **2013**, *23*, 1317–1326. [CrossRef]
17. Baylor, R.; Esper, H.; Fatehi, Y.; de Garcia, D.; Griswold, S.; Herrington, R.; Belhoussein, M.O.; Plotkin, G.; Yamron, D. *Implementing Developmental Evaluation: A Practical Guide for Evaluators and Administrators*; U.S. Agency for International Development: Washington, DC, USA, 2019.
18. Patton, M.Q. Evaluation Science. *Am. J. Eval.* **2018**, *39*, 183–200. [CrossRef]
19. Hansen, H.F. Choosing Evaluation Models. *Evaluation* **2005**, *11*, 447–462. [CrossRef]
20. Preskill, H.; Russ-Eft, D. *Building Evaluation Capacity*; Sage Publications: New York, NY, USA, 2004; ISBN 1-4833-8931-6.
21. Weiss, C.H. Have We Learned Anything New About the Use of Evaluation? *Am. J. Eval.* **1998**, *19*, 21–33. [CrossRef]
22. Patton, M.Q. *Developmental Evaluation: Applying Complexity Concepts to Enhance Innovation and Use*; Guilford Press: New York, NY, USA, 2011.
23. Patton, M.Q. *Essentials of Utilization-Focused Evaluation*; SAGE: Saint Paul, MN, USA, 2012.
24. Coryn, C.L.S.; Hattie, J.A.; Scriven, M.; Hartmann, D.J. Models and Mechanisms for Evaluating Government-Funded Research. *Am. J. Eval.* **2007**, *28*, 437–457. [CrossRef]
25. Doemeland, D.; Trevino, J. *Which World Bank Reports Are Widely Read?* The World Bank: Washington, DC, USA, 2014; pp. 1–34.
26. Wall, T.; Meadow, A.M.; Horangic, A. Developing Evaluation Indicators to Improve the Process of Coproducing Usable Climate Science. *Weather Clim. Soc.* **2017**, *9*, 95–107. [CrossRef]
27. McNie, E.C. Delivering Climate Services: Organizational Strategies and Approaches for Producing Useful Climate-Science Information. *Weather Clim. Soc.* **2013**, *5*, 14–26. [CrossRef]
28. Guido, Z.; Hill, D.; Crimmins, M.; Ferguson, D. Informing Decisions with a Climate Synthesis Product: Implications for Regional Climate Services. *Weather Clim. Soc.* **2013**, *5*, 83–92. [CrossRef]
29. VanLandingham, G.; Silloway, T. Bridging the Gap between Evidence and Policy Makers: A Case Study of the Pew-MacArthur Results First Initiative. *Public Adm. Rev.* **2016**, *76*, 542–546. [CrossRef]
30. Boschetti, F.; Cvitanovic, C.; Fleming, A.; Fulton, E. A Call for Empirically Based Guidelines for Building Trust among Stakeholders in Environmental Sustainability Projects. *Sustain. Sci.* **2016**, *11*, 855–859. [CrossRef]
31. Lacey, J.; Howden, M.; Cvitanovic, C.; Colvin, R.M. Understanding and Managing Trust at the Climate Science–Policy Interface. *Nat. Clim. Chang.* **2017**, *8*, 22–28. [CrossRef]
32. Klink, J.; Koundinya, V.; Kies, K.; Robinson, C.; Rao, A.; Berezowitz, C.; Widhalm, M.; Prokopy, L. Enhancing Interdisciplinary Climate Change Work through Comprehensive Evaluation. *Clim. Risk Manag.* **2017**, *15*, 109–125. [CrossRef]
33. Colavito, M.M.; Trainor, S.F.; Kettle, N.P.; York, A. Making the Transition from Science Delivery to Knowledge Coproduction in Boundary Spanning: A Case Study of the Alaska Fire Science Consortium. *Weather Clim. Soc.* **2019**, *11*, 917–934. [CrossRef]
34. Owen, G.; Ferguson, D.B.; McMahan, B. Contextualizing Climate Science: Applying Social Learning Systems Theory to Knowledge Production, Climate Services, and Use-Inspired Research. *Clim. Chang.* **2019**, *157*, 151–170. [CrossRef]
35. Salazar, K. Department of the Interior Secretarial Order 3289: Addressing the Impacts of Climate Change on America's Water, Land, and Other Natural and Cultural Resources, 2009. Available online: https://www.doi.gov/elips/search?query=&name=&doc_type=2408&doc_num_label=&policy_category=All&approval_date=&so_order_num=3289&so_amended_num=&chapter=&dm_prt=&archived=All&office=All&date_from%5Bdate%5D=&date_to%5Bdate%5D=&sort_by=search_api_relevance&sort_order=DESC&items_per_page=10 (accessed on 30 May 2017).
36. ACCCNRS. *Report to the Secretary of the Interior*; Advisory Committee on Climate Change and Natural Resource Science: Washington, DC, USA, 2015.
37. USGS Program Evaluation | Climate Adaptation Science Centers. Available online: https://www.usgs.gov/programs/climate-adaptation-science-centers/program-evaluation (accessed on 25 October 2021).
38. Bamzai-Dodson, A.; Lackett, J.; McPherson, R.A. *CASC Project Evaluation Survey Template*; U.S. Geological Survey Data Release: Reston, VA, USA, 2022. [CrossRef]
39. Wardropper, C.B.; Dayer, A.A.; Goebel, M.S.; Martin, V.Y. Conducting Conservation Social Science Surveys Online. *Conserv. Biol.* **2021**, *35*, 1650–1658. [CrossRef] [PubMed]
40. Wolf, C.; Joye, D.; Smith, T.; Fu, Y. *The SAGE Handbook of Survey Methodology*; SAGE Publications: London, UK, 2016.
41. Qualtrics: Provo, UT Qualtrics (Copyright 2020). Available online: https://www.qualtrics.com/ (accessed on 30 May 2017).
42. Bamzai-Dodson, A.; Lackett, J.; McPherson, R.A. *North Central and South Central Climate Adaptation Science Center Project Evaluation: Survey Data Public Summary*; U.S. Geological Survey Data Release: Reston, VA, USA, 2022. [CrossRef]

43. Ferguson, D.B.; Meadow, A.M.; Huntington, H.P. Making a Difference: Planning for Engaged Participation in Environmental Research. *Environ. Manag.* **2022**, *69*, 227–243. [CrossRef] [PubMed]
44. Arnott, J.C. Pens and Purse Strings: Exploring the Opportunities and Limits to Funding Actionable Sustainability Science. *Res. Policy* **2021**, *50*, 104362. [CrossRef]
45. Arnott, J.C.; Kirchhoff, C.J.; Meyer, R.M.; Meadow, A.M.; Bednarek, A.T. Sponsoring Actionable Science: What Public Science Funders Can Do to Advance Sustainability and the Social Contract for Science. *Curr. Opin. Environ. Sustain.* **2020**, *42*, 38–44. [CrossRef]
46. Steger, C.; Klein, J.A.; Reid, R.S.; Lavorel, S.; Tucker, C.; Hopping, K.A.; Marchant, R.; Teel, T.; Cuni-Sanchez, A.; Dorji, T.; et al. Science with Society: Evidence-Based Guidance for Best Practices in Environmental Transdisciplinary Work. *Glob. Environ. Chang.* **2021**, *68*, 102240. [CrossRef]
47. Clark, T. "We're Over-Researched Here!": Exploring Accounts of Research Fatigue within Qualitative Research Engagements. *Sociology* **2008**, *42*, 953–970. [CrossRef]
48. Young, N.; Cooke, S.J.; Hinch, S.G.; DiGiovanni, C.; Corriveau, M.; Fortin, S.; Nguyen, V.M.; Solås, A.-M. "Consulted to Death": Personal Stress as a Major Barrier to Environmental Co-Management. *J. Environ. Manag.* **2020**, *254*, 109820. [CrossRef] [PubMed]
49. Herrick, C.; Vogel, J. Climate Adaptation at the Local Scale: Using Federal Climate Adaptation Policy Regimes to Enhance Climate Services. *Sustainability* **2022**, *14*, 8135. [CrossRef]
50. Smart, J. *Planning an Evaluation: Step by Step*; Australian Institute of Family Studies: Victoria, Australia, 2020.
51. Courtney, S.; Hyman, A.; McNeal, K.S.; Maudlin, L.C.; Armsworth, P. Development of a Survey Instrument to Assess Individual and Organizational Use of Climate Adaptation Science. *Environ. Sci. Policy* **2022**, *137*, 271–279. [CrossRef]
52. Hyman, A.; Courtney, S.; McNeal, K.S.; Bialic-Murphy, L.; Furiness, C.; Eaton, M.; Armsworth, P. Distinct Pathways to Stakeholder Use versus Scientific Impact in Climate Adaptation Research. *Conserv. Lett.* **2022**, *15*, e12892. [CrossRef]

sustainability

MDPI

Article

Climate Services and Transformational Adaptation

Edward R. Carr [1,2]

[1] International Development, Community, and Environment Department, Clark University, Worcester, MA 01610, USA; edcarr@clarku.edu

[2] Humanitarian Response and Development Lab, George Perkins Marsh Institute, Clark University, Worcester, MA 01610, USA

Abstract: The Working Group II contribution to the IPCC's Sixth Assessment Report states that effective adaptation to the changing climate will require transformational changes in how people live. This article explores the potential for climate services to catalyze and foster transformational adaptation. I argue that weather and climate information are not, in and of themselves, tools for transformation. When designed and delivered without careful identification of the intended users of the service and the needs that service addresses, they can fail to catalyze change amongst the users of that information. At worst, they can reinforce the status quo and drive maladaptive outcomes. For climate services to serve as agents of transformational adaptation, the climate services community will have to change how it understands the users of these services and their needs. Building climate services around contemporary understandings of how people make decisions about their lives and livelihoods offers designers and implementers of climate services opportunities to create services that catalyze transformational adaptation. These opportunities provide examples for the wider field of adaptation to consider in its efforts to contribute to climate resilient development.

Keywords: adaptation; transformation; climate services; maladaptation; climate resilient development; risk; vulnerability; resilience

Citation: Carr, E.R. Climate Services and Transformational Adaptation. *Sustainability* **2023**, *15*, 289. https://doi.org/10.3390/su15010289

Academic Editors: Charles Herrick, Jason Vogel and Glen Anderson

Received: 7 September 2022
Revised: 9 December 2022
Accepted: 12 December 2022
Published: 24 December 2022

1. Introduction

The IPCC's Working Group II contribution to the Sixth Assessment Report offers a stark message: we have delayed action for too long for incremental changes in our systems and the ways we live in the world to deliver a just, sustainable future [1]. Pathways to a climate-resilient future require the transformation of how we live in the world.

This assessment changes the calculus of adaptation programs, projects, and interventions. Actions aimed at preserving the status quo or introducing incremental changes intended to weather coming changes in climate will, in the end, not meet the moment. Instead, these efforts must facilitate transformative changes that move people toward climate-resilient improvements in human well-being, or climate-resilient development (CRD) [1]. At the same time, a growing literature points to the very limited evidence for the efficacy of our prior adaptation efforts [2,3] and growing evidence of their maladaptive outcomes [4–8]. In short, we have not been very good at climate change adaptation when framed around preservation. To pivot adaptation toward sparking transformative changes in how people live introduces even greater uncertainties to this project.

Climate services are an interesting adaptation intervention from which to consider how to facilitate or catalyze transformative changes toward climate-resilient development. They are information that, in and of itself, is not prescriptive (though there are cases where climate services are bundled with more prescriptive interventions such as seed and fertilizer programs). Thus, the intended users of this information can choose how they use it—in part or in whole, for the purposes envisioned by the producers of the service or for completely different goals. This lack of prescriptive power is evident in a growing body of work around the outcomes and impacts of climate services programs. In the ways in which it reveals how weather and climate information are taken up and used by their intended users, this work [9–16] sheds light on what in development studies [17] is a well-trodden critique: development and adaptation experts do not fully understand the current needs of their intended users, let alone the ways in which this information might facilitate transformative changes toward CRD.

There is much we still do not know about the users of climate services and their needs [18]. However, research on the dynamics of livelihoods in the context of development and adaptation interventions, economic change, and a changing environment points us toward what to look for as we consider the use of climate services as tools for transformative change toward CRD. Specifically, when we interrogate the resilience of socio-ecologies such as those that characterize agrarian communities in West Africa, we can see opportunities to catalyze (but not direct) transformative changes. At the broadest level, these opportunities exist in terms of the ability to reduce risk and vulnerability for agrarian populations. Doing so appears to create space for innovations in livelihoods practices and transformational shifts in the identities associated with those practices [19]. At the same time, delivering information in a manner that exacerbates risk and vulnerability, including the risk of changing the existing social order in a community, can result in decisions and actions that reinforce socio-ecologies against change. Over time, this renders them more vulnerable to catastrophic failures and costly transformations [20].

For climate services to serve as agents of transformative adaptation, the climate services community will have to change the ways it understands the users of these services and their needs. After briefly describing the historical practice of climate service design, I discuss recent progress in efforts to identify climate services users and needs. I identify persistent knowledge gaps that will continue to challenge our ability to use climate services as tools for transformative adaptation. I then offer some examples of climate services that are creating space for transformative change, consider the ways in which climate services might catalyze more rapid changes toward CRD, and suggest lessons for other interventions seeking to promote transformational adaptation and CRD.

2. Designing Climate Services

The design of most climate services reflects an understanding of vulnerability as produced by exposure to the impacts of a variable and/or changing climate [15]. As a result, most early climate services were shaped by the availability of climate information and the ability to disseminate it [21,22]. Under this model, those with weather and climate information packaged and disseminated that information, allowing the recipients to do whatever they wanted with it. The design and dissemination of climate services often lacked careful consideration of who the intended users of the information were or what their needs might be.

While this was the dominant mode of climate service design until very recently, there were important exceptions, such as the initial design of Mali's Agrometeorological Advisory Program. The design of that program was shaped by Malians intimately familiar with agricultural production in the country, and who therefore understood who the users of this service were and what information they needed [23]. However, even this case of good initial design illustrates the challenges that the climate services community faced in identifying users and needs. As the drought that provided its initial impetus faded, the program was given new purposes and goals for which it had not been designed. Where once it had been well-targeted to specific users and needs, the expansion of program goals over the next two decades led to a situation where, by 2010, the program was making broad assumptions about its users and needs that were not borne out by examination [13,24,25].

More recently, a nuanced literature focused on the diversity of climate service end users and needs has emerged. Such work has lurked at the margins of the climate services community for two decades. In the early 2000s, Archer [26] and Roncoli [11,27] examined different users of climate services, whether farmers or in the forecast community. While this early work in this arena was slow to get traction, more recent development donor attention to climate services has shifted the emphasis in their design. These organizations have increasingly looked to climate services as means to address important environmental and social challenges. Refocusing climate services on the achievement of goals requires attention to those who uses the information, how they use it, and whether or not it helps them achieve their goals. Thus, the work of Roncoli and her co-authors [27] on the integration of scientific and indigenous understandings of precipitation forecasting set the stage for more recent work how to overcome persistent misperceptions about end-users in the climate services community [28–30]. Work pointing to the differences among end-users and their ability to interact with forecasts [26,31] was foundational for more recent efforts seeking to identify the specific information needs of different end-users [32–36] and facilitated the emergence of gendered and feminist approaches to understanding end-user needs [15,37]. Today, engagement with the users of climate services is integral to conversations about their impact [38,39]. At the same time, it is not controversial to suggest that climate services, as a field, struggles with the effective identification of end users and their needs [40] and that substantial, systematic research on this subject is needed to move the field forward [18].

While users and their needs have become central to the design and implementation of many climate services over the past decade, one aspect of the framing of climate services has remained constant. Whether carefully considering users and needs or not, they are framed as defensive tools for the preservation of existing ways of living the face of a changing climate, environment, and economy. Increased attention to users and needs allows us to think in more critical and nuanced ways about *whose* ways of living are preserved and protected by a particular service. However, at a time when the climate change community recognizes that transformational adaptation will be required to achieve CRD, interventions that preserve that which exists now risk becoming maladaptive. They can perpetuate practices that will become inviable over time or maintain social relations that act as barriers to significant changes in human well-being. Shifting the framing of climate services from defensive tools protecting people and livelihoods from the impacts of climate change to vehicles for the achievement of CRD requires more than just shifting the focus of climate services from the science of climate to the social and behavioral science of the intended end-user. It requires social science approaches that can identify opportunities for transformation that climate services might support or leverage.

Relatively little work in climate services has considered how they might contribute to transformational adaptation, or more broadly to CRD. Notable exceptions to this lie in work led by Hansen [38,41]. The approach in this work focuses on identifying broad relationships between the use of weather and climate information and the achievement of development goals. While this enables discussions of pathways by which climate services might, for example, address SDG 2 "Zero Hunger" [38], it does not unpack the very localized opportunities that such information leverages or the barriers that it overcomes to result in such impacts. To change climate services from efforts to hold off the bad effects of climate variability and change to vehicles for the sorts of transformation inherent to achieving development goals requires a different approach. I suggest that one productive means of understanding how climate services work as vehicles for transformational adaptation and the achievement of CRD is to understand how they intersect and interact with livelihoods, people's ways of living in the world [42–44].

3. Using Livelihoods Analysis to Design Transformational Climate Services

A starting point for transformational climate services lies in understanding not only what the intended end-users of a given climate service want, but why they want it. This requires engagement with the perceptions of individuals and the social structures that give perceptions meaning. This sort of inquiry identifies two kinds of barriers and opportunities for climate services. The first are barriers to the uptake and use of different kinds of climate information. The second are barriers to and opportunities for such interventions to catalyze transformative adaptation that aligns CRD. This is not to suggest that climate science is irrelevant to the development of transformational climate services, but that it should not be the starting point of the design and implementation of those services.

Using livelihoods to understand how climate services work is not entirely new. In an effort to strengthen drought-preparedness efforts, Roncoli and her co-authors [45] examined the livelihoods implications of a severe drought in Burkina Faso. Using the predominant framing of livelihoods at that time, one focused on material means of making a living [46–48], they examined how farmers perceptions shaped their evaluations of and predictions for agricultural seasons. Through this work, they convincingly demonstrated that the farmers in their study were not helpless victims of drought, but agents worthy of engagement when planning forecasts, famine warnings, and other forms of weather and climate service. However, this early work differs from the question at hand in two important ways.

First, what Roncoli and her co-authors were studying was coping, rather than adaptation. Their goal was to demonstrate that farmers held stores of knowledge and practice for managing shocks like drought, and that what farmers know and do should be part of conversations that had, to that point, often been limited to development, humanitarian, and meteorological organizations. They did not examine the adaptive or transformational potential of the farmers in their study because that was not their aim.

Second, their framing of livelihoods and the ways in which farmers shifted them in the context of a drought was descriptive and material in its focus. Because the aim of the research was to demonstrate the value of farmer knowledge and practice to forecasting and early warning, there was little need for discussion of the social context within which farmer perceptions could be translated into decisions and actions. This work did not engage with the ways in which making a living is inextricably intertwined with making meaning of the world and how to live in it [42–44,49,50]. However, as the Working Group II contribution to the IPCC's Sixth Assessment Report recognizes, meaning, power, and agency are critical aspects of transformation and the achievement of CRD [1].

More recent work in livelihoods studies builds on the idea that livelihoods are always both about making a living and making sense of the world. This work creates opportunities for examining issues of meaning and value central to livelihoods decisions and practices, and therefore critical to the identification of opportunities for transformative change and CRD. Livelihoods research has developed a range of theoretical approaches to the making of meaning first articulated by Bebbington [44]. These include approaches that draw on Ortner's model of "serious games" [49], Bourdieu's theory of practice [50], and Foucault's concept of governmentality [43,51]. In this article, I draw from studies that employed the latter, in the form of the Livelihoods as Intimate Government approach. These illustrate how contemporary livelihoods approaches focused on meaning *and* materiality allow for the identification of barriers to the transformational use of climate services and opportunities for weather and climate information to catalyze such transformation.

3.1. *Understanding the Transformative Potential of Climate Services through Livelihoods: LIG*

The Humanitarian Response and Development Lab (HURDL) at Clark University has employed the Livelihoods as Intimate Government approach to both evaluate the impact of climate services in sub-Saharan Africa and to inform the design of new services. Through this work, HURDL has identified nuanced reasons for the limited uptake of weather and climate services tied to social structures, power relations, and meaning [13,15,25]. At the same time, it has also identified spaces where transformational change might take root and flourish if properly supported by targeted weather and climate information [19,20].

As an approach, LIG focuses on the different understandings and experiences of the vulnerability context expressed by individuals in the same community or household. These differences speak to the understanding of different stressors, activities, and identities, providing a point of entry into the construction of meaning through livelihoods in a given place. Broadly speaking, LIG treats meaning as emerging at the intersection of three things: (1) discourses of livelihoods, which reflect local understandings of the "correct" activities to undertake and the correct way to undertake them given the challenges of the context, (2) the ways in which those discourses and understandings mobilize identity as they speak to who should conduct what activities and how they should be conducted, and (3) tools of coercion, locally-appropriate means of disciplining people to ensure they align with expectations of their identity and the discourses of livelihoods [43,51]. Methodologically, LIG employs rapid ethnographic methods, including participant observation and semi-structured interviewing. Typically, fieldwork is conducted by teams of two or more researchers, spending eight to ten weeks in a community [51].

3.2. *Climate Services as Barriers to Transformation*

A LIG analysis of the uptake and use of climate information provided by Mali's Agrometeorological Advisory Program speaks to how a well-targeted climate service might address short-term livelihoods and food security needs but over the long term hold back the sorts of transformation needed for successful adaptation. An initial assessment of the impact of the program commissioned by USAID [24] more than three decades after its launch found that the uptake of the advisories was very low and skewed toward men. Further investigation employed the LIG approach to explain this pattern of use [25]. The assessment found that the project was, on one hand, extremely well-designed for its stated purpose: addressing food availability challenges in the late 1970s and early 1980s. The advisories targeted key staple crops over which men had decision-making authority. Further, the nature of the advisories (such as providing farmers with constantly-updated information on when to plant, and what varieties to plant) meant that only the wealthiest fraction of men, those who owned both farm equipment and animal traction, could use the advisories. These, of course, were the men who would produce the most staple crop, and therefore be the audience that this program most needed to reach [13]. The assessment found that even in 2014, these men were still following the advisories [24,25].

However, the assessment also found that these advisories reinforced existing livelihoods—both their meaning and their material practices [25]. For example, by providing information that only the wealthiest, most senior men could use, the advisories reinforced the authority of these men over their households and extended families. As part of their role, these senior men are expected to make agricultural decisions for the fields of their families, most commonly the shared fields of the family. Such decisions also have implications for the fields of individual households in the concession because junior men do not want to contradict senior men. This can result in the loss of access to land and other agricultural resources. A senior man's power is not absolute. If he fails to successfully feed his family through his agricultural decisions and staple crop production, he can have his authority and status questioned or even stripped. Interestingly, the skill of these advisories, and thus their ability to productively inform on-farm decisions, has been questioned [52]. However, their accuracy might be beside the point. By providing something men could blame for faulty decisions, the advisories gave senior men a means of deflecting criticism of their decisions and therefore reduced their accountability to their households,. Reduced accountability for those with the greatest authority increases the durability of existing social structures, even under conditions of environmental stress. These structures limit women's authority and autonomy, and thus circumscribe one of the most well-understood pathways to transformative change and climate-resilient development: empowering women.

3.3. *Climate Services as Catalysts of Transformational Adaptation*

Livelihoods analysis can help us identify situations where climate services reinforce the structural causes behind observed inequities in situations where transformational change is needed. It also can help identify opportunities for climate services to catalyze transformational change. I use the term catalyze advisedly here. As noted by Schipper and her co-authors [1], CRD pathways are not prescriptive steps that one takes toward a climate resilient future. Instead, these pathways emerge from formal and informal decisions taken by individuals, households, communities, and countries. More than 80 years of formal development practice have demonstrated that prescriptive transformations tend to reflect the desires and beliefs of the "developed", who are the wealthy and powerful, and thus those with the most invested in existing economic, political, and social structures. In short, this echoes an observation about livelihoods enabled by LIG, but at much larger scales. Just as in a household livelihoods decision, a transformation of international or global structures whose means and goals are managed by the wealthiest and most powerful is unlikely to challenge the structures that grant the powerful their privileges [43]. Seen from the perspective of the powerful, transformations are likely to be transformation for others, but status quo for the wealthy and powerful. Further, many decades of development have shown us that the transformations desired by "the developed" are often not those that "the developing" would select for themselves [17], resulting in many development projects and interventions with low rates of uptake and limited impact.

If we shift our thinking from the management of transformation to the catalysis of transformation, we shift our understanding of agency and outcomes in this process. While those with resources and authority are still able to invest in certain catalysts of change, they are not able to determine the final outcome of that which they start. Catalyzing change means creating opportunities for actors to make new decisions, take up new activities, and redefine how they live in the world in terms that make sense to them. Often the outcome is not far from the goals of formal development practice. For example, where we have seen women's empowerment around the world, it has come less from donor-funded gender sensitization programs than from women who understand how to identify and leverage opportunities in their specific contexts. Numerous studies of reversals of environmental degradation have demonstrated that local knowledge of the environment often has much more to do with effective outcomes than outside technical knowledge.

If climate services are to be catalysts of transformational adaptation, they must clearly identify opportunities for weather and climate information to create the conditions within which people can act in new ways without reinforcing existing structures that act as barriers to transformational change. One example lies in a broad observation that has emerged across HURDL's work on livelihoods. A broad synthesis of livelihoods data [19] spanning more than a thousand interviews and a dozen livelihoods zones across West Africa suggests that as individuals, households, and communities experience greater security from uncertainty and locally-specific drivers of vulnerability, spaces open for transformational change. For example, in Mali and Senegal, the most food and income secure households are also the places where one is most likely to find women taking on activities or roles that do not align with expectations, such as farming a "man's crop." In the most stressed and challenged households, we see no deviation from expectations.

This difference in attitude toward innovation and potential transformation lies in the ways stressors that challenge sources of income and assets present two threats. The first is material, which in the most stressed households can manifest as existential threats. Under such circumstances, insisting that all members of the household play their roles is justified as a pathway to safety and security in a context of vulnerability. At the same time, these stressors also threaten the existing social order. In these households, men are often failing to feed their families adequately. They therefore risk loss of status and authority each season. Allowing other members of the household to take on new tasks, or to take on tasks and responsibilities that belong to men, risks demonstrating that these men and their decisions need not be at the center of livelihoods. Thus, men have an incentive to carefully enforce roles and responsibilities in their households to ward off challenges to their authority. In situations where material and social stresses converge, livelihoods can become rigid to the point of brittleness. This puts households and communities at risk of catastrophic transformations where existing livelihoods (both activities and the meanings and order behind them) are pushed past thresholds of sustainability [20]. On the other hand, in households where production, and therefore the status of the man in charge of the household, is secure, a woman farming a man's crop presents neither a material nor a social threat and is tolerated. Over time, such spaces of deviation and innovation can quietly redefine what is seen as acceptable behavior for women, junior men, or others in society, creating a pathway toward CRD.

This suggests that one way that climate services can promote transformational adaptation and pathways toward CRD is by focusing on vulnerability reduction. This is not the same thing as risk reduction. Risk requires understanding the likelihood of a hazard's occurrence. Vulnerability, on the other hand, demands we understand how that hazard impacts a person's way of living in the world. By providing information that can lower the vulnerability associated with different livelihoods, climate services can create the sorts of security and safety that allow for the sorts of innovation and transgression that can result in transformation. Such outcomes might come through increased production during average years as climate services reduce the need for inefficient hedging. Perhaps forecasts can facilitate livelihoods planning to address the impacts of excessively wet, dry, or hot years. There are any number of possible contributions climate services might make to transformational adaptation and CRD. Like the observation about vulnerability reduction above, such contributions should be identified through nuanced understandings of the current structure of activities and society and be targeted toward explicit sites where change can be catalyzed. Following this line of thinking moves us past a framing of climate services as solutions for adaptation and development challenges in and of themselves. Instead, it presents climate services as locally appropriate facilitators of transformation to CRD.

4. Conclusions

The remaining pathways toward increased CRD are transformational in character [3]. The climate change community of practice is pivoting toward CRD as a framing for climate action that moves us past the preservation of current systems, structures, and levels of well-being. For climate services to contribute to this changing understanding of climate action and its goals, we must rethink their purpose. Where once climate services were also implicitly defensive tools for the preservation of current practices and structures in the face of growing threats, today climate services should be viewed as potential catalysts of CRD. How we might do this for climate services illustrates broader principles for transformational adaptation that can be applied to all manner of adaptation interventions.

The pivot to transformational adaptation makes the ongoing attention to users and needs central not only to the long-term relevance of climate services, but also to any adaptation intervention with transformational aspirations. The climate change community requires an expansion of inquiry into the users of adaptation interventions and their needs to fill the substantial gaps in our knowledge. However, this work must not fall into the trap of an exclusive focus on understanding and preserving what currently exists in the face of change. Inquiry into the opportunities for transformation in existing systems and situations is the foundation for transformational adaptation.

A pivot toward users and needs emphasizes the potential value and importance of co-produced adaptation interventions. However, it also highlights that such co-production itself must be built on a deep understanding of users [39,53–55]. In its discussion of the structures behind observed livelihoods decisions and uses of weather and climate services, this article highlights the need for deep engagement with users of adaptation interventions that will not emerge in a single workshop, but through extended engagement and learning through the design, implementation, and monitoring of adaptation in action. The identification of transformational opportunities is fraught with micro-politics and competing interests, making everything from who participates in co-production to the means of eliciting ideas and understandings critical to the transformational potential of such activities. Co-production is not, by itself, a means of making adaptation interventions effective catalysts of CRD. The character of co-production is critical.

Finally, aligning climate services with CRD highlights the need to coordinate adaptation interventions with other efforts to create opportunities for transformational change that speak to development challenges. While an effective seasonal forecast might tell farmers what they need to plant to avoid negative outcomes, without access to seeds and appropriate farming equipment that information will not be translated into the safety and security that creates spaces of transgression and transformation. Further, even in situations where an adaptation intervention does contribute to increased safety and security, those seeking to transgress will need access to opportunity. While the women in wealthy, secure households described above *can* farm men's crops and avoid sanction or even attention, they cannot do so without access to seed, land, and farming equipment. It is only through deep engagement with the users of adaptation interventions that we will learn what opportunities they are seeking, what opportunities we can create, and the limits of different adaptation interventions on our path to CRD.

Funding: This research received no external funding.

Institutional Review Board Statement: Not applicable.

Informed Consent Statement: Not applicable.

Acknowledgments: Any time one attempts to write a piece with a broad agenda, it stands on more shoulders than can be named in an acknowledgements section. This article benefited from more than a decade of conversations about adaptation and climate services with many colleagues, including John Furlow, Sheila Onzere, Helen Rosko, Kwame Owusu-Daaku, Daniel Abrahams, Tshibangu Kalala, Jim Hansen, Cathy Vaughan, Simon Mason, Glen Anderson, Steve Zebiak, Zack Guido, Jim Buizer, Janae Davis, and Rob Goble. My thinking on adaptation was also greatly influenced by my co-authors in Chapter 18 of the Working Group II contribution to the IPCC's Sixth Assessment report, including Lisa Schipper, Siri Eriksen, Aromar Revi, Ben Preston, Luis Fernández-Carril, Bruce Glavovic, Nathalie J.M. Hilmi, Debbie Ley, Rupa Mukerji, Silvia Muylaert de Araujo, Rosa Perez, Steve Rose and Pramod Singh.

Conflicts of Interest: The author declares no conflict of interest.

References

1. Schipper, E.L.F.; Revi, A.; Preston, B.L.; Carr, E.R.; Eriksen, S.H.; Fernandez-Carril, L.R.; Glavovic, B.; Hilmi, N.J.M.; Ley, D.; Mukerji, R.; et al. Climate Resilient Development Pathways. In *Climate Change 2022: Impacts, Adaptation, and Vulnerability. Contribution of Working Group II to the Sixth Assessment Report of the Intergovernmental Panel on Climate Change*; Pörtner, H.-O., Roberts, D.C., Tignor, M., Poloczanska, E.S., Mintenbeck, K., Alegría, A., Craig, M., Langsdorf, S., Löschke, S., Möller, V., et al., Eds.; Cambridge University Press: Cambridge, UK, 2022; pp. 2655–2807. ISBN 9781009325844.
2. Sitati, A.; Joe, E.; Pentz, B.; Grayson, C.; Jaime, C.; Gilmore, E.; Galappaththi, E.; Hudson, A.; Alverio, G.N.; Mach, K.J.; et al. Climate Change Adaptation in Conflict-Affected Countries: A Systematic Assessment of Evidence. *Discov. Sustain.* **2021**, *2*, 42. [CrossRef] [PubMed]
3. IPCC. *Summary for Policymakers*; Pörtner, H.-O., Roberts, D.C., Tignor, M., Poloczanska, E.S., Mintenbeck, K., Alegría, A., Craig, M., Langsdorf, S., Löschke, S., Möller, V., et al., Eds.; Cambridge University Press: Cambridge, UK, 2022; ISBN 9781139177245.
4. Eriksen, S.; Schipper, E.L.F.; Scoville-Simonds, M.; Vincent, K.; Adam, H.N.; Brooks, N.; Harding, B.; Khatri, D.; Lenaerts, L.; Liverman, D.; et al. Adaptation Interventions and Their Effect on Vulnerability in Developing Countries: Help, Hindrance or Irrelevance? *World Dev.* **2021**, *141*, 105383. [CrossRef]
5. Atteridge, A.; Remling, E. Is Adaptation Reducing Vulnerability or Redistributing It? *Wiley Interdiscip. Rev. Clim. Change* **2018**, *9*, e500. [CrossRef]
6. Magnan, A.K.; Schipper, E.L.F.; Burkett, M.; Bharwani, S.; Burton, I.; Eriksen, S.; Gemenne, F.; Schaar, J.; Ziervogel, G. Addressing the Risk of Maladaptation to Climate Change. *Wiley Interdiscip. Rev. Clim. Change* **2016**, *7*, 646–665. [CrossRef]
7. Schipper, E.L.F. Maladaptation: When Adaptation to Climate Change Goes Very Wrong. *One Earth* **2020**, *3*, 409–414. [CrossRef]
8. Eriksen, S.H.; Nightingale, A.J.; Eakin, H. Reframing Adaptation: The Political Nature of Climate Change Adaptation. *Glob. Environ. Change* **2015**, *35*, 523–533. [CrossRef]
9. Roncoli, C. Ethnographic and Participatory Approaches to Research on Farmers' Responses to Climate Predictions. *Clim. Res.* **2006**, *33*, 81–99. [CrossRef]
10. Peterson, N.D.; Broad, K.; Orlove, B.S.; Roncoli, C.; Taddei, R.; Velez, M.-A. Participatory Processes and Climate Forecast Use: Socio-Cultural Context, Discussion, and Consensus. *Clim. Dev.* **2010**, *2*, 14–29. [CrossRef]
11. Roncoli, C.; Ingram, K.T.; Kirshen, P.; Jost, C. Burkina Faso: Integrating Indigenous and Scientific Rainfall Forecasting. *IK Notes* **2001**, *39*, 1–4.
12. Rader, M.; Kirshen, P.; Asce, M.; Roncoli, C.; Hoogenboom, G.; Ouattara, F. Agricultural Risk Decision Support System for Resource-Poor Farmers in Burkina Faso, West Africa. *J. Water Resour. Plan. Manag.* **2009**, *135*, 323–333. [CrossRef]
13. Carr, E.R.; Onzere, S.N. Really Effective (for 15% of the Men): Lessons in Understanding and Addressing User Needs in Climate Services from Mali. *Clim. Risk Manag.* **2018**, *22*, 82–95. [CrossRef]

14. Carr, E.R.; Fleming, G.; Kalala, T. *Assessing Climate Service Needs in Kaffrine, Senegal: Livelihoods, Identity, and Vulnerability to Climate Variability and Change*; USAID: Washington, DC, USA, 2015.
15. Carr, E.R.; Owusu-Daaku, K.N. The Shifting Epistemologies of Vulnerability in Climate Services for Development: The Case of Mali's Agrometeorological Advisory Programme. *Area* **2016**, *48*, 7–17. [CrossRef]
16. Owusu-Daaku, K.N. (Mal)Adaptation Opportunism: When Other Interests Take over Stated or Intended Climate Change Adaptation Objectives (and Their Unintended Effects). *Local Environ.* **2018**, *23*, 934–951. [CrossRef]
17. Chambers, R. *Whose Reality Counts?: Putting the First Last*; Intermediate Technology: London, UK, 1997.
18. Carr, E.R.; Goble, R.; Rosko, H.M.; Vaughan, C.; Hansen, J. Identifying Climate Information Services Users and Their Needs in Sub-Saharan Africa: A Review and Learning Agenda. *Clim. Dev.* **2020**, *12*, 23–41. [CrossRef]
19. Carr, E.R. Properties and Projects: Reconciling Resilience and Transformation for Adaptation and Development. *World Dev.* **2019**, *122*, 70–84. [CrossRef]
20. Carr, E.R. Resilient Livelihoods in an Era of Global Transformation. *Glob. Environ. Change* **2020**, *64*, 102155. [CrossRef]
21. Hansen, J.W. Integrating Seasonal Climate Prediction and Agricultural Models for Insights into Agricultural Practice. *Philos. Trans. R. Soc. London. Ser. B Biol. Sci.* **2005**, *360*, 2037–2047. [CrossRef]
22. Hansen, J.W.; Mishra, A.; Rao, K.P.C.; Indeje, M.; Ngugi, R.K. Potential Value of GCM-Based Seasonal Rainfall Forecasts for Maize Management in Semi-Arid Kenya. *Agric. Syst.* **2009**, *101*, 80–90. [CrossRef]
23. Moussa, A.; Traore, K. Background and Function of the Agrometeorological Advisory Program. In *Assessing Mali's Direction National de la Meteorologie Agrometeorological Advisory Program: Preliminary Report on the Climate Science and Farmer Use*; Carr, E.R., Ed.; United States Agency for International Development: Washington, DC, USA, 2014; pp. 12–18.
24. Carr, E.; Dinku, T.; Giannini, A.; Kupfer, J.; Mason, S.; Moussa, A. *Assessing Mali's Direction National de La Meteorologie Agrometeorological Advisory Program: Preliminary Report on the Climate Science and Farmer Use*; Carr, E.R., Ed.; United States Agency for International Development: Washington, DC, USA, 2014.
25. Carr, E.R.; Onzere, S.; Kalala, T.; Owusu-Daaku, K.N.; Rosko, H. *Assessing Mali's l'Agence Nationale de La Météorologie's (Mali Meteo) Agrometeorological Advisory Program: Final Report on the Farmer Use of Advisories and the Implications for Climate Service Design*; United States Agency for International Development: Washington, DC, USA, 2015.
26. Archer, E.R.M. Identifying Underserved End-User Groups in the Provision of Climate Information. *Bull. Am. Meteorol. Soc.* **2003**, *84*, 1525–1532. [CrossRef]
27. Roncoli, C.; Ingram, K.; Kirshen, P. Can Farmers of Burkina Faso Use Seasonal Rainfall Forecasts? *Pract. Anthropol.* **2000**, *22*, 24–28. [CrossRef]
28. Nidumolu, U.; Lim-Camacho, L.; Gaillard, E.; Hayman, P.; Howden, M. Linking Climate Forecasts to Rural Livelihoods: Mapping Decisions, Information Networks and Value Chains. *Weather Clim. Extrem.* **2018**, *27*, 100174. [CrossRef]
29. Loboguerrero, A.M.; Boshell, F.; León, G.; Martinez-Baron, D.; Giraldo, D.; Recaman Mejía, L.; Díaz, E.; Cock, J. Bridging the Gap between Climate Science and Farmers in Colombia. *Clim. Risk Manag.* **2018**, *22*, 67–81. [CrossRef]
30. Guido, Z.; Zimmer, A.; Lopus, S.; Hannah, C.; Gower, D.; Waldman, K.; Krell, N.; Sheffield, J.; Caylor, K.; Evans, T. Farmer Forecasts: Impacts of Seasonal Rainfall Expectations on Agricultural Decision-Making in Sub-Saharan Africa. *Clim. Risk Manag.* **2020**, *30*, 100247. [CrossRef]
31. Roncoli, C.; Orlove, B.S.; Kabugo, M.R.; Waiswa, M.M. Cultural Styles of Participation in Farmers' Discussions of Seasonal Climate Forecasts in Uganda. *Agric. Hum. Values* **2011**, *28*, 123–138. [CrossRef]
32. Nkiaka, E.; Taylor, A.L.; Dougill, A.; Antwi-Agyei, P.; Fournier, N.; Bosire, E.N.; Konte, O.; Lawal, K.A.; Mutai, B.; Mwangi, E.; et al. Identifying User Needs for Weather and Climate Services to Enhance Resilience to Climate Shocks in Sub-Saharan Africa. *Environ. Res. Lett.* **2019**, *14*, 123003. [CrossRef]
33. Carr, E.R.; Abrahams, D.; De la Poterie, A.T.; Suarez, P.; Koelle, B. Vulnerability Assessments, Identity and Spatial Scale Challenges in Disaster-Risk Reduction. *Jàmbá J. Disaster Risk Stud.* **2015**, *7*, a201. [CrossRef]
34. Henriksson, R.; Vincent, K.; Archer, E.; Jewitt, G. Understanding Gender Differences in Availability, Accessibility and Use of Climate Information among Smallholder Farmers in Malawi. *Clim. Dev.* **2021**, *13*, 503–514. [CrossRef]
35. Diouf, N.S.; Ouedraogo, I.; Zougmoré, R.B.; Ouedraogo, M.; Partey, S.T.; Gumucio, T. Factors Influencing Gendered Access to Climate Information Services for Farming in Senegal. *Gend. Technol. Dev.* **2019**, *23*, 93–110. [CrossRef]
36. Buckland, S.F.; Campbell, D. An Assessment of Factors Influencing Awareness, Access and Use of Agro-Climate Services among Farmers in Clarendon, Jamaica. *Geoforum* **2021**, *126*, 171–191. [CrossRef]
37. Carr, E.R.; Fleming, G.; Kalala, T. Understanding Women's Needs for Weather and Climate Information in Agrarian Settings: The Case of Ngetou Maleck, Senegal. *Weather Clim. Soc.* **2016**, *8*, 247–264. [CrossRef]
38. Hansen, J.; List, G.; Downs, S.; Carr, E.R.; Diro, R.; Baethgen, W.; Kruczkiewicz, A.; Braun, M.; Furlow, J.; Walsh, K.; et al. Impact Pathways from Climate Services to SDG2 ("Zero Hunger"): A Synthesis of Evidence. *Clim. Risk Manag.* **2022**, *35*, 100399. [CrossRef]
39. Vincent, K.; Archer, E.; Henriksson, R.; Pardoe, J.; Mittal, N. Reflections on a Key Component of Co-Producing Climate Services: Defining Climate Metrics from User Needs. *Clim. Serv.* **2020**, *20*, 100204. [CrossRef]
40. Porter, J.J.; Dessai, S. Mini-Me: Why Do Climate Scientists' Misunderstand Users and Their Needs? *Environ. Sci. Policy* **2017**, *77*, 9–14. [CrossRef]

41. Hansen, J.; Hellin, J.; Rosenstock, T.; Fisher, E.; Cairns, J.; Stirling, C.; Lamanna, C.; van Etten, J.; Rose, A.; Campbell, B. Climate Risk Management and Rural Poverty Reduction. *Agric. Syst.* **2019**, *172*, 28–46. [CrossRef]
42. Scoones, I. Livelihoods Perspectives and Rural Development. *J. Peasant Stud.* **2009**, *36*, 171–196. [CrossRef]
43. Carr, E.R. Livelihoods as Intimate Government: Reframing the Logic of Livelihoods for Development. *Third World Q.* **2013**, *34*, 77–108. [CrossRef]
44. Bebbington, A. Capitals and Capabilities: A Framework for Analyzing Peasant Viability, Rural Livelihoods and Poverty. *World Dev.* **1999**, *27*, 2021–2044. [CrossRef]
45. Roncoli, C.; Ingram, K.; Kirshen, P. The Costs and Risks of Coping with Drought: Livelihood Impacts and Farmers' Responses in Burkina Faso. *Clim. Res.* **2001**, *19*, 119–132. [CrossRef]
46. Chambers, R.; Conway, G. *Sustainable Rural Livelihoods: Practical Concepts for the 21st Century*; Institute of Development Studies: Sussex, UK, 1992.
47. Farrington, J.; Carney, D.; Ashley, C.; Turton, C. Sustainable Livelihoods in Practice: Early Applications of Concepts in Rural Areas. In *Natural Resource Perspectives*; ODI: London, UK, 1999.
48. Carney, D. Implementing the Sustainable Livelihoods Approach. In *Sustainable Rural Livelihoods: What Contribution Can We Make?* Department for International Development: London, UK, 1998; ISBN 9781861920829.
49. Jakimow, T. Serious Games in Livelihood Analysis: Reflections from the Case of Agricultural Wage Labourers in Andhra Pradesh. *J. Dev. Stud.* **2012**, *48*, 1274–1287. [CrossRef]
50. Sakdapolrak, P. Livelihoods as Social Practices—Re-Energising Livelihoods Research with Bourdieu's Theory of Practice. *Geogr. Helv.* **2014**, *69*, 19–28. [CrossRef]
51. Carr, E.R. From Description to Explanation: Using the Livelihoods as Intimate Government (LIG) Approach. *Appl. Geogr.* **2014**, *52*, 110–122. [CrossRef]
52. Mason, S.J.; Giannini, A.; Dinku, T. Assessment of Climate Science Used by Direction Nationale de La Météorollogie Du Mali for Provision of Agrometeorological Information to the Rural Community. In *Assessing Mali's Direction Nationale de la Météorologie Agrometeorological Advisory Program: Preliminary Report on the Climate Science and Farmer Use of Advisories*; USAID: Washington, DC, USA, 2014; pp. 19–34.
53. Vincent, K. Development Geography I: Co-Production. *Prog. Hum. Geogr.* **2022**, *46*, 890–897. [CrossRef]
54. Vincent, K.; Daly, M.; Scannell, C.; Leathes, B. What Can Climate Services Learn from Theory and Practice of Co-Production? *Clim. Serv.* **2018**, *12*, 48–58. [CrossRef]
55. Bremer, S.; Wardekker, A.; Dessai, S.; Sobolowski, S.; Slaattelid, R.; van der Sluijs, J. Toward a Multi-Faceted Conception of Co-Production of Climate Services. *Clim. Serv.* **2019**, *13*, 42–50. [CrossRef]

sustainability MDPI

Review

Key Considerations for the Use of Nature-Based Solutions in Climate Services and Adaptation

Arsum Pathak [1,*], Laura E. Hilberg [2], Lara J. Hansen [2,*] and Bruce A. Stein [3]

1 National Wildlife Federation, Austin, TX 78752, USA
2 EcoAdapt, Bainbridge Island, WA 98110, USA
3 National Wildlife Federation, Washington, DC 20006, USA
* Correspondence: pathaka@nwf.org (A.P.); lara@ecoadapt.org (L.J.H.)

Abstract: Nature-based solutions (NbS) involve the reliance on natural or nature-based systems to enhance community resilience through delivering both climate adaptation and mitigation outcomes. While NbS do not necessarily represent new "technology" or methods, the intentional incorporation of these approaches into climate adaptation and mitigation efforts is often considered novel, particularly within the climate services sector where interventions have historically prioritized structural infrastructure approaches. NbS can offer an effective replacement for or complement to such traditional infrastructure approaches. Additionally, natural and nature-based systems can respond to climate change in a manner that engineered solutions often cannot, providing long-term holistic adaptation and mitigation success with additional benefits to ecosystem services such as improved air and water quality, carbon sequestration, outdoor recreation, and biodiversity protection. The incorporation of NbS as a core component of climate services increases the likelihood of adoption and effective implementation, ensuring greater long-term effectiveness for both communities and the natural systems on which they depend. This article supports the adoption and effective implementation of NbS by climate service providers through presenting a set of seven "key considerations" for their use in community-based adaptation. These key considerations are based on a review of work in the field to date, both within the United States and globally. Although these key considerations were developed in support of US adaptation planning applications (specifically, the US Climate Resilience Toolkit), they have global relevance.

Keywords: climate services; nature-based solutions; vulnerability assessment; climate adaptation; resilience; co-production

Citation: Pathak, A.; Hilberg, L.E.; Hansen, L.J.; Stein, B.A. Key Considerations for the Use of Nature-Based Solutions in Climate Services and Adaptation. *Sustainability* **2022**, *14*, 16817. https://doi.org/10.3390/su142416817

Academic Editors: Mohammad Valipour, Charles Herrick, Jason Vogel and Glen Anderson

Received: 30 September 2022
Accepted: 13 December 2022
Published: 15 December 2022

Publisher's Note: MDPI stays neutral with regard to jurisdictional claims in published maps and institutional affiliations.

1. Introduction

As the pace and scale of climate change and its impacts become increasingly evident across the United States and globally, there is a growing need for robust and reliable climate services, which can be defined as the provision of climate information for use in decision making [1]. Climate data from across multiple sensors and observation platforms document increasing climate variability, including changes in temperature and precipitation patterns, more extreme weather events, and rising sea level. The number of billion-dollar weather and climate disasters in the United States has doubled from about five events per year in 1990–1999 to close to 13 events annually in 2010–2019, with losses exceeding 900 billion USD in these last 10 years [2]. These climatic changes and their associated impacts are affecting our water systems, biodiversity, food supply, and health with far-reaching consequences for US communities and ecosystems [3]. As these climate impacts become even more disruptive in the future, the approaches to mitigate these risks will require novel thinking that moves away from business-as-usual strategies such as traditional structural solutions (e.g., levees, sea walls, and stormwater drainage channels). In the face of accelerating climate change, the limitations of such hard infrastructure approaches—high costs, limited lifetime, risks of maladaptation—are becoming all too evident [4]. Such static structures, which are often designed to standards based on past climatic conditions, may not be able to keep up with the increasing climate variability and, accordingly, may have escalating maintenance costs. A lack of community or ecological co-benefits and, in many cases, negative or maladaptive consequences can make them an unsuitable adaptation solution in certain circumstances.

The role of nature, on the other hand, is receiving increasing attention for coping with growing climate risks, and the International Union for Conservation of Nature (IUCN) has published several reports designed to define and operationalize the concept of nature-based solutions (NbS) [5,6]. The concept of NbS builds on the framework of "ecosystem services", which, over the past few decades, has emerged as an important approach for understanding, documenting, and valuing the varied contributions of nature to people [7,8]. By providing protective benefits, NbS nest within the "regulation of environmental processes" category of ecosystem services as defined by the Intergovernmental Panel on Biodiversity and Ecosystem Services [8]. Nature-based solutions can offer effective approaches for addressing climate vulnerabilities and reducing risks through replacing or complementing traditional infrastructure approaches [5]. Additionally, natural and nature-based systems can respond to changing climatic conditions in a manner that engineered solutions often cannot, offering long-term holistic adaptation and mitigation outcomes [4,9,10].

Nature-based solutions are rapidly becoming a core component of what US climate service providers can offer to communities in support of their adaptation and resilience planning and implementation efforts. This review is, therefore, intended to help climate service providers understand the conceptual basis for NbS and explore a set of key considerations for the broader application of NbS in community-based adaptation and resilience planning. Specifically, the aim of this review article is to support increased or improved adoption and implementation of NbS in climate services to effectively reduce climate change impacts. This is accomplished through two main objectives. First, the article presents seven key considerations designed to guide climate service providers in the US through the process of incorporating NbS into community adaptation planning. Second, the article highlights factors required to be considered by climate service providers to help them embed these considerations across different levels and users of the climate service process. These key considerations stemmed from literature reviews, organizational expertise, and discussions with adaptation practitioners, and were developed in support of the US Climate Resilience Toolkit [11]. Although our focus here is on US applications and adaptation planning, these key NbS considerations have global relevance.

In general, climate services tend to focus on making climate information more available and accessible for decision makers at all levels, as well as filling data gaps as they arise and demand is identified. One of the greatest challenges for the uptake of NbS is their limited integration into existing climate services and use by climate service providers. Therefore, among the first steps in better integrating NbS will be to broaden the information scope of climate services and climate service providers to share NbS-relevant data and options in support of adaptation. Existing climate services frameworks already offer avenues through which the integration of NbS could be explored at each level of the climate service process. For instance, each component of the Global Framework for Climate Services proposed by Hewitt et al. [12]—users, user interface platform, climate services information system, observations and monitoring, and research, modeling, and prediction—can incorporate multiple nature-based considerations. Starting with ensuring that natural systems are included in observation and monitoring to designing research and models that can generate NbS-relevant projections is essential to creating a climate services information system that not only considers NbS (and natural systems more broadly) in its process but can also provide easily discoverable information (through a user interface platform) relevant to NbS so that users can apply the outputs to NbS implementation. Across all of this is the need to increase capacity of users and service providers at each level in order to expand skills and knowledge related to NbS.

As climate service providers support partners in developing adaptation strategies, there are many opportunities to incorporate NbS, including as part of hazard mitigation/risk reduction, restoration, and infrastructure development, and ensuring the sustainability of ecosystem services (e.g., water quality and quantity, and carbon sequestration) [9,10,13]. Indeed, NbS may often be able to address multiple climate stressors (temperature, flooding, drought, and sea level rise) in support of multiple local goals with less directed long-term management. For example, floodplain protection and/or management is an NbS that incorporates well into hazard mitigation planning to address flood impacts. Similarly, integration of vegetation buffers (e.g., forests and riparian habitat) around and through communities can serve to reduce flood risk, ameliorate thermal stress, and support ecosystem services (water quality, air quality, wildlife, and recreation), as well as aquifer recharge. Furthermore, NbS can level the playing field by offering more affordable, sustainable adaptation solutions, making them a particularly useful option for climate service providers to share with partner communities.

2. What Are Nature-Based Solutions

"Nature-based solutions" (NbS) refer to the use of natural systems and processes to deliver a variety of environmental benefits, especially for climate adaptation/resilience and climate mitigation goals. These strategies can range from planting trees and installing rain gardens in urban areas to provide shade and reduce stormwater flow to restoring rivers and floodplains to reduce flood risks and improve water quality. NbS include a broad range of strategies, from conservation of intact natural systems and restoration of priority ecosystems to the use of engineered systems designed to mimic natural system functions [10]. They also include nonstructural solutions such as open-space preservation through buyouts and easements. Nature-based strategies can complement structural solutions to form hybrid or "green/gray" systems for climate adaptation and risk reduction.

The International Union for the Conservation of Nature (IUCN) defines NbS as " ... actions to protect, sustainably manage and restore natural and modified ecosystems in ways that address societal challenges effectively and adaptively, to provide both human well-being and biodiversity benefits" [5].

This broad conception of NbS encompasses or intersects with several related terms and approaches, as noted above. This includes terms that completely overlap with the NbS concept (e.g., natural defenses, natural infrastructure, and ecosystem-based adaptation), as well as terms that are a subset of NbS tailored to a specific concern (e.g., green infrastructure (for stormwater management) and natural climate solutions (for carbon sequestration)). Preferred terminology may depend on the specific sector, community, or location. For instance, the term natural infrastructure often resonates with planners and policymakers accustomed to working with traditional gray infrastructure; accordingly, this NbS-related term is often used in federal and state policies and funding authorizations. The framing of NbS can also influence public perception and policy choices, which can lead to both ambiguity and overly narrow conceptualizations, thereby sometimes enabling practices with negative consequences for biodiversity and people [14,15]. The seven key considerations presented below are based on a broad conception of NbS and designed to reduce such ambiguity and overly narrow characterizations in order to promote the effective incorporation of the concept in climate services and adaptation more generally.

3. Key Considerations for Incorporating Nature-Based Solutions

In 2020, the IUCN published a collaboratively developed global standard for the design and application of NbS, offering a common framework for increasing the scale and impact of these approaches while seeking to avoid inconsistent and ungrounded applications of the concept [6]. Although the IUCN standard reflects a significant advance in mainstreaming NbS globally, the structure of that standard (eight criteria and 28 associated indicators) is highly conceptual and may not meet the practical needs of US climate service providers and their local community adaptation clients. Building on that standard, this paper proposes seven "key considerations" for the use of NbS (Figure 1) that can be embedded in existing climate services, products, and frameworks at the local and regional levels. Through incorporating these considerations, the field of climate services can expand the scope and relevance of its support for community-based adaptation clients and other users.

Figure 1. Key considerations for use of nature-based solutions.

3.1. Recognize Natural Systems and Processes as Critical Infrastructure

Climate service providers working through an adaptation planning process with communities typically focus first on project scoping, which includes identifying key community assets and critical infrastructure (e.g., schools, hospitals, emergency services, power plants and other utilities, and levees and seawalls). Because damage to or loss of these structures and the services they provide would have significant impacts on the health and safety of the community, these structures and services are typically priorities for protection from climate-related hazards. Although they are often not included within inventories of critical community assets, natural systems such as forests, rivers, floodplains, tidal wetlands, and coral reefs also provide crucial benefits and services to human communities, including protection from natural hazards such as flooding, erosion, and extreme heat [4,10,16–19]. The essential ecosystem services provided by natural systems also include other social, economic, and cultural benefits such as freshwater supplies, improved air and water quality, provisioning of food and other resources, pollination services, and recreational opportunities, as well as the nonmaterial benefits (e.g., cultural, spiritual, and aesthetic value) provided by natural ecosystems [5,20].

Many community benefits and ecosystem services provided by natural systems cannot be easily replaced by engineered structures, which can be costly to build and maintain and provide fewer additional benefits compared to functioning natural systems. For instance, living shoreline projects use natural techniques to stabilize shorelines, providing wave attenuation that buffers storm surge while also supporting birds, marine life, and local recreational opportunities, often costing less than conventional shoreline armoring techniques [21–23]. Some benefits provided by natural systems may also be difficult or impossible to replace once the natural system is degraded or lost. For instance, large-scale loss of forest cover can also result in significant alterations in local or regional hydrology, with resulting implications for community water supplies and water quality [24]. In many instances, the protective functions and other ecosystem services provided by these systems are likely to become even more critical as the climate changes, exacerbating coastal erosion, extreme heat events, water shortages, biodiversity loss, and other stressors impacting community safety and wellbeing [17].

For climate service providers, working with a community to identify critical natural systems/assets and features is essential for supporting climate adaptation, including the design and implementation of NbS. Practitioners can draw from a range of information sources to identify the natural systems and processes that comprise a community's natural assets, such as existing planning documents, local inventories, remote sensing, and community knowledge. Employing a range of methods results in a comprehensive understanding of natural assets that extends beyond designated parks and recreational infrastructure (e.g., piers and nature trails) to also include wetlands and waterways, riparian buffers and floodplains, urban tree canopies, and wildlife habitat corridors. During this process, it is important for climate service providers to help communities explore the full range of co-benefits and ecosystem services that natural assets can provide (e.g., climate regulation, storm surge protection, and job opportunities), as these may not have been previously discussed or identified as valuable by community members. For instance, a flood-prone community losing mangroves to development or expanding shrimp aquaculture may not be appropriately valuing the role of mangroves in flood and erosion control and as a nursery habitat that supports recreational fishing. The process of identifying natural systems and the critical services they provide to the community can also serve as an opportunity for community members to come together and build relationships that will support collaborative planning and implementation of specific NbS projects within the community. For climate service providers, the discussions that occur in these settings can illuminate community values and interests, paving the way for meaningful stakeholder engagement.

3.2. *Consider Climate Impacts on Priority Natural Resources*

The ability of natural systems to respond to disturbances by withstanding or recovering from the disruption, as well as past exposure leading to existing adaptation to such stresses, is one aspect of why NbS are attractive to communities as part of their adaptation planning [10,25]. For example, transport and deposition of sediments from higher up in the watershed results in accretion of soils in downstream wetlands and estuaries. This may provide enough additional land to replace erosional loses after storms or match land loss due to sea level rise [25,26]. However, climate change is likely to challenge ecosystems, making even intact, disturbance-adapted systems vulnerable to rapidly changing conditions and climate extremes. Ecosystems that have been altered or degraded by human land uses or activity are generally even more vulnerable to climate impacts, as anthropogenic stressors reduce the natural adaptive capacity of those systems to respond to and cope with change [10]. For example, urban encroachment and upstream dams can limit the natural movement of sediment into tidal marshes, preventing natural accretion and reducing their resilience to storms and sea level rise [27]. As a result, the vulnerability of human communities to the impacts of climate stressors and extreme climate events cannot be considered in isolation from how natural systems are affected by climate change. In doing so, the protective benefits and ecosystem services provided to those communities by natural systems could be overestimated. Evaluating both natural systems and human community vulnerabilities together allows for identifying where existing and intact natural systems are more likely to continue to deliver protective benefits and services, as well as where this may not be the case, requiring either additional assistance to restore ecosystem functioning or non-NbS solutions. Therefore, it is essential that climate services are inclusive of data and resources that also explore the implications of climate change for these natural systems.

The process of understanding how natural ecosystems are likely to be affected by climate change is typically accomplished through assessing the vulnerability of these resources [28]. A vulnerability assessment can support community adaptation planning through the following approaches:

- Identifying which priority natural assets are most and least likely to be impacted by current and projected climate conditions;
- Understanding why priority assets are vulnerable to inform the identification of possible adaptation actions that can reduce vulnerabilities and climate risks;
- Determining when and how to most effectively implement adaptation actions targeted toward priority natural assets.

Numerous approaches exist to assess the climate-related vulnerabilities and risks to species and ecological systems [28]. Just as all adaptation is local, selecting the right vulnerability assessment process is also dependent on the locality (e.g., resources present, detail required, data availability, and resource availability). When considering the climate change vulnerability of natural systems, climate service providers should seek to use the most ecologically relevant climate variables (which often may involve extremes rather than averages) and multiple future scenarios. Relying on ecologically relevant facts can provide the most accurate picture of ecosystem responses, but can be complex and may not necessarily be supported by the most widely accessible climate datasets. Ultimately, understanding the components of the vulnerability of ecosystems their associated species is essential to creating and implementing adaptation strategies that will successfully benefit human communities.

3.3. Consider Equity Implications in the Design and Application of NbS

Natural assets can be important to communities, offering valuable options for reducing climate risks and enhancing their overall wellbeing and resilience. However, NbS can also magnify existing inequities and/or create new challenges within a community. Historically, natural features such as parks, nature trails, and green spaces have benefitted predominantly white and more affluent communities [29]. Even when natural infrastructure is prioritized in low-income communities and communities of color, historical disinvestment and underinvestment in those same communities can make natural infrastructure projects less equitable. In Baltimore, Maryland, several smaller natural infrastructure projects installed by nongovernmental organizations and community groups have been predominantly located in areas with higher African-American populations that are less likely to have larger, city-funded projects. As compared to large-scale city-funded projects in neighboring communities, these non-city-led projects are limited to small rain gardens or micro-bioretention facilities [30].

Certain nature-based approaches such as creation of green spaces and floodplain acquisitions can also create new challenges and increase risks for socially vulnerable populations. Creating green spaces and other urban greening programs can increase housing costs and property values with several factors such as location (e.g., distance from downtown), scale, and function affecting whether a place gentrifies, risking displacement of the community members these strategies are intended to benefit [31]. Nonstructural solutions such as flood buyouts also commonly benefit whiter, wealthier, and more urban communities, although lower-value properties and properties owned by communities of color are more likely to accept and be bought out than higher-value properties through these programs [32].

Designing and implementing equitable NbS efforts are crucial to break the patterns of existing and historic inequities and build the adaptive capacity of socially vulnerable and marginalized communities. Natural and nature-based features, when prioritized in at-risk, socially vulnerable communities, can effectively address climate risks for these communities, as well as substantively contribute to an improvement in quality of life for community residents. Strategies such as inclusive and collaborative planning, partnership with tribal, indigenous, and other natural resource-dependent groups, and meaningful outreach and education can support representative NbS planning and implementation.

The effectiveness of climate products and services hinges on how well they center the needs of different users, particularly underrepresented and marginalized groups. This will require climate service providers to engage diverse stakeholders (e.g., community-based organizations) and co-produce products (e.g., risk assessments and adaptation action plans) that address user needs. Adaptation options driven by equitable and inclusive climate services will not only expand the usability of climate services by decision makers and public agencies, but also lead to more just outcomes in climate-resilient communities.

3.4. Ensure That NbS Yield Net Positive Biodiversity Benefits

Nature-based solutions should enhance biodiversity value and yield net biodiversity benefits, for instance, through incorporating site-specific designs and materials. Biodiversity includes multiple levels of biological organization—from genes and species to ecosystems—and each level of organization can be understood as consisting of three major components: composition, structure, and function [33]. NbS, depending on the type of project, may rely on one or more of these levels or components, for instance, a particular species (i.e., composition), habitat (i.e., structure), or biological processes (i.e., function). Strategies such as tree plantings using native and climate-resilient species, beaver reintroductions to restore wetlands and riparian areas, and living shorelines that use oysters and marsh grasses to stabilize coasts both offer protective benefits and strengthen long-term ecosystem resilience.

Natural ecosystems and native species are already under stress from a variety of anthropogenic sources, which, in many instances, has significantly reduced their natural adaptive capacity. Climate change is adding another layer of threats to already sensitive species or degraded systems, sometimes directly (e.g., higher temperatures and increasing drought) and, at other times, by exacerbating existing stresses. Addressing the species and ecosystem impacts of climate change, through targeted adaptation actions and climate-smart conservation practices [34], is essential to sustaining NbS functions. In doing so, it is important to seek net positive biodiversity and ecological outcomes, which not only slow ecological deterioration but also achieve actual ecological enhancement in value and function. Understanding climate change impacts on biodiversity, both spatially and temporally, will influence these decisions. In certain cases, priority habitats (e.g., sites containing the sole remaining populations of endangered species) will require an immediate emphasis. Similarly, areas where the effects of climate change are likely to be buffered and, therefore, hospitable for the lasting survival of particular species, also known as climate change refugia, may be useful to protect and prioritize.

Climate services offer an opportunity to embed biodiversity conservation outcomes in NbS design and implementation efforts. The recently released IPBES Values Assessment [35] highlights the current dominant, yet narrow, focus on short-term profits and economic growth when valuing nature in decision making as a key driver of the global biodiversity crisis. Climate service providers can help broaden this focus by bringing a holistic understanding of ecological values and services in the design and implementation of NbS. This can be achieved by embedding a combination of biodiversity data, local ecological knowledge, and multiple stakeholders throughout the generation and provision of climate service products. Additionally, continuous monitoring and evaluation of NbS to account for ecosystem uncertainties and climate impacts can mitigate unintended consequences, as well as inform future efforts to enhance the functionality and connectivity of ecosystems to achieve net biodiversity enhancement [36,37].

3.5. Seek to Protect or Restore Critical Natural Infrastructure

Intact natural systems are themselves at risk of climate change. Along the Gulf of Mexico, coastal ecosystems such as beaches and dune systems, offering protective benefits to nearby communities, are susceptible to erosion and conversion to open water due to sea level rise and saltwater intrusion [38]. Similarly, rangelands in Arizona, which offer habitat for an array of wildlife species, are experiencing mass mortality events due to more frequent and severe periods of droughts and climate change-fueled wildfires [39].

Protecting and restoring critical natural infrastructure will be essential adaptation strategies to help ensure that such systems will continue to provide ecosystem services and community benefits. This can involve prioritizing the protection of intact natural systems, restoration of degraded systems, incorporating nature-based features in engineered systems, and/or integration of natural (green) and engineered (gray) approaches in hybrid infrastructure. Protecting existing biodiversity and still extant natural systems should be a priority, but restoring the composition, function, or structure of already degraded ecological systems is becoming increasingly important for achieving adaptation and mitigation outcomes through NbS. In an era of rapid climate change, ecological restoration should not be viewed solely as a return to prior or historical states, but rather in the context of sustaining ecological function under current and future conditions. The International Standards for the Practice of Ecological Restoration [40] provides such a framework for guiding the development and implementation of ecological restoration projects. Ultimately, protection and restoration efforts will need to be taken in the light of broader climatic changes where the focus lies not just on preservation and restoration to historical conditions (i.e., managing for persistence), but one that is simultaneously open to anticipating and actively facilitating ecological transitions (i.e., managing for change) [34].

Climate services will need to integrate and embed ecological knowledge, as well as natural resource expertise to provide the required context for nature-based adaptation decisions. This includes efforts such as risk mapping and impact modeling for vulnerable ecosystems based on long-term climate and ecological datasets. Such integrated data serves as a useful climate service product for ecological and natural resource scientists and managers, planners, and representatives from federal, state and local agencies (e.g., fish, wildlife, and parks departments), conservation organizations, tribal and indigenous groups, equity-centered organizations, and others to share technical expertise, datasets, and knowledge of the region's natural resources.

3.6. *Give Natural Features and Processes Space to Function*

By nature, most intact ecosystems are dynamic and possess at least some ability to respond to change over time [25]. For example, coastal dune systems that are unrestricted by the presence of roads or development are constantly shifting as wind and waves move the sand, allowing them to naturally migrate inland as sea levels rise [41,42]. Rivers, wetlands, forests, and grassland systems all also have the ability to respond to environmental changes, through varied mechanisms such as sediment accretion (or erosion) and shifts in vegetation communities, among others [43–45]. In general, the systems that offer the most significant protective benefits to human communities are often those that have evolved to cope with a wide range of conditions and/or rapid fluctuations in environmental conditions, and so are well adapted to absorb the impacts of extreme weather and other climate-related hazards without significant degradation to the system or surrounding areas. However, in many places, human land uses have altered or constrained natural systems, preventing them from absorbing and responding to change and increasing exposure of surrounding communities to climate-related hazards such as flooding. For example, floodplains are well equipped to capture and hold excess stormwater, allowing it to be absorbed slowly and preventing downstream flooding [46]. Unfortunately, floodplains are often highly valued for development, and, where this occurs, the flood protection and erosion control benefits of this system are lost for surrounding natural and human communities, as well as those that lie downstream [47,48]. Furthermore, as climate change increases the frequency and severity of extreme precipitation events, inappropriate siting of new infrastructure and development is likely to cause even greater risks or create new hazards in areas that were not previously vulnerable to flooding.

Well-functioning natural systems also include complex processes that operate across a variety of spatial and temporal scales, such that what happens in one area may be inextricably tied to the functions of adjacent systems or more distant locations. For example, streamflow volume and water quality in a given stream reach is heavily dependent on the surrounding land uses, both upstream and in neighboring upland and riparian areas that absorb and filter runoff [49,50]. As a result, successful implementation of NbS designed to address flood risk or water quality issues would need to consider both the impacts of climate change and the hoped-for benefits of NbS at the watershed scale and not just the community scale.

Climate service providers supporting communities in expanding the use of NbS for hazard reduction and other co-benefits must think in terms of larger ecosystem scales and processes in order to ensure that natural systems have the space that they need to function. This becomes even more important as the climate changes and natural systems are responding to more extreme conditions, which may necessitate protection, restoration, or creation of systems or consideration of natural processes that extend across larger areas. In contrast to non-NbS approaches that require human intervention for modifications or adjustments (as well as constant maintenance), NbS designed at appropriate spatial and temporal scales have the potential to respond to changing conditions while still delivering desired ecosystem services.

3.7. Integrate NbS into Existing Planning Processes

The easiest way to include NbS is to simply adopt these measures as part of existing planning processes. This includes integrating climate-smart solutions into legally required land-use planning efforts, such as comprehensive or general plans, multi-hazard mitigation plans, community/neighborhood plans, and utility plans, as well as climate action plans that may or may not be required in any given jurisdiction. NbS as presented in the examples in this paper, can be offered by climate service providers as components of these plans that reduce climate risk, with the added potential benefits of reduced long-term maintenance cost, less direct management, possible autonomous improvement, and the support of associated ecosystems and their services.

To those ends, there is opportunity, heretofore underutilized, to include NbS as key climate change adaptation elements in traditional local planning processes. For example, aspects of each of the many elements of a local comprehensive or general plan (e.g., housing, transportation, public facilities, and environment) are vulnerable to climate change [51], and there is opportunity to incorporate NbS into planning for each of these sectors to reduce those vulnerabilities.

While NbS can be incorporated into almost any local planning process, there are also opportunities to build local capacity and uptake of NbS through the participants in those processes. This may require engaging multi-solution climate service provides, rather than just hard infrastructure-focused engineering firms, in order to focus on developing more holistic solutions that include NbS. At the same time, local community desires to include NbS could increase climate service provider awareness, and NbS-interested stakeholders included in local planning processes could increase community ability to implement NbS. For example, including natural resource managers and environmental justice stakeholders with natural system interests will help to identify opportunities for NbS that can support the needs of both nature and local communities.

As previously mentioned, climate services may also need to include additional information to support fully integrating NbS into local planning. For example, the data needed to understand the scope of local impacts and vulnerabilities may require different components (e.g., stream flow and timing, soil temperature, and species composition) at different scales (e.g., watershed and seasonal/decadal) with different thresholds (e.g., extremes and timing/phenology).

4. Conclusions

Given the vital role of climate services in delivering support to local entities in their efforts to develop effective community-based adaptation plans, ensuring that NbS are fully integrated into climate service offerings will be essential for achieving successful adaptation and mitigation outcomes. The seven "key considerations" outlined above are designed to guide the inclusion of NbS into local planning processes, resulting in improved adoption of climate adaptation actions that are sustainable for both communities and the ecosystems around them. In addition, designing and implementing NbS on the basis of these key considerations can offer adaptation solutions that may be less expensive and less fragile than comparable gray infrastructure options.

Climate services are meant to provide useful and usable climate information in a timely and tailored manner to support adaptation. To date, climate services have been envisioned to focus on delivery of climate data (e.g., temperature, precipitation, and sea level rise), socioeconomic data, vulnerability assessments, and guidance to assist users (individuals and decision makers). However, existing climate service frameworks can be improved and informed by inclusion of nature-based adaptation solutions in general and incorporation of these NbS-specific key considerations. With growing interest in and need for the use of natural and nature-based approaches for climate risk reduction, it is necessary to more explicitly embed NbS within the framework of any climate service to ensure the intended benefits of these services to all users.

Author Contributions: Conceptualization, A.P., L.E.H., L.J.H. and B.A.S.; writing, A.P., L.E.H., L.J.H. and B.A.S. All authors have read and agreed to the published version of the manuscript.

Funding: This project was funded by the Climate Resilience Fund (Grant #17755/17757-2021-4), in support of the US Climate Resilience Toolkit.

Institutional Review Board Statement: Not applicable.

Informed Consent Statement: Not applicable.

Data Availability Statement: Not applicable.

Acknowledgments: The authors gratefully acknowledge Patty Glick and Jessie Ritter (National Wildlife Federation), who contributed to a white paper from which this manuscript developed [11]. The authors also acknowledge Ned Gardiner (National Oceanic Atmospheric Administration), Kim Rhodes (Fernleaf), and Darren Long (Climate Resilience Fund), who provided input and support for our work on the development of these NbS key considerations.

Conflicts of Interest: The authors declare no conflict of interest.

References

1. Hewitt, C.D.; Stone, R. Climate Services for Managing Societal Risks and Opportunities. *Clim. Serv.* **2021**, *23*, 100240. [CrossRef]
2. Smith, A.B.U.S. *Billion-Dollar Weather and Climate Disasters, 1980–Present*; Accession 0209268; NOAA National Centers for Environmental Information (NCEI): Asheville, NC, USA, 2020.
3. Reidmiller, D.R.; Avery, C.W.; Easterling, D.R.; Kunkel, K.E.; Lewis, K.L.M.; Maycock, T.K.; Stewart, B.C. (Eds.) *Impacts, Risks, and Adaptation in the United States: The Fourth National Climate Assessment, Volume II*; U.S. Global Change Research Program: Washington, DC, USA, 2018.
4. Sutton-Grier, A.E.; Wowk, K.; Bamford, H. Future of Our Coasts: The Potential for Natural and Hybrid Infrastructure to Enhance the Resilience of Our Coastal Communities, Economies and Ecosystems. *Environ. Sci. Policy* **2015**, *51*, 137–148. [CrossRef]
5. Cohen-Shacham, E.; Walters, G.; Janzen, C.; Maginnis, S. *Nature-Based Solutions to Address Global Societal Challenges*; International Union for Conservation of Nature: Gland, Switzerland, 2016.
6. International Union for Conservation of Nature (IUCN). *IUCN Global Standard for Nature-Based Solutions: A User-Friendly Framework for the Verification, Design and Scaling up of NbS: First Edition*, 1st ed.; International Union for Conservation of Nature: Gland, Switzerland, 2020.
7. Millennium Ecosystem Assessment. In *Ecosystems and Human Well-Being: General Synthesis*; Island Press: Washington, DC, USA, 2005.
8. Intergovernmental Science-Policy Platform on Biodiversity and Ecosystem Services (IPBES). *Global Assessment Report on Biodiversity and Ecosystem Services of the Intergovernmental Science-Policy Platform on Biodiversity and Ecosystem Services*; IPBES: Bonn, Germany, 2019.
9. Depietri, Y.; McPhearson, T. Integrating the grey, green, and blue in cities: Nature-based solutions for climate change adaptation and risk reduction. In *Nature-based Solutions to Climate Change Adaptation in Urban Areas*; Springer: Cham, Switzerland, 2017; pp. 91–109.
10. Glick, P.; Powell, E.; Schlesinger, S.; Ritter, J.; Stein, B.A.; Fuller, A. *The Protective Value of Nature: A Review of the Effectiveness of Natural Infrastructure for Hazard Risk Reduction*; National Wildlife Federation and EcoAdapt: Washington, DC, USA, 2020.
11. Pathak, A.; Glick, P.; Hansen, L.; Hilberg, L.; Ritter, J.; Stein, B. *Incorporating Nature-Based Solutions into Community Climate Adaptation Planning*; NOAA Climate Smart Communities Series, Volume 5; National Wildlife Federation: Washington, DC, USA, 2022.

12. Hewitt, C.; Mason, S.; Walland, D. The global framework for climate services. *Nat. Clim Change* **2012**, *2*, 831–832. [CrossRef]
13. Seddon, N.; Chausson, A.; Berry, P.; Girardin, C.A.J.; Smith, A.; Turner, B. Understanding the value and limits of nature-based solutions to climate change and other global challenges. *Phil. Trans. R. Soc. B* **2020**, *375*, 20190120. [CrossRef] [PubMed]
14. Osaka, S.; Bellamy, R.; Castree, N. Framing "nature-based" solutions to climate change. *WIREs Clim. Change* **2021**, *12*, e729. [CrossRef]
15. Melanidis, M.S.; Hagerman, S. Competing narratives of nature-based solutions: Leveraging the power of nature or dangerous distraction? *Environ. Sci. Policy* **2022**, *132*, 273–281. [CrossRef]
16. Small-Lorenz, S.L.; Stein, B.A.; Schrass, K.; Holstein, D.N.; Mehta, A.V. *Natural Defenses in Action: Harnessing Nature to Protect Our Communities*; National Wildlife Federation: Washington, DC, USA, 2016.
17. Martin, T.G.; Watson, J.E.M. Intact ecosystems provide best defence against climate change. *Nat. Clim Change* **2016**, *6*, 122–124. [CrossRef]
18. Storlazzi, C.D.; Reguero, B.G.; Cole, A.D.; Lowe, E.; Shope, J.B.; Gibbs, A.E.; Nickel, B.A.; McCall, R.T.; van Dongeren, A.R.; Beck, M.W. *Rigorously Valuing the Role of US Coral Reefs in Coastal Hazard Risk Reduction*; US Geological Survey Open-File Report 2019-1027; US Geological Survey: Reston, VA, USA, 2019.
19. Rezaie, A.M.; Loerzel, J.; Ferreira, C.M. Valuing natural habitats for enhancing coastal resilience: Wetlands reduce property damage from storm surge and sea level rise. *PLoS ONE* **2020**, *15*, e0226275. [CrossRef]
20. Dahl, T.A.; Theiling, C.H.; Echevarria, W. *Overview of Levee Setback Projects and Benefits*; US Army Corps of Engineers ERDC/CHL CHETN-VII-17: Vicksburg, MS, USA, 2017.
21. Gittman, R.K.; Peterson, C.H.; Currin, C.A.; Joel Fodrie, F.; Piehler, M.F.; Bruno, J.F. Living shorelines can enhance the nursery role of threatened estuarine habitats. *Ecol. Appl.* **2016**, *26*, 249–263. [CrossRef]
22. Narayan, S.; Beck, M.W.; Wilson, P.; Thomas, C.; Guerrero, A.; Shepard, C.; Reguero, B.G.; Franco, G.; Ingram, C.J.; Trespalacios, D. *Coastal Wetlands and Flood Damage Reduction: Using Risk Industry-Based Models to Assess Natural Defenses in the Northeastern USA*; Lloyds Tercentenary Research Foundation: London, UK, 2016.
23. Powell, E.J.; Tyrrell, M.C.; Milliken, A.; Tirpak, J.M.; Staudinger, M.D. A review of coastal management approaches to support the integration of ecological and human community planning for climate change. *J. Coast Conserv.* **2019**, *23*, 1–18. [CrossRef]
24. Ellison, D.; Morris, C.E.; Locatelli, B.; Sheil, D.; Cohen, J.; Murdiyarso, D.; Gutierrez, V.; Noordwijk, M.; van Creed, I.F.; Pokorny, J.; et al. Trees, forests and water: Cool insights for a hot world. *Glob. Environ. Change* **2017**, *43*, 51–61. [CrossRef]
25. Spalding, M.D.; McIvor, A.L.; Beck, M.W.; Koch, E.W.; Möller, I.; Reed, D.J.; Rubinoff, P.; Spencer, T.; Tolhurst, T.J.; Wamsley, T.V.; et al. Coastal ecosystems: A critical element of risk reduction. *Conserv. Lett.* **2014**, *7*, 293–301. [CrossRef]
26. Morris, J.T.; Sundareshwar, P.V.; Nietch, C.T.; Kjerfve, B.; Cahoon, D.R. Responses of coastal wetlands to rising sea level. *Ecology* **2002**, *83*, 2869–2877. [CrossRef]
27. Peteet, D.M.; Nichols, J.; Kenna, T.; Chang, C.; Browne, J.; Reza, M.; Kovari, S.; Liberman, L.; Stern-Protz, S. Sediment starvation destroys New York City marshes' resistance to sea level rise. *Proc. Natl. Acad. Sci. USA* **2018**, *115*, 10281–10286. [CrossRef] [PubMed]
28. Glick, P.; Stein, B.A.; Edelson, N.A. *Scanning the Conservation Horizon: A Guide to Climate Change Vulnerability Assessment*; National Wildlife Federation: Washington, DC, USA, 2011.
29. Wolch, J.R.; Byrne, J.; Newell, J.P. Urban green space, public health, and environmental justice: The challenge of making cities 'just green enough'. *Landsc. Urban Plan.* **2014**, *125*, 234–244. [CrossRef]
30. University of California, Davis. Is Green Infrastructure a Universal Good? Available online: https://www.caryinstitute.org/storymap-green-infrastructure-universal-good (accessed on 12 September 2022).
31. Rigolon, A.; Németh, J. Green gentrification or 'just green enough': Do park location, size and function affect whether a place gentrifies or not? *Urban Stud.* **2020**, *57*, 402–420. [CrossRef]
32. Elliott, J.R.; Brown, P.L.; Loughran, K. Racial inequities in the federal buyout of flood-prone homes: A nationwide assessment of environmental adaptation. *Socius* **2020**, *6*, 237802312090543. [CrossRef]
33. Noss, R.F. Indicators for monitoring biodiversity: A hierarchical approach. *Conserv. Biol.* **1990**, *4*, 355–364. [CrossRef]
34. Stein, B.A.; Glick, P.; Edelson, N.; Staudt, A. *Climate-Smart Conservation: Putting Adaption Principles into Practice*; National Wildlife Federation: Washington DC, USA, 2014.
35. Intergovernmental Science-Policy Platform on Biodiversity and Ecosystem Services (IPBES). *Summary for Policymakers of the Methodological Assessment of the Diverse Values and Valuation of Nature of the Intergovernmental Science-Policy Platform on Biodiversity and Ecosystem Services*; IPBES secretariat: Bonn, Germany, 2022.
36. Kumar, P.; Debele, S.E.; Sahani, J.; Rawat, N.; Marti-Cardona, B.; Alfieri, S.M.; Basu, B.; Basu, A.S.; Bowyer, P.; Charizopoulos, N.; et al. An overview of monitoring methods for assessing the performance of nature-based solutions against natural hazards. *Earth-Sci. Rev.* **2021**, *217*, 103603. [CrossRef]
37. Hoffman, J.R.; Hansen, L.J. *Moving from Faith-based to Tested Adaptation Process and Approach: How Will We Know We're Adapting? NOAA Climate Smart Communities Series, Volume 4, Adaptation Insight*; NOAA Technical Report NOS 1: Silver Spring, MD, USA, 2022.
38. Sweet, W.V.; Hamlington, B.D.; Kopp, R.E.; Weaver, C.P.; Barnard, P.L.; Bekaert, D.; Brooks, W.; Craghan, M.; Dusek, G.; Frederikse, T.; et al. *Global and Regional Sea Level Rise Scenarios for the United States: Updated Mean Projections and Extreme Water Level Probabilities along US Coastlines*; NOAA Technical Report NOS 1: Silver Spring, MD, USA, 2017.

39. Briske, D.D.; Joyce, L.A.; Polley, H.W.; Brown, J.R.; Wolter, K.; Morgan, J.A.; McCarl, B.A.; Bailey, D.W. Climate-change adaptation on rangelands: Linking regional exposure with diverse adaptive capacity. *Front. Ecol. Environ.* **2015**, *13*, 249–256. [CrossRef]
40. McDonald, T.; Gann, G.D.; Jonson, J.; Dixon, K.W. *International Standards for the Practice of Ecological Restoration–Including Principles and Key Concepts*; Society for Ecological Restoration: Washington DC, USA, 2016.
41. Hesp, P.A.; Walker, I.J. Coastal dunes. In *Treatise on Geomorphology, Vol. 11 Aeolian Geomorphology*; Academic Press: San Diego, CA, USA, 2013; pp. 328–355.
42. Leo, K.L.; Gillies, C.L.; Fitzsimons, J.A.; Hale, L.Z.; Beck, M.W. Coastal habitat squeeze: A review of adaptation solutions for saltmarsh, mangrove and beach habitats. *Ocean Coast. Manag.* **2019**, *175*, 180–190. [CrossRef]
43. Gasith, A.; Resh, V.H. Streams in mediterranean climate regions: Abiotic influences and biotic responses to predictable seasonal events. *Ann. Rev. Ecol. Syst.* **1999**, *30*, 51–81. [CrossRef]
44. Stromberg, M.R.; Corbin, J.D.; D'Antonio, C. (Eds.) *California Grasslands: Ecology and Management*; Univ of California Press: Berkeley, CA, USA, 2007.
45. Swanson, M.E.; Franklin, J.F.; Beschta, R.L.; Crisafulli, C.M.; DellaSala, D.A.; Hutto, R.L.; Lindenmayer, D.B.; Swanson, F.J. The forgotten stage of forest succession: Early-successional ecosystems on forest sites. *Front. Ecol. Env.* **2011**, *9*, 117–125. [CrossRef]
46. Colloff, M.J.; Lavorel, S.; Wise, R.M.; Dunlop, M.; Overton, I.C.; Williams, K.J. Adaptation services of floodplains and wetlands under transformational climate change. *Ecol. Appl.* **2016**, *26*, 1003–1017. [CrossRef]
47. Mahmoud, S.H.; Gan, T.Y. Urbanization and climate change implications in flood risk management: Developing an efficient decision support system for flood susceptibility mapping. *Sci. Total Environ.* **2018**, *636*, 152–167. [CrossRef]
48. Ntelekos, A.A.; Oppenheimer, M.; Smith, J.A.; Miller, A.J. Urbanization, climate change and flood policy in the United States. *Clim. Change* **2010**, *103*, 597–616. [CrossRef]
49. Hasan, S.S.; Zhen, L.; Miah, M.G.; Ahamed, T.; Samie, A. Impact of land use change on ecosystem services: A review. *Environ. Dev.* **2020**, *34*, 100527. [CrossRef]
50. Richardson, C.J.; Flanagan, N.E.; Ho, M.; Pahl, J.W. Integrated stream and wetland restoration: A watershed approach to improved water quality on the landscape. *Ecol. Eng.* **2011**, *37*, 25–39. [CrossRef]
51. Hansen, L.J.; Nordgren, S.J.; Mielbrecht, E.E. *Climate Change Adaptation through Local Comprehensive Planning: Guidance for Puget Sound Communities*; EcoAdapt: Bainbridge Island, WA, USA, 2016.

sustainability

MDPI

Article

Young People Are Changing Their Socio-Ecological Reality to Face Climate Change: Contrasting Transformative Youth Commitment with Division and Inertia of Governments

Alfredo Pena-Vega [1,*], Marianne Cohen [2], Luis Manuel Flores [3], Hervé Le Treut [4], Marcelo Lagos [5], Juan Carlos Castilla [6], Aurora Gaxiola [6] and Pablo Marquet [6]

1 Laboratoire d'Anthropologie Politique, EHESS-CNRS UMR 8177, 75006 Paris, France
2 Geography and Planning Department, UR Mediations, Faculty of Letters, Sorbonne Université, 75005 Paris, France
3 Facultad de Educación, Pontificia Universidad Católica de Chile, Santiago 8331150, Chile
4 Institut Pierre-Simon Laplace, Membre de l'Académie des Sciences, 75005 Paris, France
5 Instituto de Geografía, Pontificia Universidad Católica de Chile, Santiago 8331150, Chile
6 Facultad de Ciencias Biologicas and Centro de Cambio Global, Pontificia Universidad Católica de Chile, Santiago 8331150, Chile
* Correspondence: penavega@ehess.fr

Abstract: This paper contributes to a critical re-reading of the notion of climate services. It does so by problematizing the discontinuity between young people's commitment to climate change, and the lack of a common vision regarding climate policy among governments. In this essay, youth commitment is characterized in terms of participation in the Global Youth Climate Pact (GYCP, 2015–2022). Here, young people share projects from their own high schools and communities and participate in a citizen consultation. Most projects have achieved a good success score, increasing over the years, especially for those carried out in emerging and developing countries. Some of them were presented at the COPs. In contrast, a textual analysis of intended nationally determined contributions (INDC) illustrates divergent understandings of the Paris Agreement and exemplifies the poor results of governmental climate diplomacy. This study establishes the need to closely monitor early warning signs of climate change in conjunction with high schools and school communities. The initiatives of young people are building a civic and planetary awareness for climate change in contrast with governmental division and inertia. In this sense, climate services, directed to young people, could contribute to design a sustainable future. We approach the practices, attitudes, and commitments of young people from the angle of cooperation rather than a moral vision of responsibility. Particularly, we propose a dialogical link between the treatment of climate issues and its effects on the constitution of networks, notably as they relate to practices of action, that is, the way in which distinct groups of young people develop relationships with their environments, organize themselves, and act and transform reality.

Keywords: young people; involvement; climate services; knowledge; awareness; transformation action

Citation: Pena-Vega, A.; Cohen, M.; Flores, L.M.; Le Treut, H.; Lagos, M.; Castilla, J.C.; Gaxiola, A.; Marquet, P. Young People Are Changing Their Socio-Ecological Reality to Face Climate Change: Contrasting Transformative Youth Commitment with Division and Inertia of Governments. *Sustainability* **2022**, *14*, 15116. https://doi.org/10.3390/su142215116

Academic Editors: Charles Herrick, Jason Vogel and Glen Anderson

Received: 31 July 2022
Accepted: 27 October 2022
Published: 15 November 2022

1. Introduction

The environmental challenges that humanity is facing in the 21st century, and climate change in particular, require a radical transformation in the way we dwell in, inhabit, and understand nature. Responses given by the governments are still below requirements, despite the mobilization of part of the international community [1], including researchers, citizens, and particularly, young people. A complex and integrative vision of the world is needed to tackle climate change issues, including the cultural diversity between countries [2,3] and generations. There is an urgent need to change the way we prepare the next generation of scientists and social leaders to effectively deal with the problems of the Anthropocene [4]. Unfortunately, there is a shortfall in the way we effectively teach subjects such as climate change in the classroom due to a variety of causes, ranging from ideology, inadequate training, and state or country level ordinances, for example [5–7]. This is a cause for concern, considering the growing disconnect between young people and nature [8].

To what extent are climate services an appropriate method to undertake such issues? According to the analysis of the 27 volumes published by the Climate Services Journal (2016–2022, Appendix A), climate services seek to provide understandable climate data and scenarios to facilitate decision-making by individuals, governments, economic interests, and public sector actors. Climate services often entail the co-production of indicators that make sense to stakeholders, considering their perceptions and knowledge and evaluating the relevance of communication tools (web service, mapping, application). However, the climate services enterprise needs to be challenged to better support interdisciplinary, action-directed educational efforts. As an original contribution, this notion has not yet been applied for education purposes nor addressed to young people. Our approach is, therefore, innovative, seeking to increase awareness among a social category thus far under-addressed in the climate services literature and research. In this sense, our project adopts an action-based research approach to inform future leaders who will have to cope with climate change. Co-construction is at the heart of the pedagogical approach we advocate.

In the first section, this paper reports the case study of the Global Youth Climate Pact (GYCP hereafter) project and the methodology used to monitor and assess its efficiency and effectiveness, compared to the inertia and division of governments. The second section traces how GYCP was driven by young people's growing concern about climate change, the transformation that the GYCP project generates in students, in their awareness of the problem and in their eagerness to become actors of change. The third section compares this transformative pathway with the fragmentation of views and inaction of the governments, highlighted by a mapping of the discourse underlying the contribution made by governments to the Paris agreement. The conclusion puts this experiment in perspective with those conducted in the context of climate services.

2. Presentation of the Study Case and the Methodology

2.1. The GYCP Project: An Urgent Need for an Active and Participative Pedagogy

The GYCP project began in 2014 as a response to the critical lack of involvement from young generations and insufficient consideration of the human dimensions of climate change in the COP meetings and reports. Developed in 30 countries (Figure 1), it involves more than 12,000 young people and prioritizes their awareness and continuing education. The question was whether young people were aware of the challenges and opportunities that a pathway toward a low-carbon model entails and ready to join the world debate on the ways to reach it. In this context, we contacted schools and science teachers in different countries to carry out a pilot experience of reflecting on climate change in the school and in their localities. This reflection process led to the organization of projects by the students and required further knowledge to support the work of both teachers and students. Scientists have a leading role to play in this process. As agents of knowledge dissemination, they interacted with students to explore scientific results, issues and uncertainties about global warming, and helped build their pilot projects.

Our approach differs from traditional pedagogy, which tends to treat young people as mere receptacles of adult knowledge. In this context, it is critical that climate change is understood by young people and not merely explained to them. This new paradigm needs to tackle creative complexity, particularly in the field of education [9–11]. It points to the need for a new teaching strategy that embraces the interconnected nature of planetary society, where fundamental notions such as uncertainty, interdependency, and nonlinearity become embodied knowledge, thus, aligning the increasing commitment of young people with their ownership of scientific knowledge. This issue is opening a new field of research [10,12–15] and action–research projects such as the Global Youth Climate Pact (http://www.globalyouthclimatepact.eu/ (accessed on 30 July 2022).

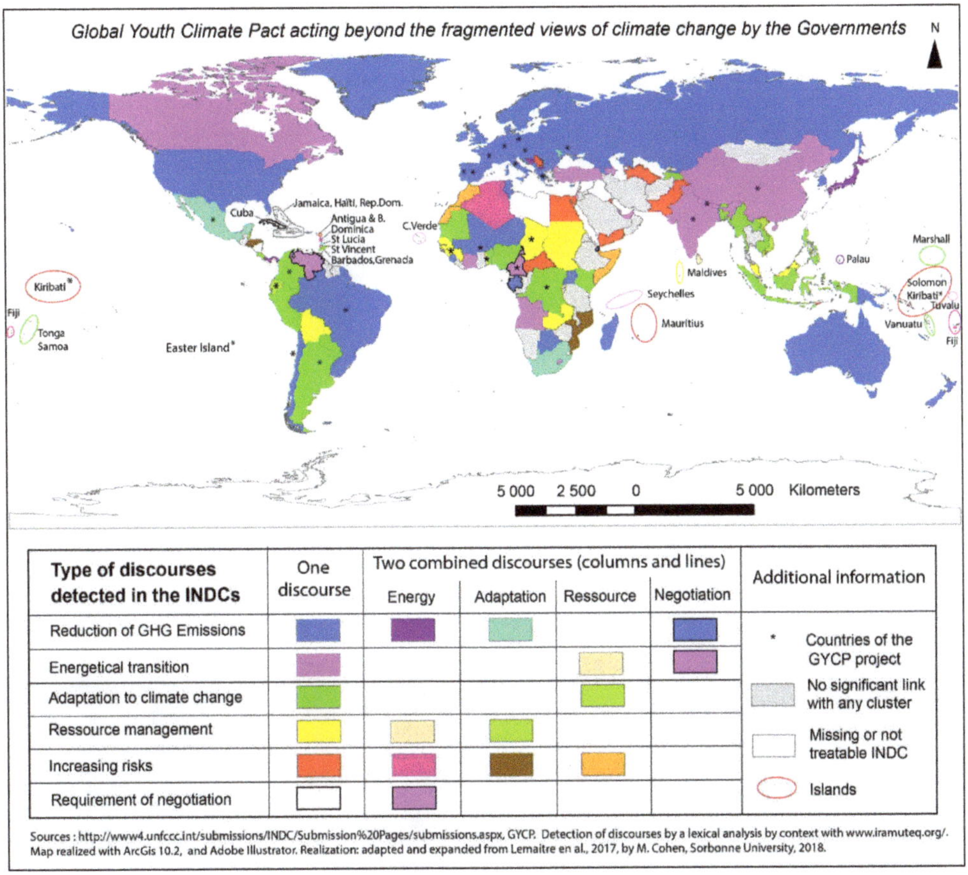

Figure 1. Map of the discourse underlying the intended nationally determined contributions (INDC) to the Paris Agreement and locations of the GYCP projects. Realization: Marianne Cohen (Material English, French and Spanish corpus: INDC https://unfccc.int, accessed on 30 July 2020, location of GYCP projects: http://www.globalyouthclimatepact.eu, accessed on 30 July 2022; Method: lexical analysis by context of the 3 corpus with IraMuTeQ free software; Mapping: ArcgisPro© software.

2.2. *Young People, at the Crossroads of Knowledges and Cultural Diversity*

Contrary to current stereotypes of young people [16,17], the GYCP experience demonstrates that they are not disengaged from the ongoing climatic crisis. They are eager to understand and find ways to generate actions in the face of governmental inaction. They desire to be agents of change now and in the future. Drawing upon this awareness, one can conclude that there is a great opportunity to propose action-oriented pedagogical experiences that nurture this interest and transform it into ways of knowing and creating citizens that can fit into a carbon neutral world by 2050. This requires, however, rethinking the dominant discipline-oriented teaching paradigm and working instead toward problem-oriented strategies that allow and enable contributions from all disciplines, integrating and crossing knowledges and enhancing the open-mindedness of all the actors of the teaching system [18,19]. In particular, it is critical to develop an open-minded learning by doing and to create learning institutions that serve human interactions. In this sense, the GYCP experience points toward the opportunity and need for a radical re-reading of the notion of climate services (Appendix A).

2.3. Methodology Used to Monitor and Assess the GYCP Project

Several methods were used to monitor the projects and assess the growing awareness of the students and to co-construct a common parlance and vision while respecting cultural diversity. During the large meetings that brought together the delegations of all the countries, we used crowdsourcing to question a large panel of participants dynamically, including those who were physically present. This method, implemented with the help of an external service provider specializing in this technique, made it possible to obtain significant statistical results. The main question asked referred to the degree of sensitivity of young people to the impact of climate change. Three editions followed one another, in 2015 during the COP 21 with 600 participants, in 2017 with 918 participants, and in 2019 during the COP24 in Madrid with 300 participants.

During the development of the project in each country, we used focus groups to capture and characterize the experience of specific groups of students. The focus groups made it possible to bring to light deep-seated questions on the part of the young people and stimulate the appropriation of the knowledge related to climate change, its integration in their cultural reality, and the elaboration of action-oriented projects.

We also sought to evaluate the success of 53 projects by developing a composite index, taking into account the focus of the projects, their ability to be disseminated in the wider youth community, and the level of concrete realization they achieved. Each of these criteria were scored from 1 to 4 (Table 1), providing a readable indicator for evaluating the projects and comparing their level of success.

Table 1. Multi-criteria assessment of the success of projects.

Indexes	Focus	Dissemination	Effectiveness
1	General	Only one class	Meeting
2	Accurate but limited	More high school class	Community
3	Accurate and concerning more people	More high schools	Community, decision makers
4	Targeting	Community	Implementation of concrete experiments

Finally, we compared the accomplishment of the young people who participate in the GYCP project with governmental inertia and fragmentation of national policies. With this aim, we overlaid the location of all the GYCP projects with a map highlighting the diverse ways governments understood their engagement to the Paris Agreement. This map was produced by analyzing the intended contributions of 191 countries with an automatic method (Appendix B).

3. The GYCP Project, a Process from Consciousness to Action

3.1. A Growing Youth Concern and Engagement

According to our first (2015) crowdsourcing survey during the COP21 with 600 young people, more than 89% of respondents said that they and their families were concerned about climate change. While it is not surprising that individuals attending a COP would be highly focused on climate change issues, this is nevertheless a high proportion, especially when considered in light of pervasive, widespread stereotypes of a youth that is uninformed and uninterested in climate issues [16,17]. We differentiated the answers according to the origin of the young people: developed countries (European origin), emerging countries (China, India, Brazil, Colombia, Chile), and developing countries (Guinea, Burkina-Faso, Lebanon, Nepal). Thus, while 87% of young people from developed countries said they were concerned, 100% of young people from emerging countries and 94% of those from developing countries did so. The level of concern is higher in emerging and developing countries and lower among young people from developed countries, perhaps because they perceive climate change as a distant threat, removed from their lives both in space and time. It could be argued that at that time (2015), in the minds of people and including young people, climate change risks were perceived as non-personal, about the future, other places, and other species (plants and animals, not humans) [1].

Comparing these results with the 2017 survey on perceived impacts, we also see a shift. Indeed, among the 918 exchanges of views that generated the most interest, 41.5% believe that extremely negative impacts will be felt in terms of rising sea levels, land flooding, and the disappearance of some cities (Venice) or island countries (Kiribati, Maldives). In 2017, the proportion of the risk of natural disasters was twice as low (21%), on par with concern about a disruption of the seasonal cycle.

During the COP25 held in Madrid in 2019, this concern was shared by 15.4% about one or more of these disasters. In the same year, the second most important concern (24%) was about the lack of water resources, desertification, and the increased risk of famine, a result similar to 2017. A smaller proportion of exchanges (18.9%) revealed concern about the reduction in biodiversity, more than twice as many as in 2017. When asked about the impact of climate change on their way of life, 300 young people from 8 European and Latin American countries gave a clear answer: more than 90% of the responses from young people from Europe and Latin America agreed that global warming would have negative or very negative consequences. There is a convergence of views compared to the 2015 survey, the negative or very negative impacts of global warming are, all things considered, identical in terms of percentages. In 2019, unlike in 2015, young Europeans were proportionally more likely than young Latin Americans to perceive negative impacts; they were no longer seen as a distant threat by 96.5% of them [20].

We take these figures as an illustration that shows that students from emerging countries were more likely to experience, in their own lives, the impact of climate change in the year 2015, whereas the concern about climate change is nowadays shared by all students whatever country they are living in.

3.2. *Young People May Be the Wellspring of Socio-Ecological Change*

Concrete examples are given by the diagnosis established by learners in contrasting geographical contexts (Figure 1). Projects carried out under the GYCP came from young people living all over the world: central France, Colombia, Brazilian Amazonia, rain forest in central Africa, semi-arid northern Chile, Easter Island, and northern Argentina, and all highlight a multi-level ecological deficit. From their diagnosis, they set up specific action projects and contribute to socio-ecological change in their territories. During this process, scientists provide their knowledge, and learners contextualize it to elaborate a diagnosis and further an action plan, teachers being the linchpin.

A key point, brought in by the students' projects, is how ancestral knowledge may contribute to thinking about the future under climate change, particularly in territories where the memory of the past has a great importance in knowledge transmission. For example, in the Puyanawa indigenous community of Brazilian Amazonia (Appendix C), a link was established between the students and the elders' knowledge, and the diagnosis and action project were translated into the indigenous experience. Other projects by young Pygmies belonging to Bantu, Nilotic, and Sudanese communities, living in the Congo Basin Forest, were engaged in combating deforestation as "forest gatekeepers". In Easter Island, students were engaged in the reactivation and reinterpretation of the Rapa Nui techniques of "rock gardens" as a way of enhancing the sustainability of agriculture and addressing water shortages affecting the territory (Appendix C). From this insular experience, learners suggest rethinking "our planet like an island in the middle of the Universe". Other proposals are oriented on innovative technologies without abandoning the native heritage. In the Chincolco agricultural school in central Chile, learners are elaborating on a multi-objective hydroponic technology to adapt to water shortages (Appendix C), while students in the Azapa Valley of Northern Chile are reflecting on improving carbon sequestration in their local vegetation and wetlands. Young Colombians, on the other hand, are developing an agroecology book to bring alternatives to rural populations facing climate change, thus, becoming actors in one of the most important mitigation and adaptation strategies for climate change, known as nature-based solutions [18,21,22]. In a French rural territory and in northern Argentina, scholars are rethinking the carbon footprint of their school canteen or their city and bringing a proposal at the regional level.

This demonstrates how each contextualized experience may facilitate or compel socio-ecological change when it is built upon a place-based reality anchored in experience, including the most recent projects designed during the health context of the pandemic. All these examples are creating virtuous circles in which young people are at the source of socio-ecological change, bridging the gap between traditional knowledge, local experiences, innovative solutions, and transformative change.

In the process of building projects with young people, we did not try to transpose the UN concepts. They were free to choose their own words, and what was important was creativity. The frequency of key words in the titles of the projects shows the way in which young people appropriate their "climate reality". Thus, among the titles of the 54 projects, 10 mention environmental issues, 9 citizenship and awareness, and the same number education or resources. According to a focus group, these notions underlie the notion of "transformation" that is dear to the hearts of the young people. The projects related to the climate issue as such occupy only the fifth position with eight efforts directly focusing on climate change. The notion of adaptation is not mentioned, and mitigation is rarely put forward (six projects), and this is the case regardless of the level of development of the countries where the young people live. Project success indicators show that most projects achieved a good score, increasing over time (average score from 4.4 in 2015 to 9.8 in 2021, decreasing to 7.2 in 2021 due to the COVID-19 pandemic). The level of success of the projects was inversely proportional to the level of development of the countries, the average score being 4.9 in developed countries, 6.6 in emerging countries, and 8 in developing countries. This shows that young people in countries where the effects of climate change are most dramatic are the most involved in finding local solutions (see Table 2).

Table 2. Results of the Multi-criteria assessment of projects.

Title of the Project	Country	Type	Year	Focus	Dissemination	Effectiveness	Score
Environmental education	China	Developed	2015	1	1	1	3
Environmental awareness	Italy	Developed	2015	1	1	1	3
Gardens in the school	Poland	Developed	2015	4	1	1	6
Environmental awareness	Romania	Developed	2015	4	1	1	6
Citizenship and environment	Ukraine	Developed	2015	2	1	1	4
Biosorbent	Colombia	Developing	2016	1	1	1	3
Crossroads Climate Program	Guinea	Developing	2016	3	4	3	10
Climate change global—ocean	Kiribati	Developing	2016	2	4	2	8
Waste management in Kathmandu	Nepal	Developing	2016	3	4	2	9
Stone gardens from the ancestors of Rapa Nui	Rapa-Nui	Developing	2016	4	4	4	12
CO_2 outside Melle city	France	Developed	2016	1	1	1	3
Education environment	Brazil	Emergent	2016	1	1	1	3
Planetary citizenship, sustainability	Brazil	Emergent	2016	1	1	1	3
Carbon sequestration in Azapa Valley soils	Chile	Emergent	2016	3	4	4	11
Climate change monitoring—Chiloé Island	Chile	Emergent	2016	1	1	2	4

Table 2. *Cont.*

Title of the Project	Country	Type	Year	Focus	Dissemination	Effectiveness	Score
Environment social service	Colombia	Developing	2017	4	4	1	9
Eco Blanket and spherical panel	Colombia	Developing	2017	1	1	1	3
Higher Education Organized for the Prevention of Environmental Damage	Colombia	Developing	2017	4	4	3	11
Fight against global warming and environmental education	Congo	Developing	2017	3	4	3	10
Educobien's student alliance	Venezuela	Developing	2017	4	3	1	8
Human regulation of ecosystems in the Chizé forest.	France	Developed	2017	1	1	1	3
CO_2 Footprint in Santiago's High Schools	Chile	Emergent	2017	1	1	1	3
Anthropic disasters in the O'Higgins region	Chile	Emergent	2017	3	1	1	5
Hydroponic system for all	Chile	Emergent	2017	4	4	4	12
Domestic wastes compacter	Colombia	Developing	2018	3	4	1	8
Land rescue	Colombia	Developing	2018	1	3	1	5
One million trees for the world	Colombia	Developing	2018	1	4	1	6
Walking with the Frailejones	Colombia	Developing	2018	1	1	1	3
Carbon footprint, San Pedro de Jujuy	Argentina	Emergent	2018	4	2	2	8
Puyanawa: Traditional Knowledges and the challenges changes	Brazil	Emergent	2018	3	4	2	9
Compost generation from organic matter produced in food specialties.	Chile	Emergent	2018	3	2	1	6
Environmental awareness	Morocco	Developing	2019	4	1	2	7
To reduce the carbon footprint of the school canteen	France	Developed	2019	3	1	1	5
Nature in the city	Spain	Developed	2019	4	4	3	11
Raising awareness of the climate problem	Spain	Developed	2019	3	1	1	5
Climate change and rural youth	Chile	Emergent	2019	1	1	1	3
Bocashi, a natural fertilizer	Chile	Emergent	2019	1	3	1	5
The impact of our footprints	Chile	Emergent	2019	3	3	1	7
Coyhaique, an acid city?	Chile	Emergent	2019	4	4	1	9
Coyhaique submerged	Chile	Emergent	2019	3	4	1	8
Water explorers in Patagonia	Chile	Emergent	2019	3	4	1	8
Social research on the harvesting of wild fruits in Balmaceda	Chile	Emergent	2019	2	1	1	4
The school garden of the 21st century.	Chile	Emergent	2019	3	3	1	7
Natural fertilizers	Chile	Emergent	2019	3	4	1	8

Table 2. *Cont.*

Title of the Project	Country	Type	Year	Focus	Dissemination	Effectiveness	Score
Biodiversity and deforestation	Cameroon	Developing	2020	4	4	3	11
Gayer—reforestation	Madagascar	Developing	2020	3	4	2	9
Reforestation	Mali	Developing	2020	3	4	1	8
Youth and the environment	Rwanda	Developing	2020	4	4	3	11
Waste management	Panama	Developing	2021	3	4	1	8
Reforestation with native trees	Chile	Emergent	2021	4	4	2	10
Rainwater harvesting	Chile	Emergent	2021	4	3	2	9
Understanding Antarctica to understand the climate	Chile	Emergent	2021	3	1	1	5
Education climate change	India	Emergent	2021	3	1	1	4

3.3. *Youth Acting for the Adaptation to Climate Change*

These experiences are all based on the desire " to act now", but what is really acting? Contrary to the statement of Claudia Gorr [23], young people who participate in the GYCP do not agree that "citizens cannot do anything to mitigate climate change". This common vision also emerges from the claim for political action raised by the growing youth mobilization against climate change. In the GYCP, participant's drive to take action rests on their engagement and on a collective and reflexive participation primarily oriented towards adaptation to climate change and, secondarily, towards its mitigation. For half of them, "I, us, young people" can act against climate change; it is up to them. A considerable proportion think that it requires the involvement of all the inhabitants of the planet, with only a few thinking that governments have to play the major role. Contrary to other studies, the experience associated with the projects developed by students raised awareness of the importance of becoming involved in the struggle against one of the most important problems facing society, climate change. The experience of the GYCP provides information that appears to counter the conventional wisdom that young people do not take seriously the issues related to climate change. Nevertheless, there are clearly barriers to overcome in order to change the way we explain complex and multifactorial problems, such as climate change. This requires finding new forms of explanation. This problem could be addressed by enhancing a deep understanding by teaching through actions and by bringing scientists to schools to monitor the progress of projects and provide scientific evidence, which has the important co-benefit of fostering an appreciation for science and science-based actions.

4. GYCP Is at Odds with the Division and Inertia of the Governments

The actions of youth seem at odds with the *climate paradox*: the more acute the consequences of climate change, the more divided and paralyzed are decision-makers [1,24,25]. Due to their engagement in local actions, such as those illustrated through the GYCP, young people provide hope and scope for action. Their projects were presented at the different COPs. In Paris, during the COP21 in 2015, we presented 21 proposals from 10 countries in a side-event to the organizers. The same happened at COP 22 in Marrakech in 2016 and at COP 23 in Bonn. In Katowice at COP 24, in 2018, posters reporting on the different projects (Appendix C) received a warm welcome from the assembled authorities, including ministers of the environment, the UN youth representative, the European Commission environment representative, and mayors of large cities. In Madrid, we were invited by the president of the COP to present our proposal for the future (six items: education) [26–28], valuing forests, reduced carbon footprint, biodiversity, resources, water and soil, sustainable agriculture, protection of the oceans, and nature in the city (see Appendix D), but this ended with a report being presented to her presidency coordinator. Should we see in this kind but short-lived reception an illustration of the distance between the concrete commitment of young people and the inaction of political leaders and a justification for young people's lack of trust in decision-making processes at national and international levels? In Glasgow, the young people were not admitted to the conference for health reasons, but they did participate in a citizen's consultation, of which we made ourselves the spokespersons (Appendix E). From a report on the degradation of ecosystems, the growing role of social networks and a global crisis, they proposed among the solutions, acceleration of decision-maker actions, awareness among young people, and change in the dominant modes of consumption [29].

On the other hand, government discourse underlying their contributions to the Paris Agreement INDCs are revealing a divided world and explain the inaction on the part of governments (Figure 1). This division is shaped by strategic alliances (e.g., UNFCCC negotiation groups, OECD, ASEAN, Francophonie, Commonwealth ...) beyond the "South-North" divide. As previously described, these discourses were evaluated by means of an automatic text analysis of the INDCs of 191 countries, first applied to the English corpus [30] and further to French, Spanish and English corpus separately, synthetized and mapped (Appendix B). Regions highly vulnerable to climate change have a fragmented view on the issue (Africa, island countries) that weakens their political force in the negotiations [31]. Developed countries focusing their INDC on the "Reduction of GHG emissions" constitute a more homogenous group strengthened by the efficiency of their negotiation groups. Emergent countries have different visions, either adopting the dominant discourse on the "Reduction of GHG emissions" (Chile, Brazil) or a composite discourse in line with the vulnerability of part of their territory (Mexico, South Africa). Inequalities are not enough counterbalanced by the financial aid brought to low-emitting countries to cope with climate change while a low proportion of countries focusing their INDC on the "Reduction of GHG emissions" and on the "Energetic transition" implemented a carbon policy, illustrating a disconnect between rhetoric and decision-making [20]. This may explain the disappointing results obtained five years after the Paris Agreement since CO_2 emissions have continued to rise between 2015 and 2019 by 4.88%, in 8 out of 10 countries emitting nearly 70% of total carbon dioxide. The 2020 numbers show a reverse trend, related to the effects of the COVID-19 pandemic (www.carbonproject.com, accessed on 30 March 2022). While countries adopting the discourse on the reduction in GHG emissions moderately decreased their CO_2 emissions (−1.8%), those focusing their INDC on energetic transition experienced a very strong increase (7.3%) (https://edgar.jrc.ec.europa.eu/overview.php?v=booklet2020 (accessed on 30 March 2022)) due to their high coal consumption, the exact opposite of the rhetoric behind their INDC. Beyond this geographical fragmentation, this inaction can also be interpreted by the low political benefit expected from the drastic measures needed, which will only materialize after 2050, given the inertia of greenhouse gases in the atmosphere. The distance between the INDCs and the actions, 7 years after the Paris Agreement, suggests a climate diplomacy made of magic words, very far from or even the exact opposite of the reality of political decisions. In this sense, the map of the discourse underlying the INDCs of the governments in 2015 was premonitory of the weak results obtained through COP26, which was, however, decisive. Far from the acceleration desired by young people, a policy of small steps has prevailed, leaving fears that the objectives of the Paris Agreement are out of reach.

5. Conclusion: Youth Lessons beyond the Cop26 and Renewing Climate Services

As we are writing, the growing youth claim to act against climate change is another expression of their awareness and need for action. Involvement of the younger generation should go beyond wishful thinking or goodwill, and advance toward mobilization, such as the movement launched by Greta Thunberg. Young people of the GYCP involved in action-oriented projects conceived to change their territory are currently acting beyond the fragmentation between countries while nevertheless building their projects in recognition of cultural specificities. This demonstrates the importance of climate change education through active teaching methods but also the need to integrate them as qualitative indicators of countries' commitments, as proposed by the GYCP during the COP25. Young people are opening new avenues that can bring a change to us all. Unfortunately, in Glasgow, decision-makers once again demonstrated their lock-in to short-sighted geopolitical divisions, contrasting with the promising narratives employed in their intentional contributions to the Paris Agreement. Our hope is thin that the voices of the young people were heard by their representatives, but despite this, the GYCP will continue its transformative action. Young people´s call for actions to mitigate and adapt to climate change and ultimately to foster multilateral cooperation is becoming stronger. It can provide the required additionality to cross the tipping point and help to address collective action problems associated with climate change, fostering an increase in countries´ level of ambition in their CO_2 abatement commitments.

In this sense, the GYCP has great potential as a test bed for activities that could contribute to a renewal within the enterprise of climate services by deepening their objectives and applying them to a promising segment of society. Our approach, while inspired by certain aspects of the concept of climate services, is a rather radical re-reading of it on several points. Our partners are underprivileged high school students from the public sector. The researchers and associated teachers are volunteers. The depth of the co-construction of knowledge is far from both traditional pedagogy and the co-construction carried out within the framework of climate services, which aims at transforming knowledge and climate data into "useful" indicators for policy and economic activity.

From a social experimentation point of view, we solicit the collective intelligence of young people by suggesting a reflexive involvement and a conscious commitment from the elaboration of action and experimentation projects. The impact on educational orientations is reflected above all in the way we deal with the complexity of an essentially transversal subject. Our approach is the opposite of a utilitarian, top-down orientation. It is the group of students who, after a local diagnosis, suggest a contextualized and problematized project, according to a bottom-up approach. The question is not what researchers can solve in climate-vulnerable regions but what the "good actions" are that these young people can propose to obtain useful results. This suggests the benefits that may accrue through integration of place-based, ancestral knowledge in climate services. All these principles are the basis of a pedagogical book for teachers [18].

Appropriately configured, climate services may be a tool that can help young generations to cope with the dangers and perspectives linked with the increasing variability and change in climate; perhaps serving as a mechanism to reinforce actions undertaken by the students, similar to forest gatekeepers, stone-gardens, or agroecology. The question raised by this experience is how far climate services can support young people in their efforts to design a sustainable climate reality.

Finally, considering our results, we would like to highlight a line of thought, part of which is borrowed from Michael H. Glantz's book: Climate Affairs [32]). We have described how our project is based on three fundamental principles: reflexive knowledge, awareness raising, and the importance of socio-anthropology in the climate system. Although there are sophisticated models of global warming scenarios with increasingly advanced technological means to scrutinize our Earth, there are still a significant number of people who ignore the evidence of climate change. However, there are several ways of integrating a socio-anthropological dimension to climate problematization. Some are tactical, others are strategic. On the tactical level, the socio-anthropological dimension can be considered in climate discussions when it directly, visibly, and largely, influences issues of societal change. On the other hand, a strategic focus on climate disruption leads to an overemphasis on the long-term climate change issue at the expense of the shorter-term socio-anthropological dimension. However, a multidimensional approach that encompasses both tactical and strategic concerns in time and space integrates the socio-anthropological condition into the overall complex problem of global warming. "We are in a world faced with the difficulties of global thinking, which are the same as the difficulties of complex thinking" [32,33].

Author Contributions: Conceptualization, A.P.-V., L.M.F., P.M. and M.L.; methodology: A.P.-V., L.M.F., M.C. and A.G.; software: M.C.; validation, A.P.-V., P.M. and J.C.C.; formal analysis, A.P.-V.,

M.C., L.M.F. and P.M.; investigation, A.P-V., L.M.F., P.M., A.G., M.L., J.C.C. and H.L.T.; resources, A.P.-V.; data curation, P.M.; writing—original draft preparation, A.P.-V., M.C., L.M.F. and P.M.; writing—review and editing, A.P-V. and M.C.; supervision, A.P-V.; project administration, A.P.-V. and L.M.F.; funding acquisition, A.P.-V. and L.M.F. All authors have read and agreed to the published version of the manuscript.

Funding: This research was funded by French Ministry of Ecology, Region nouvelle Aquitaine. Chilean Ministry of Education, the Explore Program of Conicyt and the Centro de Cambio Global and Catedra Arauco en Etica Ambiental, F. Ciencias Biologicas, U. Catolica de Chile. We also received support from the Centro de Modelamiento Matemático (CMM), ACE210010 and FB210005, BASAL funds for centers of excellence from ANID-Chile and Grant ACE210006, AG and PM acknowledge support from Grant BASAL FB210006 to the Instituto de Ecología y Biodiversidad (IEB), AG acknowledges support from Projecto Anillo ACT192027.

Institutional Review Board Statement: Ethical review and approval were waived for this study because crowdsourcing has been done anonymously via a web platform.

Informed Consent Statement: Individual informed consent statement was unnecessary due to the use of the crowdsourcing methodology. Hundreds of students contribute simultaneously and anonymously to the discussions via the web platform Synthetron (http://synthetron.com accessed on 10 August 2022), identified only by their country.

Data Availability Statement: Data are not available due to our agreements with our external service provider.

Acknowledgments: We acknowledge students and teachers who participated in the GYCP project, the company who realized the crowdsourcing, Master students from Sorbonne Université who contributed to the INDC analysis (supervised by MC), students of Sciences-Po (Poitiers) who analyzed the information collected in Rapa-Nui (Supervised by APV) and Guillermo Marini, associated professor at the Universidad Católica de Chile, contributed to the revision of the English language.

Conflicts of Interest: The authors declare no conflict of interest.

Appendix A. Meta-Analysis of the Literature on Climate Services

In order to define climate services, we listed and analyzed the titles of 212 articles published in the journal Climate Services, published by Elsevier, between 2016 and 2022 (https://www.sciencedirect.com/journal/climate-services (accessed on 10 August 2022)). Additionally, we used Google Scholar to search for articles associating the key-word climate services with "youth", "education", or "high school".

We found that 68% of the 212 titles focused on a region, a country, or a level of development; 36% were applied to developed countries; and 32% to developing countries. The remaining 31% had a global or a general approach. Less than half (43%) of the titles focused on the benefits of climate services on the economy, with agriculture being the main sector (25%). Another significant proportion (21%) referred to the contribution of climate services to risk management (floods, droughts, heat waves, etc.) without focusing economic stakeholders. Only one title out of 5 referred to the need for co-construction of indicators and to the perception and knowledge of stakeholders. A higher proportion (36%) referred to the description of climate data, projects or climate services in general, i.e., following the point of view of the researchers. No article associated the term climate service with youth, education, or high school in this corpus, neither on Google Scholar.

The lexical analysis, by context with IRaMuTeQ 0.7 free software, differentiated six types of titles of equal importance. More information on this software is given in Appendix B. The words and illustrative variables best related to the types are listed in the Table A1, along with the chi^2 value that assess the strength of their relationship with the type. Only one type referred to a disadvantaged social category, namely, rural smallholders, preferentially published in volume 20 (year 2020). The others referred to market and methodological issues of data analysis and simulations that differ in time. The word citizen was used in only one title, and the words youth, student, or education were not used.

Six types of titles of Climate Services were published during the period 2016–2022, analysis with IRaMuTeQ 0.7 software

Table A1. Results of the lexical analysis by context of the titles of *Climate Services* journal.

Market		**Smallholder**		**Research Framework**		**Simulation**					
						Season		**Country**		**Region**	
Type 1	**Qhi²**	**Type 2**	**Qhi²**	**Type 3**	**Qhi²**	**Type 4**	**Qhi²**	**Type 5**	**Qhi²**	**Type 6**	**Qhi²**
Forms (words)											
Market	35	smallholder	36	Framework	13	prediction	44	scenario	29	West	19
Uptake	17	agro	21	Process	11	seasonal	17	national	29	Multi	14
EU	17			Enable	11					Africa	13
Engagement	17										
Illustrative variables											
Volume 17	11	Volume 20	10	volume 22	4	Volume 27	5	Year 2016	10	Volume 13	5
						Year 22	4	Volume 1	9	Volume 11	5
								Volume 4	9		

Appendix B. Automatic Text Analysis of the INDCs of 191 Countries

In order to explore the human dimension of climate change at the global scale, we applied a lexical analysis by context to 191 intended nationally determined contributions (INDCs) to the Paris Agreement (COP21). Furthermore, [34] assessed the expected effects of state commitments on greenhouse gas emissions.

We first downloaded the INDCs and gathered them in three corpuses, according to their language: English (558,641 words); French (66,026 words), and Spanish (45,065 words). We then formatted these three corpuses, deleted tables of numbers and graphs, corrected the spelling, and homogenized technical notations (for example CO_2 eq).

We further processed each corpus using an R interface for lexical analysis by context with IRaMuTeQ 0.7 free software. Among many techniques, this type of lexical analysis is easily reproducible and well adapted to highlight the differences of views in a large corpus [35]. First used for literary analysis [36], it was applied to political or sociological purposes (i.e., [37]).

The software cut the corpus into basic context units containing about 200 characters, further grouped in context units, and then in significant statement classes using a descending hierarchical analysis. Each cluster was characterized by its own vocabulary according to the Chi-square test. Combined with a careful reading of the text, this analysis made it possible to understand the linguistic particularities of the different discourses. We detected six main types of discourses and then mapped out the significantly-linked discourse by countries, according to the Chi-squared test, by using the geographical information system (ArcGIS 10.2©).

Appendix C. Three Examples of Projects presented in the COP24 in Katowice in 2018)

Global Youth Climate Pact

Puyanawa: Traditional Knowledges and the Challenges Changes

UFRJ

UnB

Authors: Jósimo Constant;

Introduction/Problem

Our people Puyanawa (people of the frog) we are an indigenous ethnic group that we are located in the State of Acre, north of Brazil, before the arrival of the rubber explorers, we inhabited the banks of the Juruá River. According to the elders, we derive from the junction of the frog and the leaf. This work aims to identify a series of problems that we are facing because of climatic factors. It will be taken in relevance the memory of the oldest, the oral transmission of the traditional stories of the own people.

Justification

Despite all the mishaps that we have experienced over time in the hands of the settlers, we are working hard so that our traditional customs are preserved. The relationship with nature to us indigenous peoples is something symmetrical and relevant, we Puyanawa, we still master many practices, tactics and techniques of our ancestors. It is these knowledge of the ancients that made us go through many difficulties, continue to seek in nature all the elements necessary for maintaining the group and preserving the environment.

Considerations

Colonization was the main negative factor to break with our customs, cultural and environmental ties. But we are working hard for what remains of our customs are maintained, especially our mother tongue which is seriously threatened by the reduced number of speakers. We are monitoring, raising awareness to promote the sustainable management of each resource extracted from our environment. We are seriously concerned about the terrible and sad environmental catastrophes that are destroying the world.

Methodology

The methodology used is translated into the indigenous experience itself, but the memory of the elders will be taken into account. In addition, the ethnographic method will be used, which I propose through this method to carry out a dense description of indigenous traditional knowledge together with the other members of the community and the effective participation of students from the Itxuwãy Rabuy Puyanawa school, indigenous medical practices and impacts environmental and their consequences in our indigenous territory.

Results

The remarkable project is being the gateway to many other things, especially in opening the minds of the younger ones, so that they become more interested in traditional knowledge. Lately our community has been working to rescue the traditional language and culture, but we still need a deeper, dense and long-term work as deforestation, fishing and predatory hunting have increased. In this way, sustainable management and traditional knowledges have been important for the preservation of our land.

References

CONSTANT, Jósimo. History, memory, traditional knowledge and challenging climate change under the Puyanawa indigenous perspective.

Contact: josimo.constant@gmail.com

Figure A1. Puyanawa: Traditional knowledges and Challenge changes. Source: Constant Josimo (UFRJ), GYPC.

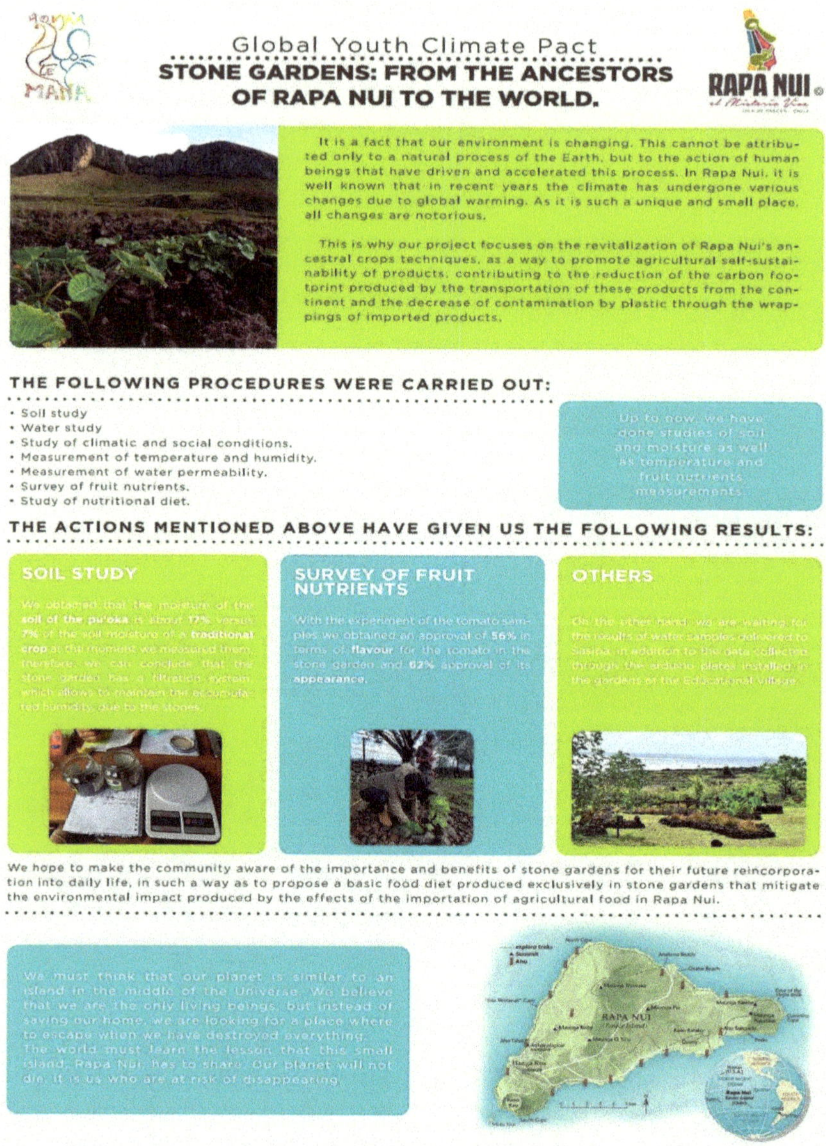

Figure A2. Stone gardens: from the ancestors of Rapa-Nui to the world. Source: Young people, High school Aldea, Rapa Nui, GYPC.

Figure A3. Hydroponics for all. Source: Young people, High school Cordillera, GYCP.

Appendix D. The Proposals of Action in the Face of the Climate Emergency presented in COP25 in Madrid in 2019)

Table A2. Working group, GYCP.

Topics/Workshops	Findings	Arguments	Propositions	Future Impacts
Education	Lack of practical and theoretical knowledge to generate sustainable actions. The youth remain without answers or concrete actions.	The mission of the program is to advise public and private institutions in the creation of an educational system based on ecology through transversal projects.	A program on four pillars: —Environment —Engagement —Empowerment —Efficiency	Implement age-appropriate environmental and sustainability activities.
Valuing forests, less carbon footprint	Massive deforestation of large virgin forests such as the Amazon, the Congo Basin Forest, or the Indonesian forests, among others.	The magnitude of the consequences of pollution from large industries incompatible with a sustainable future.	Creating positive rewards, not necessarily economic, for responsible and environmentally friendly countries; the greening of urban spaces	Consider in the near future a legislation that will have to control the countries involved in large-scale deforestation
Biodiversity, Resources, Water and soils	Concerned about a sixth mass extinction, that biodiversity, ecosystems and our own existence are in danger.	Deforestation causes the destruction of ecosystems, loss of animal habitats, and soil infertility.	—raise public awareness: to know the biodiversity —Created: a labeling system to rate the ecological footprint	Reforestation program to restore ecosystems and boost biodiversity
Sustainable agriculture	At least 1300 tons of food are wasted each year; more than 800 million people do not have enough to eat	Indifference of consumers [ourselves!] as to the origin of food products.	The preservation of a sustainable agroecology in which the relationship between producer and consumer will be reciprocal	to make the school a privileged place for balanced eating habits
Protection of the Oceans	Our oceans are a complex system that connects us as neighbors to all nations. We are in a state of environmental emergency with respect to the conservation status of our oceans.	Our governmental and commercial systems and the absurd over-consumption of unnecessary goods have overloaded the capacity of the oceans.	-Impose a maximum rate of imports and exports from each nation. -Nations must commit to reducing their carbon emissions	Raise awareness among students and the community at large about the causes and effects of climate change on the oceans.
Nature in the city	To build a project that starts from their anchorage in the local social environment and that crosses the fight against global warming and the defense of biodiversity [38].	This is a long-term local action project. Over the course of 5 years, students will record the sounds in the Villaverde area.	-Creation of complete ecosystems We want to revitalize this area with our Anillo Verde project, plant enough vegetation to isolate the Anillo Verde.	We want to create a biome composed of local and preferably endangered fauna and flora.

Appendix E. Listen to the Youth! Presented in the COP26, in Glasgow (2021)

Figure A4. Listen to the Youth. Source: Survey post-COVID-19, 380 people.

References

1. Rogelj, J.; Den Elzen, M.; Höhne, N.; Fransen, T.; Fekete, H.; Winkler, H.; Schaeffer, R.; Sha, F.; Riahi, K.; Meinshausen, M. Paris Agreement climate proposals need a boost to keep warming well below 2 °C. *Nature* **2016**, *534*, 631–639. [PubMed]
2. Adger, W.N.; Barnett, J.; Brown, K.; Marshall, N.; O'Brien, K. Cultural dimensions of climate change impacts and adaptation. *Nat. Clim. Chang.* **2013**, *3*, 112.
3. Gupta, J.; Arts, K. Achieving the 1.5 °C objective: Just implementation through a right to (sustainable) development approach. *Int. Environ. Agreem. Politi-Law Econ.* **2018**, *18*, 11–28.
4. Cooke, J.; Araya, Y.; Bacon, K.L.; Bagniewska, J.M.; Batty, L.C.; Bishop, T.R.; Burns, M.; Charalambous, M.; Daversa, D.R.; Dougherty, L.R.; et al. Teaching and learning in ecology: A horizon scan of emerging challenges and solutions. *Oikos* **2020**, *130*, 15–28. [CrossRef]
5. Plutzer, E.; McCaffrey, M.; Hannah, A.L.; Rosenau, J.; Berbeco, M.; Reid, A.H. Climate confusion among U.S. teachers. *Science* **2016**, *351*, 664–665. [PubMed]
6. Plutzer, E.; Hannah, A.L. Teaching climate change in middle schools and high schools: Investigating STEM education's deficit model. *Clim. Chang.* **2018**, *149*, 305–317.
7. Hannah, L.; Roehrdanz, P.R.; Bahadur, K.C.B.; Fraser, E.D.G.; Donatti, C.I.; Saenz, L.; Wright, T.M.; Hijmans, R.J.; Mulligan, M.; Berg, A.; et al. The environmental consequences of climate-driven agricultural frontiers. *PLoS ONE* **2020**, *15*, e0228305.
8. Soga, M.; Gaston, K.J. Extinction of experience: The loss of human–nature interactions. *Front. Ecol. Environ.* **2016**, *14*, 94–101.
9. Morin, E. *On Complexity*; Hampton Press: Cresskill, NJ, USA, 2008.
10. Montuori, A. Créativité et complexité en temps de crise. *Communications* **2016**, *95*, 179–198.
11. Bocchi, G.; Cianci, E.; Montuori, A.; Trigona, R.; Nicolaus, O. Educating for creativity. *World Futures* **2014**, *70*, 336–369.
12. Busch, K.C.; Román, D. Fundamental climate literacy and the promise of the next generation science standards. In *Teaching and Learning about Climate Change*; Routledge: London, UK, 2017; pp. 120–134.
13. Henderson, M.; Selwyn, N.; Aston, R. What works and why? Student perceptions of 'useful' digital technology in university teaching and learning. *Stud. High. Educ.* **2017**, *42*, 1567–1579.
14. Hestness, E.; McGinnis, J.R.; Riedinger, K.; Marbach-Ad, G. A study of teacher candidates' experiences investigating global climate change within an elementary science methods course. *J. Sci. Teach. Educ.* **2011**, *22*, 351–369. [CrossRef]

15. Ojala, M.; Bengtsson, H. Young people's coping strategies concerning climate change: Relations to perceived communication with parents and friends and proenvironmental behavior. *Environ. Behav.* **2019**, *51*, 907–935. [CrossRef]
16. Shea, N.A.; Mouza, C.; Drewes, A. Climate change professional development: Design, implementation, and initial outcomes on teacher learning, practice, and student beliefs. *J. Sci. Teach. Educ.* **2016**, *27*, 235–258. [CrossRef]
17. Han, H.; Wuk Ahn, S. Youth mobilization to stop global climate: Narratives and impact. *J. Sustain.* **2020**, *12*, 4127. [CrossRef]
18. Pena-Vega, A. *Les 7 Savoirs Nécessaires à L'éducation au Changement Climatique. Comment Les Jeunes S'engagent Pour L'urgence Climatique (The 7 Skills Needed for Climate Change Education. How Young People Are Getting Involved in the Climate Emergency)*; Atlantique Editions: Poitiers, France, 2021.
19. Giordan, A. Les grandes régulations du corps humain. In *Relier Les Connaissances, Journées Thématiques Conçues et Animées par Edgar Morin*; Morin, E., Ed.; Seuil: Paris, France, 2000; pp. 185–197.
20. Wolf, J.; Moser, S.C. Individual understandings, perceptions, and engagement with climate change: Insights from in-depth studies across the world. *Wiley Interdiscip. Rev. Clim. Chang.* **2011**, 2, 547–569.
21. Cohen-Shacham, E.; Walters, G.; Janzen, C.; Maginnis, S. (Eds.) *Nature-Based Solutions to Address Global Societal Challenges*; International Union for Conservation of Nature (IUCN): Gland, Switzerland, 2016. [CrossRef]
22. Kabisch, N.; Frantzeskaki, N.; Pauleit, S.; Naumann, S.; Davis, M.; Artmann, M.; Haase, D.; Knapp, S.; Korn, H.; Stadler, J.; et al. Nature-based solutions to climate change mitigation and adaptation in urban areas: Perspectives on indicators, knowledge gaps, barriers, and opportunities for action. *Ecol. Soc.* **2016**, *21*, 39. [CrossRef]
23. Gorr, C. Changing Climate, Changing Attitude? *Mus. Soc. Issues* **2014**, *9*, 94–108.
24. Griscom, B.W.; Adams, J.; Ellis, P.W.; Houghton, R.A.; Lomax, G.; Miteva, D.A.; Schlesinger, W.H.; Shoch, D.; Siikamäki, J.V.; Smith, P.; et al. Natural climate solutions. *Proc. Natl. Acad. Sci. USA* **2017**, *114*, 11645–11650. [CrossRef]
25. Stoknes, P.E. Rethinking climate communications and the "psychological climate paradox". *Energy Res. Soc. Sci.* **2014**, *1*, 161–170. [CrossRef]
26. Balundė, A.; Perlaviciute, G.; Truskauskaite-Kuneviciene, I. Sustainability in Youth: Environmental Considerations in Adolescence and Their Relationship to Pro-environmental Behavior. *Front. Psychol.* **2020**, *11*, 582920. [CrossRef]
27. Shutaleva, A.; Martyushev, N.; Nikonova, Z.; Savchenko, I.; Abramova, S.; Lubimova, V.; Novgorodtseva, A. Environmental Behavior of Youth and Sustainable Development. *Sustainability* **2021**, *14*, 250. [CrossRef]
28. Shutaleva, A.; Nikonova, Z.; Savchenko, I.; Martyushev, N. Environmental Education for Sustainable Development in Russia. *Sustainability* **2020**, *12*, 7742. [CrossRef]
29. Figueres, C.; Schellnhuber, J.H.; Whiteman, G.; Rockström, J.; Hobley, A.; Rahmstorf, S. Three years to safeguard our climate. *Nat. News* **2017**, *546*, 593.
30. Lemaitre, T.; Louis-Lucas, T.; Sampere, J.; Cohen, M. *Quels Discours Sur le Changement Climatique y a-t-il Derrière l'Accord de Paris?* Report ITE-Sorbonne; Sorbonne University: Paris, France, 2017. Available online: https://www.su-ite.eu/ (accessed on 15 July 2022).
31. Hsu, A.; Moffat, A.S.; Weinfurter, A.J.; Schwartz, J.D. Towards a new climate diplomacy. *Nat. Clim. Chang.* **2015**, *5*, 501–503.
32. Glantz, M. A political view of CO_2. *Nature* **1979**, *280*, 289–290.
33. Morin, E. *Penser Global. L'humain et Son Univers*; Robert Laffont: Paris, France, 2015.
34. Le Treut, H.; Boucher, O.; Criqui, P.; Benveniste, H.; Bellassen, V.; Bréon, F.-M.; Séférian, R. Groupe Interdisciplinaire Sur Les Contributions Nationales (Interdisciplinary Group on National Contributions): Working Paper. Preliminary Version, IPSL Climate Modelling Center. Scientific Note N° 32. Paris, France. 2015. Available online: https://cmc.ipsl.fr/images/publications/scientific_notes/GICN_working_paper2.pdf (accessed on 10 August 2022).
35. Kronberger, N.; Wagner, W. Keywords in context: Statistical analysis of text features. In *Qualitative Researching with Text, Image and Sound: A Practical Handbook for Social Research*; Bauer, M.W., Gaskell, G., Eds.; Sage: London, UK, 2000; pp. 299–317.
36. Chaves, M.M.N.; dos Santos, A.P.R.; dos Santosa, N.P.; Larocca, L.M. Use of the software IRAMUTEQ in qualitative research: An experience report. In *Computer Supported Qualitative Research*; Springer: Cham, Switzerland, 2017; pp. 39–48.
37. Reinert, M. Une méthode de classification descendante hiérarchique: Application à l'analyse lexicale par contexte (A top-down hierarchical classification method: Application to lexical analysis by context). *Cah. L'analyse Données* **1983**, *8*, 187–198.
38. Winkler, H.; Höhne, N.; Cunliffe, G.; Kuramochi, T.; April, A.; de Villafranca Casas, M.J. Countries start to explain how their climate contributions are fair: More rigour needed. *Int. Environ. Agreem. Politics Law Econ.* **2018**, *18*, 99–115.

sustainability

MDPI

Article

Exploring the Climatic Potential of Somo's Surf Spot for Tourist Destination Management

Anna Boqué Ciurana [1,*], Melisa Ménendez [2], María Suárez Bilbao [2] and Enric Aguilar [1]

[1] Centre for Climate Change, C3, Geography Department, Universitat Rovira i Virgili, Vila-seca, 43480 Tarragona, Spain; enric.aguilar@urv.cat
[2] Environmental Hydraulics Institute (IHCantabria), Universidad de Cantabria, 39005 Santander, Spain; melisa.menendez@unican.es (M.M.); maria.suarezbilbao@unican.es (M.S.B.)
* Correspondence: anna.boque@urv.cat

Abstract: Surfing is one of the most popular activities in coastal tourism resorts. However, the sport depends strongly on the met-ocean weather conditions, particularly on the surface wind-generated waves that reach the coast. This study provides examples of how users' needs and user perspectives are considered by climate data specialists to develop needed, highly useful information addressing human and social needs. In this vein, the climate analysis of such data can provide input on the expected length of a surfing season, according to the surfer's level of expertise. In addition, other water sports, such as SUP Wave and windsurfing, among others, might be indicated when surfing conditions are not optimal. Finally, the safety of surfers and other tourists who venture into the sea is also dependent on those conditions. We collaborated with the surfing community to define a series of indices for quantifying surfing days (SD), surfing days stratified by surfers' skills (SDS), alternate offers (AOs), and surfers' and swimmers' safety (SuS and SwS). These are of general applications but require wind and wave data at a very fine scale as the input. To illustrate the potential of our indices, we applied them to the Somo beach (Cantabria, Spain). We downscaled a global wave hindcast dataset covering a 30-year period to a spatial resolution of 100 m to obtain wave-surfing information at Somo's surf spot. The results confirmed Somo's status as a year-round surf spot, with SD values of 229.5 days/year and monthly values between 22 days/month and 16 days/month. SDS showed different seasonal peaks according to the surfers' skills. Beginners' conditions occurred more often in the summer (18.1 days/month in July), intermediate surfers' conditions appeared in the transitional seasons (14.1 days/month in April), and advanced and big-wave riders in the winter (15.1 days/month in January and 0.7 days/month, respectively). The AO index identified the SUP wave values of 216 days/year. Wind water sports presented values of 141.6 days/year; conversely, SUP sports were possible on only 7.4 days/year. SuS and SwS identified different seasonal hazard values, decreasing from the winter, autumn, and spring to minimum values in the summer.

Keywords: resilience; wave climate; tourism management; surfing; climatology; decision making; climate service; sustainability; adaptation

Citation: Boqué Ciurana, A.; Ménendez, M.; Suárez Bilbao, M.; Aguilar, E. Exploring the Climatic Potential of Somo's Surf Spot for Tourist Destination Management. *Sustainability* **2022**, *14*, 8496. https://doi.org/10.3390/su14148496

Academic Editors: Charles Herrick, Jason Vogel and Glen Anderson

Received: 18 May 2022
Accepted: 8 July 2022
Published: 11 July 2022

1. Introduction

Climate services are defined as the provision of climate information to help individuals and organizations make climate-resilient decisions. The World Climate Conference-3 (WCC-3), organized in 2009 by the World Meteorological Organization, established the Global Framework for Climate Services (GFCS) [1]. Climate data and information are transformed into customized products to provide decision makers in climate-sensitive sectors with better information to adapt to climate variability and change [2]. The goal of climate services is to provide access to scientific knowledge and, thereby, to reduce vulnerability and create opportunities to promote innovation, business opportunities, and employment, highlighting the importance of involving users in developing climate services [3]. Research has revealed [4] that peer-reviewed literature on the availability and use of climate services in the operations and management of tourism is scarce, and that a need exists for a new generation of specialized climate information products that can enhance climate risk management amongst tourism suppliers. Adaptation to climate change is becoming more urgent, but the wealth of knowledge that informs adaptation planning and decision making is currently not being used to its full potential [5]. In this context, climate services can provide valuable information that can help society enhance resilience, survival, and even prosperity in the face of climate risk [6].

Climate assessment for recreation and tourism have increasingly become dynamic research topics, especially in the age of the anthropogenic climate crisis [7]. Coastal destinations can offer different tourist activities in the same territory and all of them are influenced by meteo-climatic conditions to a specific degree [8]. We assert that there is a need to explore the climatic viability of different activities. By doing so, the development of climate services with tailored climate information about particular destinations can shed light on system changes.

The results of this research, specifically all the information generated with the indicators, imply an improved capacity for destination managers to promote particular destinations. This can lead to a destination being promoted in a more resilient way, not only by knowing which season is better for a specific level of surfing but also by knowing the viability of offering complementary activities. Thus, destination managers can plan tourist offers better and can be prepared to adapt activities when surfing is not possible. This will lead to investing in resources, from hiring staff to planning surfing championships, that will be planned more efficiently and sustainably. Definitively, using this information will enable destination managers to apply informed climate-resilient actions in their sector.

The present research bridges the gap between users and producers of climate information in line with our previous study, in which surfers and surf companies identified which meteorological and climatological information they need access to for better surfing experiences [9]. The new contacts that were gained through the survey conducted in the previous study helped the researchers of this study refine its focus.

Climate index application and validation for tourism is a complicated topic and presents several challenges [10–12]. In this context, the significance of this study is the need to transform meteo-oceanic data into information that can assist decision making in coastal destinations that need sustainable development. As coastal tourist destinations can offer different activities, we focus on surfing, one of the water activities that is offered at several destinations. Following the scientific literature, we have identified a gap in this specific activity and a need to develop a climate service that addresses it. Therefore, this research aims to contribute to the development of a specific climate service for surfing by considering specific users' needs and also by developing high-resolution meteo-oceanic data. The paper's primary objective is to present a set of climate indices for surfing destinations, taking as its experimental area the well-known Spanish surf spot of Somo (see the next section for details). With our analyses, we achieve two secondary objectives: (1) to obtain a downscaled dataset of wave data and (2) to describe with climate data the surfing potential of Somo's surf spot. As our results will specifically define the surfing potential of the spot, this information will assist surfing destination managers in promoting climate-resilient pathways for sustainable development in surfing tourism. In this regard, we intend to contribute modestly to the achievement of the various UN 2030 Agenda Sustainable Development Goals (SDGs), namely (3) good health and well-being, (8) decent work and economic growth, (12) responsible consumption and production, (13) climate action, (14) life below water, and (15) life on land.

2. Literature Review

Several authors have defended the idea [13] that climate change communication and user engagement can work as a tool to anticipate climate change. The visual communication of climate information is one of the cornerstones of climate services; thereby, the characteristics that make a climate service self-explanatory rely on the type of representation used. In this context, guidance on the climate information published by official bodies should adopt a consistent approach, with a clear narrative that describes the transition from science to guidance [14]. The form in which climate services information is needed for the required end-user decisions requires careful thought, including appropriate communication of the associated uncertainties using best practices and experiences from related sectors [15].

Numerous authors have discussed the importance of climate [16], weather [17–21], and extreme weather [22–24] in the establishment and choice of tourism destinations. Outdoor recreation is strongly and increasingly affected by climate change and its impacts present marked seasonal and geographical variations that determine its viability [25]. In the past, the Tourism Climate Index (TCI) [26,27] has been used in suitability analyses. Several studies calculated this index to determine the climatic comfort conditions for tourism in different areas [28,29]. Specific research has focused on exploring the state of weather and climate information for tourism and explored sustainable tourism and the grand challenge of climate change [30,31]. Regarding the idea of the TCI, other studies have developed the Holiday Climate Index (HCI) [13,14] and computed it, in a reshaped formulation, for beach and urban destinations with climate data downscaled dynamically [31]. Other studies [8] have proposed the co-creation of specific indices for each specific activity/destination. One such study described indices for beach and snow tourism [32], while others developed indices for skiing [33,34], and still others have focused on surfing [35]. Sports tourism, based either on attending a sports event or on practicing the sport, has experienced considerable growth in the last several decades. Surfing as a tourist activity has traditionally been labeled as sports tourism [36] or nautical, maritime, or marine tourism [37]. Most recently, researchers defined it as 'blue tourism', a concept intimately related to the blue economy and the blue growth strategy [38]. Blue tourism highlights the sea as the central resource for leisure and recreation activities and leisure and tourism industries [39,40].

Surf and surfing tourism affect the environment and depend on its preservation and there is a concern regarding not only the quality of the activity but also its sustainability. New research has ranked Cape Town beaches in terms of sustainability by using surf-tourism-related indicators [41]. Similarly, other authors have used the Driving Forces-Pressures-State-Impacts-Responses (DPSIR) framework to propose indicators to measure human activities affecting surf breaks [42]. Similarly, it has been affirmed that surf breaks are finite, valuable, and vulnerable natural resources that not only influence community and cultural identities but are also a source of revenue and provide a range of health benefits [43]. Despite this, surf breaks lack recognition as coastal resources and, therefore, the associated management measures required to maintain them. It has also been recognized that conserving biodiversity and ecosystem services requires diverse models that empower communities to act steward of such resources and also to benefit from them. They investigate the potential of surfing resources and the consciousness of surfing communities as beacons of environmental and marine biodiversity preservation. In fact, the sustainable management of these resources ensures their ability to provide for the character, economy, and development of coastal communities worldwide [44]. Valencia et al. [45] studied how surfing tourism's effects are perceived by local residents; the results of their research have implications for surf tourism management at the destination.

Fox et al. [46] focused their research on recreational ocean users, specifically surfers, and how their blue space activities may inform the understanding of ocean processes and human–ocean interconnections. They presented novel insights about the opportunities for integrating ocean sustainability strategies through blue space activity mechanisms and coastal community engagement. They defined the surfing social-ecological system adapted from McGinnis et al. [47] and demonstrated how the human (social) and ocean (ecological) systems provide opportunities for interactions between surfers (users) and waves (resource units), producing ocean literacy understanding and awareness.

Another aspect that has an impact on the perception and development of surf is the safety of the practitioners. Mindes [48] analyzed hazards perceptions among surfers in

Southern California. Rip currents are a primary mechanism associated with dangerous situations [49] and have been the focus of beachgoer education and awareness strategies [50]. Surfers and lifeguards often utilize rip currents to expedite their journey across the surf zone [51]. Attard et al. [52] found that 63% of surfers believe they have saved a swimmer's life. The enjoyability and safety of the surfing experience are enhanced when the right information is communicated in the right way. Boqué et al. [9] surveyed surfers in Spain to explore which meteorological and climatological information they find necessary for a better surfing experience.

De Andrés et al. [53], who studied surfers' balance during surfing activity between competitive surfers and non-competitive surfers in Somo, in collaboration with Escuela Cantabra de Surf and Somo Surf Center, defended that surfing in training and competition is characterized by a great variability of environmental factors such as different sizes and breaking shapes of the waves and changing weather conditions. Nevertheless, there are limitations and possibilities for the world surfing reserves [54] that can be assessed by surfing climatology and surfing forecasts [9].

3. Study Area, Data and Methods

3.1. Study Area

The pilot area of the Somo surf spot is part of the municipality of Ribamontán al Mar Municipality. Ribamontán al Mar is located on the northern shore of the Iberian Peninsula in the Cantabria region (Figure 1) close to its capital of Santander. It hosts Spain's first surfing school, established in 1991. Ribamontán al Mar (declared in 2012 as a World Surfing Reserve, the first in Spain and the second in Europe) is a pioneering territory in its commitment to surfing tourism through its Surfing Competitiveness Plan (2009–2014) and in promoting territorial balance through the competitiveness of destinations, international projection, specialization of tourism products, and deseasonalization [54].

The area is characterized by an oceanic climate, specifically Cfb, in the Köpen Climate Classification [55]. The Cfb type is defined as being temperate mesothermal, without a dry season, and with a mild summer. Using monthly values, the annual thermometric regime is regular, with the highest average values in August and the lowest in January. Precipitation is significant even in the drier months [56]. Wind variations are present throughout the year. Northwest and southeast winds dominate in the winter. In the spring, northerly winds usually blow and then shift to a northeasterly direction in the summer. High-intensity winds are more frequent in the winter and at the end of autumn [57].

3.2. Data and Methods

Data for our analysis were obtained after applying the high resolution downscaled ocean waves (DOW) approach [58,59] to the global ocean waves hindcast [60] data. This hindcast is a historical hourly wave reconstruction generated with the WAVEWATCH III model [61], using the atmospheric forcing from the Climate Forecast System Reanalysis (CFSR) global reanalysis from 1979 to 2010 [62] and extended to the present by CFSv2 [63] with a ~0.2° resolution. GOW2 has global coverage with a spatial resolution of $0.5° \times 0.5°$ and a resolution of $0.25° \times 0.25°$ in zones near the coast. The DOW approach is a global framework to downscale waves to coastal areas, which takes into account a correction of open sea significant wave height (directional calibration). The approach combines numerical models (dynamical downscaling) and mathematical tools (statistical downscaling). First, a regional hindcast is numerically simulated with the Simulating Waves Nearshore (SWAN) model using high-resolution winds from the Cantabrian domain of downscaling winds (a 3 km historical reconstruction from global CFSR reanalysis) and the GOW2 spectral data as the boundary conditions.

Then, the DOW Cantabria database is used, which is based on regional waves as initial conditions for waves in the contours of high-resolution numerical domains, at ~100 m resolution.

Figure 1. Somo surf spot location: global and local context.

Our methodological approach (Figure 2) used significant wave height (*Hm*0), peak period (*Tp*), wind speed (*Ws*), and wind direction (*Wd*) downscaled climate data from DOW in the Somo surf spot in the definition of a climate service for the management of *surfing destinations*. In addition, using *Hm*0 and *Tp* as input from DOW, we computed the wave energy flux (*We*) with the following formula [64]:

$$\begin{gathered} We = Hm0^2 * Tp \\ We = wave\ energy\ flux \\ H_{m0} = significant\ wave\ height \\ Tp = peak\ wave\ period \end{gathered} \quad (1)$$

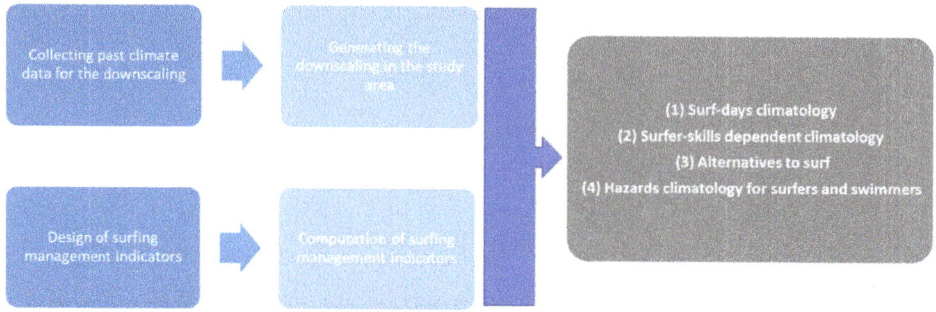

Figure 2. Development workflow of the climate service for surfing destination management.

We designed the surfing management indicators by combining the variables previously described and constraining hourly data to daylight time (obtained through the R package suncalc, https://cran.r-project.org/web/packages/suncalc/suncalc.pdf) when surfing activity was concentrated. We obtained (1) a daily surf climatology, (2) a surfer-skill climate indicator, (3) an index for alternatives to surfing, and (4) a hazard climate indicator for surfers and swimmers.

Surfing climatology yields the number of expected surfing days per year, i.e., days when, following Espejo et al. [65] and Boqué et al. [35], $Hm0 \geq 0.5$, $Tp \geq 6$, and $Ws < 20$. Days that do not meet these requirements are considered non-surfing times. For these periods, we described and indexed combining *Hm*0 and *Ws* to suggest to surfers and surf schools the best surf-related alternatives (e.g., other water sports), according to the state of the wind and the sea. We considered a surf-related activity to be any activity requiring the use of a board. We grouped them as (1) Stand Up Paddle Surf (SUP) activities, for which waves are not required, e.g., SUP yoga, SUP Pilates on board, or a water polo match using surfboards [66]; (2) SUP activities that require waves and are similar to surfing—called SUP Wave; and (3) sports such kitesurfing, in which wind speed is the key element [67]. These activities and their optimal values of *Hm*0 and *Ws* are shown in Table 1.

Table 1. Alternative surf activity definition.

Alternative Surf Activity		
Categorization	**Conditions Required**	**Explanation**
SUP/SUP yoga/SUP Pilates/Surf polo better than surfing	$Ws < 10$ $Hm0 < 0.5$	Waves are not high enough for surfing, but wind conditions allow the practice of other related activities
SUP Wave	$Ws < 20$ $Hm0 > 0.5 \leq 1.5$	Significant wave height and wind speed will probably make SUP Wave possible
Kite surfing, windsurfing, wing better than surfing	$Ws > 20$	Wind speed is too extreme for surfing but is suitable for other related activities

The second index (Table 2) categorizes the *Hm*0 values as different surf-skill levels (i.e., beginner, intermediate, advanced, or big wave rider). The values of the different intervals are an adaptation of Hutt et al. [68], who defined the maximum and minimum values of wave height according to the surfers' skills. We also combined these values for the peak period following the thresholds suggested by Espejo et al. [65].

Table 2. Surfing skill-oriented climatology definition.

Surfing Skill-Oriented Climatology		
Categorization	**Conditions Required**	**Explanation**
Beginner/Longboard/Fatty boards	$Hm0 \geq 0.5 < 0.9$ $Tp \geq 6$	Small waves useful for beginners, longboarders, or fatty board riders
Intermediate	$Hm0 \geq 0.9 < 1.5$ $Tp \geq 6$	Wave height is useful for intermediate surfers (in green waves) but also for beginners in white water
Advanced	$Hm0 \geq 1.5 < 3$ $Tp \geq 6$	Wave height is so high that the surfers require advanced skills to arrive at the peak zone and to surf
Big wave rider	$Hm0 \geq 3$ $Tp \geq 6$	Wave height is suitable only for big wave riders and tow-in surfers

To compute these two monthly indices from hourly observations, we used our own formula as follows:

$$I_m = \frac{(\sum obs_{crm})}{\sum obs_m} n_m \quad (2)$$

where I_m (Equation (2)) corresponds to the monthly indicator for a specific month and expresses the number of complete days that meet a set of given conditions, regardless of how they are distributed within the month; obs_{crm} is the number of hourly observations that meet the required conditions; obs_m is the total number of observations per month; and n_m is the number of days in that month (e.g., 31 in January, 28/29 in February, etc.).

For the hazard indicator, we followed Attard et al. [52], who demonstrated that surfers do well in locations that can be hazardous to swimmers. In line with Attard's approach [52], we used *Hm*0, *Ws*, *Wd*, and *We*, according to formula II. Following Koon et al. [69], Mazzone [70], Whitcomb [71], and Miloshis et al. [72], we computed hazard scores for intermediate surfers, the third general degree established by the surfing Spanish federation framework, and intermediate swimmers, according to the classification of the Real Federación Española de Natación achieving the level fry 2. As swimmers' and surfers' interactions with the ocean are intrinsically different, we defined specific cut-off points for each, as reflected in Table 3, and attribute values from 0 to 4 to each condition to create a composite index that can take values between 0 and 10. Maximum values (10) relate to hazardous conditions; minimum values (0) relate to conditions without hazards.

Table 3. Hazard management: surfers' versus swimmers' definition.

Hazard Management: Surfers versus Swimmers Definition				
Variable Based	**Conditions Required (Swimmers)**	**Value (Swimmers)**	**Conditions Required (Surfers)**	**Value (Surfers)**
Wind-based	$Ws < 10$ Wd = all directions	0	$Ws < 15$ Wd = all directions	0
	$Ws \geq 10 < 15$ Wd = onshore	1	$Ws \geq 15 < 20$ Wd = all directions	3
	$Ws \geq 10 < 15$ Wd = offshore	2	$Wd \geq 20$ Wd = all directions	4
	$Ws \geq 15 < 20$ Wd = onshore	1	NA	NA
	$Ws \geq 15 < 20$ Wd = offshore	3	NA	NA
	$Ws \geq 20$ Wd = all directions	4	NA	NA

Table 3. *Cont.*

Hazard Management: Surfers versus Swimmers Definition				
Variable Based	Conditions Required (Swimmers)	Value (Swimmers)	Conditions Required (Surfers)	Value (Surfers)
Significant wave-height-based	$Hm0 > 0.5 < 0.9$ $Hm0 \geq 0.9 < 1.5$ $Hm0 > 1.5$	1 2 3	$Hm0 > 1.5 > 3$ $Hm0 \geq 3$ NA	1 2 NA
Wave energy flux-based	$We < 45$ $We \geq 45 < 100$ $We \geq 100 < 1000$ $We \geq 1000$	0 1 2 3	$We \geq 500 < 1000$ $We \geq 1000$ NA NA	1 4 NA NA

We obtained each daily hazard indicator by selecting the maximum hourly value of the hazard score per day. These values were packaged (1) in the form of calendars and in graphical time series where maximum monthly values are shown, as we will present in Section 4.

For SD, SDS, and AO, we represent the monthly values as boxplots, and we also show the annual values in a graphical time series to observe the evolution for the 1985–2015 period. For all sets of indicators, the Mann–Kendall test was calculated to explore the trends. For SuS and SwS, we represent the annual mean of the monthly mean of the daily maximum value in the time series.

4. Results

4.1. Surf Climatologies

Figure 3 presents the monthly climatology of the expected surfing days computed from 1985–2015 at the Somo surf site. The annual number of expected surfing days was 229.5. The highest monthly value corresponded to July (22 days), followed by August (21.7 days/month) and June (21 days/month). Lower values corresponded to November (16.3 days/month), February (16.9 days/month), December (17.8 days/month), and April (17.9 days/month). The winter months (December, January, and February) showed larger interquartile ranges.

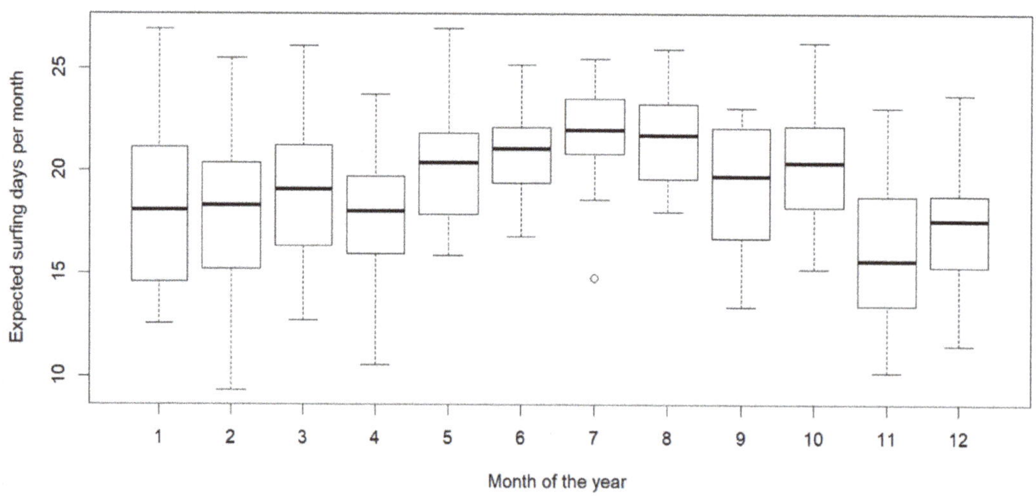

Figure 3. Expected distribution of surfing days per month, Somo, 1985–2015.

Figure 4 shows the evolution of the annual SD for the 1985–2015 period. The SD annual values ranged from 247.8 days (the year 2015) to 206.19 days (the year 2010). The plot shows the variation of the annual SD between the years; the standard deviation corresponded to 10.09 days.

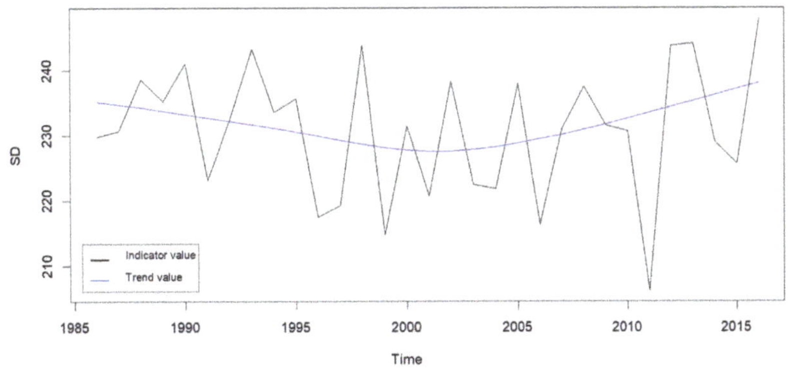

Figure 4. Evolution and trend of annual surfing days; reference period is 1985–2015 in Somo.

Figure 5 adds the consideration of the surfer's skill level. Our results showed that, depending on the practitioner's skills, the season shifted from summer to winter, opening the door to the deseasonalization of tourist resorts. In this regard, the peak number of the expected days for the beginners clustered again in the summer: June (17.3 days/month), July (18.19 days/month), and August (17.2 days/month). By contrast, intermediate surfers should expect to find a larger number of optimal days in the transition seasons, with peaks in April (14.4 days/month) and September (13.4 days/month). Finally, advanced surfers and big wave riders will find better conditions in the winter. For advanced surfers, the expected days peaked in January (15.1 days) and December (12.3 days/month). Big wave riders should expect <1 day/month, concentrated throughout the period of the November–April semester and peaking in January (0.7 days/month).

Figure 6a–d show the SDS annual evolution and trend for the 1985–2015 period. The maximum SDS were detected on surfing days for intermediate surfers at 167.02 days (in 2011), followed by beginners with 157.36 days (in 1985), 108.21 days (in 1986) for advanced surfers, and 10.02 days (in 2014) for big wave riders. The minimum SDS annual values were ranked from big wave riders with 0 days (in 1992), advanced surfers with 43.16 days (the year 2010), beginners with 94.94 days (the year 2011), and intermediates with 114.5 days (in 1989). The standard deviation ranged from 2.19 days (big wave riders) to 17.41 days for advanced surfers. The case for intermediates was 11.89 days and for beginners was 16.2 days.

4.2. *Alternative Offer*

Days when environmental conditions do not favor surfing might still be suitable for alternative water sport activities (Figure 7a–c). From the series of activities considered in Section 3, in the case of the Somo surf spot, the surf activity offered most frequently was SUP Wave (216 days/year); specifically, July (22.7 days/month) had the largest number of expected days. Kitesurfing was the alternative surf activity offered second most frequently (141.6 days/year), and the spring and summer months presented the lowest values for expected kitesurfing days per year, linked with summer's calm winds. SUP yoga (7.4 days/year) was the alternative that offered lower possibilities, which indicates that if the activity needs to be promoted, it should probably ubicate in rivers next to the main surf spot. SUP Wave and kitesurfing seemed to be complementary, as when there is so much wind to practice SUP Wave, there is enough wind to practice kitesurfing, wing, or windsurfing. The high values for these wind activities were present specifically in autumn and winter: November (15.8 days/month), December (16 days/month), and January (16.8 days/month). A good period for practicing SUP Wave is during the spring and summer, and at the beginning of autumn: May (21 days/month), June (22 days/month), July (22.7 days/month), August (22.4 days/month), and September (19.6 days/month).

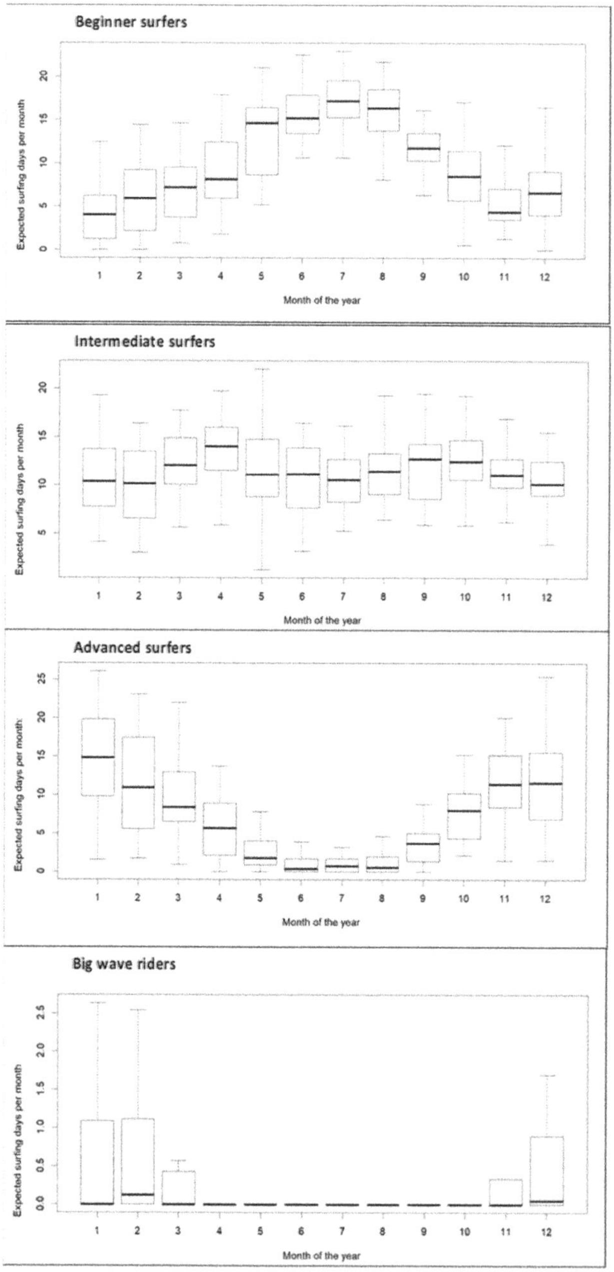

Figure 5. Expected distribution of surfing days per month sorted by surfer's skill level; reference period is 1985–2015 in Somo.

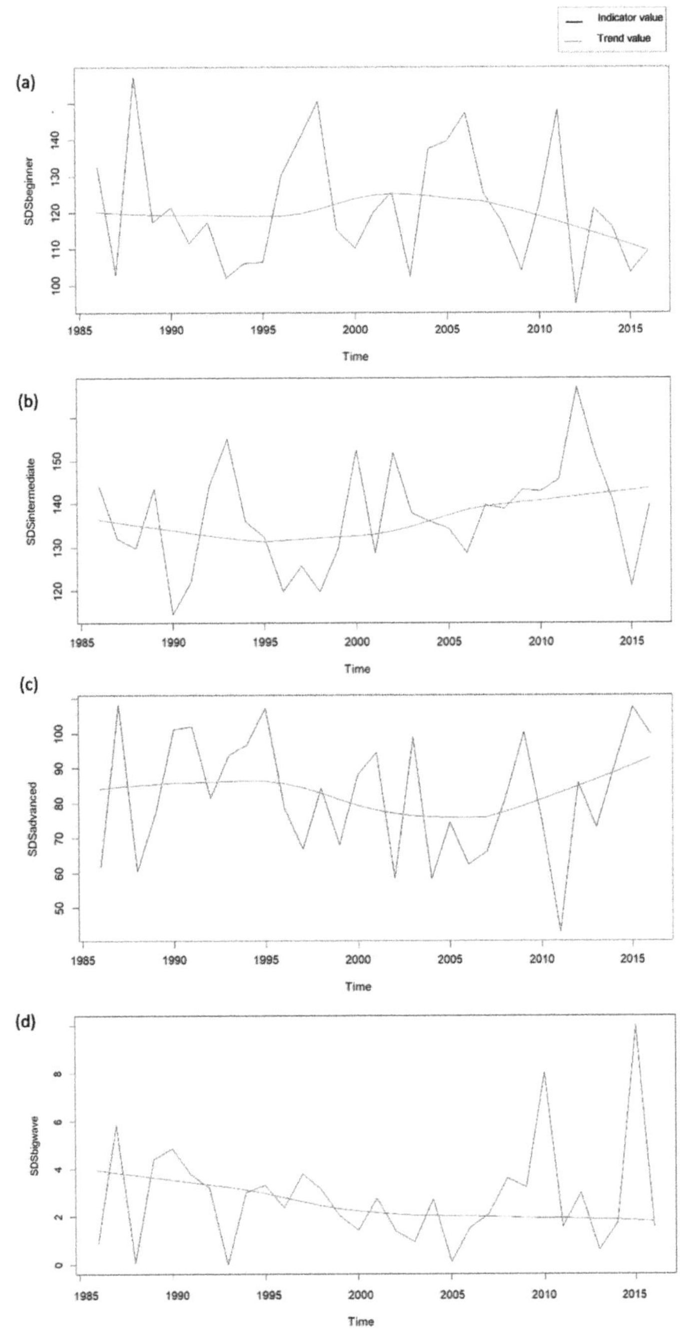

Figure 6. (**a**) Evolution and trend of annual surfing days for beginner surfers; reference period is 1985–2015 in Somo. (**b**) Evolution and trend of annual surfing days for intermediate surfers; reference period is 1985–2015 in Somo. (**c**) Evolution and trend of annual surfing days for advanced surfers; reference period is 1985–2015 in Somo. (**d**) Evolution and trend of annual surfing days for big wave riders; reference period is 1985–2015 in Somo.

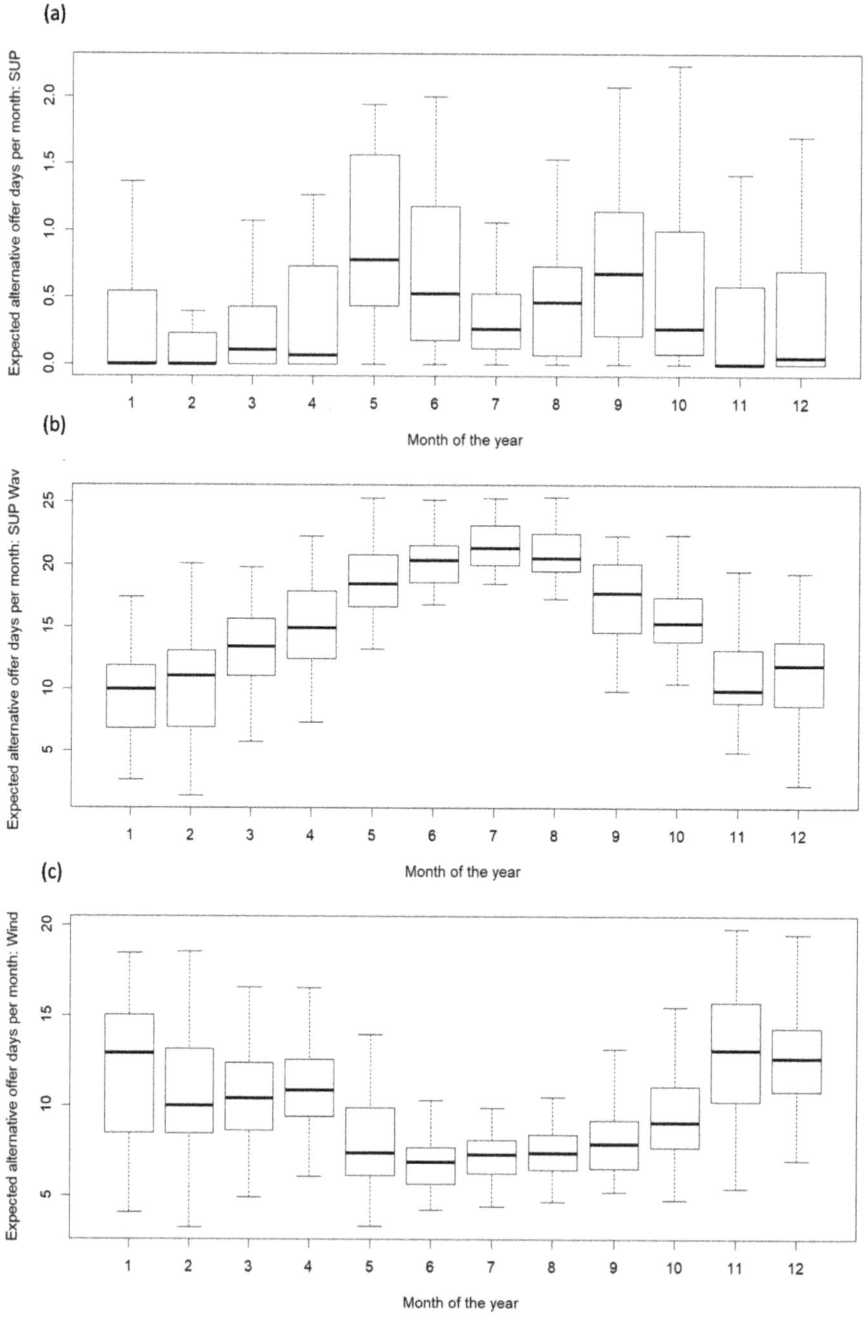

Figure 7. (**a**) Expected distribution of alternative offer monthly days for SUP-related sports; reference period is 1985–2015 in Somo. (**b**) Expected distribution of alternative offer monthly days for SUP Wave sport; reference period is 1985–2015 in Somo. (**c**) Expected distribution of alternative offer monthly days for wind-related sports, i.e., windsurfing, kitesurfing, wing surfing; reference period is 1985–2015 in Somo.

Figure 8a–c shows the annual AO evolution and trend for the 1985–2015 period. The Mann–Kendall test denoted the absence of a trend in the data. For the annual AO values, SUP-related activities presented the lowest values of annual days: a minimum of 3.35 days in 1986 and a maximum days of 11.81 days in 1997. SUP Wave presented a maximum of 207.74 annual days in 2001 and a minimum of 165.23 days in 1993. Wind and water sports such as windsurfing, wing surfing, or kitesurfing presented high maximum annual values in 2010, corresponding to 138.71 days, and lower values were in 1998, corresponding to 102.89 days.

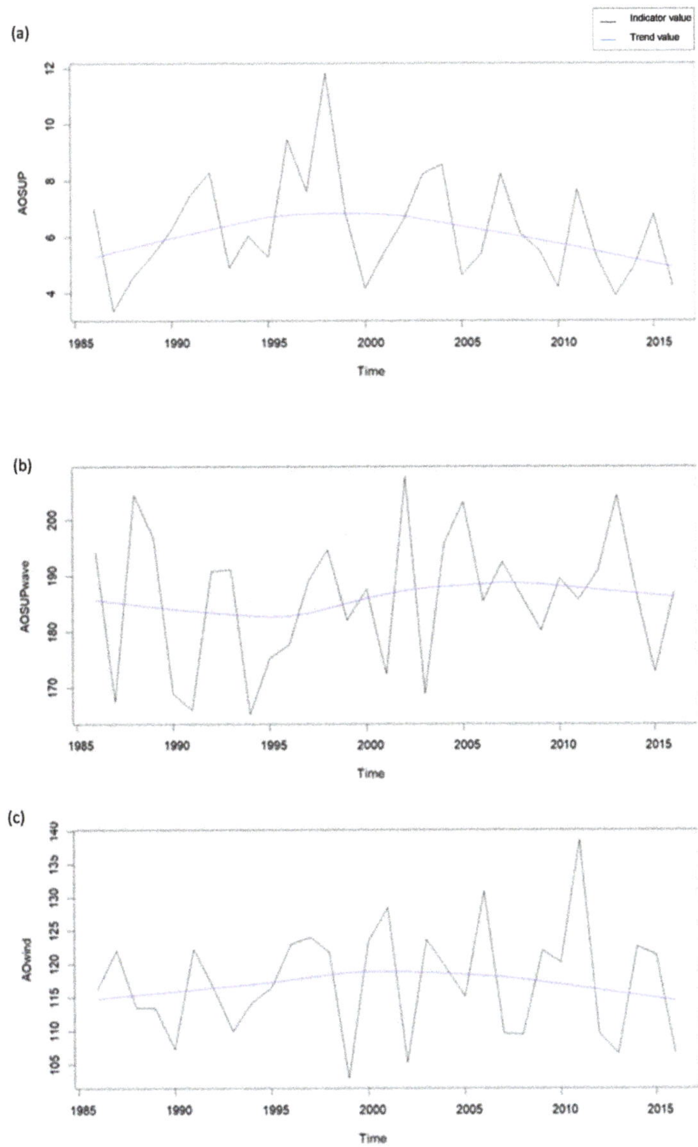

Figure 8. (**a**) Evolution and trend of annual alternative offer days for SUP-related sports; reference period is 1985–2015. (**b**) Evolution and trend of annual alternative offer days for SUP Wave sport; reference period is 1985–2015. (**c**) Evolution and trend of annual alternative offer days for wind-related sports, i.e., windsurfing, kitesurfing, wing surfing; reference period is 1985–2015.

4.3. Hazards Management for Surfers and Swimmers

As expected, the results showed that, in the coordinates of the Somo surf spot, the hazard score was higher for swimmers than for surfers (Figure 9). The maximum possible values were 10 for both swimmers and surfers, and even so, at any time of the studied period, a score of 10 was reached. The scores for surfers were always lower than those for swimmers (Figure 9). Higher hazard values were present in the winter, autumn, and spring; lower values corresponded to the summer season. After analyzing higher scores for surfers versus swimmers year round, we found the following values: January (4.1 vs. 7.3), February (4.2 vs. 7.3), March (3.9 vs. 7), April (3.7 vs. 6.7), November (4.4 vs. 7.8), and December (3.9 vs. 7).

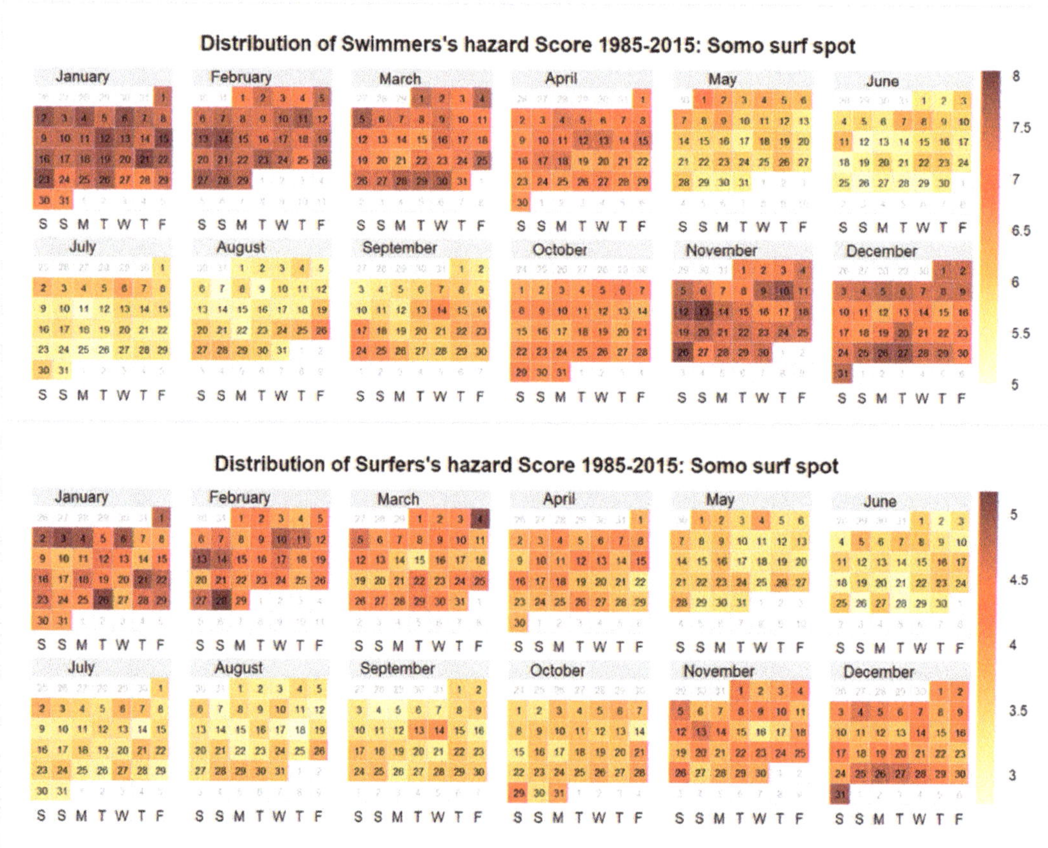

Figure 9. Distribution of swimmers' and surfers' hazard score, 1985–2015: Somo surf spot.

Figure 10a,b presents the evolution and trend of the annual values of SwS and SuS for the 1985–2015 period. The highest values for SwS and SuS were in 2014 (a score of 9.21 vs. 7.07) and the lowest happened in 1987 (a score of 7.32 vs. 4.25).

The Mann–Kendall test denoted the absence of significant trends in the series of all the indicators, characterized by interannual variability.

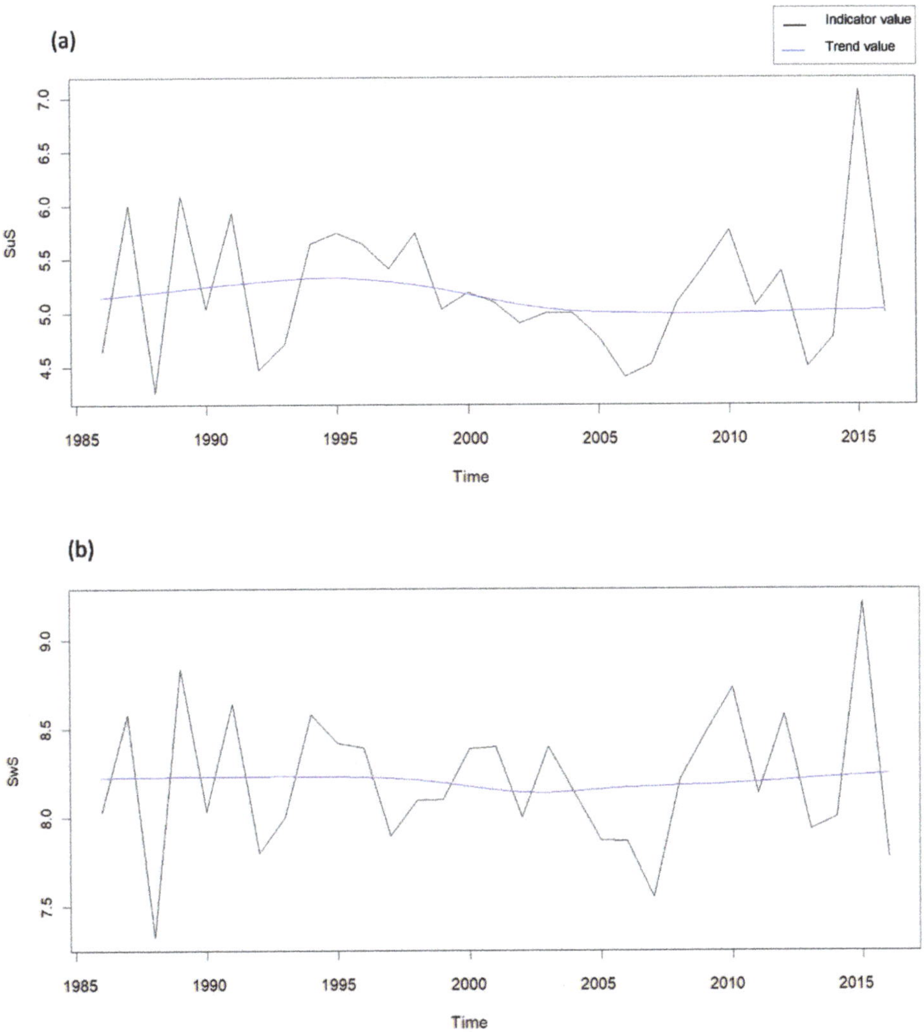

Figure 10. (**a**) Evolution and trend of annual maximum SuS from 1985 to 2015 in Somo. (**b**) Evolution and trend of annual maximum SwS.

5. Discussion

As described in Section 3, surfing days were computed considering peak period (*Tp*), significant wave height (*Hm*0), wind direction (*Wd*), and wind speed (*Ws*) parameters. The highest values in the summer will probably be linked to the period of calm winds in the area. Nevertheless, the months in the winter that presented lower values will probably present high values in other spots of the east of the beach where the wind speed is not as high as in this region due to orientation and exposure factors. These results improved those of Boqué et al. [35], who calculated expected surfing days without considering wind direction and wind speed, basing their calculations only on buoy data information from *Puertos del Estado* and *Instituto Marinha Portugal*.

As Scarfe et al. [73] suggested, we have developed a surfing wave climatology intended as an information resource for surfing management. Espejo et al. [65] developed a global index for analyzing surfing climatic potential, but the horizontal spatial resolution of ocean data was coarser than ours. Espejo et al. [65] based their analysis on a global scale, while we focused on the local scale by utilizing downscaled data with a hybrid method. Tausía [74] studied the surfing conditions in the Somo surf spot with a slightly coarser spatial resolution of 100 m, focusing on the numerical simulation of the physical processes that affect surfing waves.

Advanced surfers had a higher number of expected days per month from October to April. Intermediate surfing days per month had fewer fluctuations year round. As suggested by Hutt [68], surf breaks were classified according to surfing skills. In this sense, we followed Barlow et al. [75], who examined the effect of wave conditions and surfer ability on performance and the physiological response of recreational surfers. Hence, by combining climatic conditions and surfing levels as defined by Hutt [68], we see that we can contribute to the knowledge about expected surfing days by considering surfers' skills. Thus, we have more evidence about how different sizes of waves are associated with the balance of surfers during surfing activities, which will depend on surfers' skills as De Andrés et al. [53] stated.

These results provide important insights into demonstrating the different capacities for offering water-related activities for a specific territory. In some cases, lectures on the deseasonalization of the tourist activity are supported by the offer of other kinds of tourist products. Peñas de Haro [76], defended deseasonalizing sun and beach tourism in Mallorca, which is typically concentrated in the summer months. The deseasonalization proposal is based on the offer of surfing and body surfing activities, as these activities are possible when sun and beach climatic requirements are not in their best conditions. Martín et al. [77] also presented a proposal for the diversification of products in consolidated tourist destinations, giving special mention to the possibility of promoting Costa del Sol as a surfing destination. Even so, these studies did not specifically analyze climate data to determine the exact climatology of the products that can diversify the tourist offer, which is one of the aims of our study.

Regarding the hazard information from swimmers, as stated by Short et al. [78], rip currents and beach hazards have an impact on public safety and have implications for coastal management. We believe that surfers and lifeguards can assist swimmers in a hazardous situation and that swimmers should have lessons on rip current escape strategies [72]. In the event that a swimmer does not know how to escape from a rip current, surfers and lifeguards, who know how rip currents work [50], can perform a rescue [51]. Surfers possess this ability because they usually use rip currents to arrive at the surfing waiting-area zone for surfing [50]. Therewith, we consider in which moments surfers present the highest hazard score because, in that situation, they are not going to be able to rescue swimmers. During these times, lifeguards should check on both surfers and swimmers. Based on climatic conditions, our results reveal the difference between swimmers' and surfers' hazards, and thus, this information can assess lifeguards' decision making related to which periods are better for assisting only swimmers and which are important for assessing the safety of both swimmers and surfers. In Somo, lifeguards are only present during the summer months; therefore, this information can be of value when deciding whether to extend the period of lifeguards' presence if required.

6. Conclusions and Perspectives

León et al. [79] explained that the tourism sector is recognized as being highly vulnerable to climate change, and research supporting destinations to enhance their resilience capacities is still considered scarce. As Bradshaw [80] found, a review of the related tourism literature raises awareness of surfing as a sport, tourism, and innovation opportunities for policymakers in the context of a highly entrepreneurial country, highlighting the benefits that surf tourism offers for sustainable growth and positioning surf tourism as an innovative product.

Our research represents an advance in the knowledge of (1) the expected surfing conditions, (2) the expected surfing conditions related to surfers' skills, (3) the expected conditions for alternative surf offers, and (4) the expected hazard conditions and their differences for surfers and swimmers. Our case is applied in Somo's surf spot but the general framework can work as a model for other specific surfing destinations, specifically sandy beaches. Surfing destinations with point breaks and estuaries propagations of swell should follow another approach; nevertheless, surfing management indicators can be applied in the same way.

Following Borne [81], who defended the functions of academic and more-popular literature within different language games, academic accounts can seem turgid, dense, and overcomplicated, while popular media may sometimes be seen as repeating banal and superficial observations. However, the scope for surfing-related authors to seek to bridge the gap between scholarship and surfing culture is exceedingly broad. For this reason, we developed specific indicators and represented them to assist surfing destination managers to be better prepared to make climate-smart decisions as recommended by the Global Framework for Climate Services [2]. In this vein and following Kumar et al. [82], who explored how the visualization and communication of the forecast support the end users' decision making, our graphics in the results section are designed to be simple and easy to interpret for surfing destination managers, surf schools, and surfers, among others.

Our results contribute to the blue economy knowledge, as Spinrad [83] highlighted that the new blue economy is realized as the commercialization of value-added data, information, and knowledge about the marine environment. The economic benefits are enabled by dramatic improvements in observational capabilities and the development of predictive models. Increases in the volume, diversity, and quality of data, as well as more skillful methods of forecasting and nowcasting, make possible the production of products and services enhancing traditional components of the blue economy.

Surf tourism development provides economic opportunities to residents in coastal destinations, yet it has also been criticized for associations with gentrification, pollution, and inequality. The pandemic exacerbated existing sustainability challenges by accelerating development near surf breaks in Bocas del Toro, Panama. Mach [84] also found that there is an urgent need for stakeholders in surf communities, and particularly surf tourism business owners, to cooperate to preserve surf experiences that are vital to residents' mental and physical health and well-being as well as attractiveness as a surf tourism destination. As Mach et al. [85] explained, we defend the idea that surfing tourism deserves a more significant place in funding initiatives, discussions, and research related to fostering sustainable development from ocean resources in the rapidly changing world.

Our research can modestly contribute to Spain's goals for its Sustainable Tourism Strategy 2030. This is because, in 2019, the general guidelines of the Sustainable Tourism Strategy were presented, but surfing tourism was not mentioned.

This study presents a foundation for surfing climate service surfing. Future work will apply our indices to other surf spots and will validate the predictability of the indices. In addition, more indicators can be generated to assess surfing activities if more variables are added; an example is wetsuit recommendations if seawater temperature is analyzed. The present study has focused on surf tourism, but the methodology can be applied to other outdoor and sport-tourism-related activities following Silva et al. [86] and other dimensions of adventure tourism [87].

As surfers have their experiential standards for the surfability of particular places and conditions, and following Hutt et al. [68], research can affirm that, depending on surfing skills, surfers will be able to perform in specific meteo-oceanic conditions or not. The general idea is that the advanced surfers can surf in all conditions when they are not adverse. Conversely, beginning surfers cannot perform in all situations. Nevertheless, when high waves that are beneficial for advance surfers occur, beginners may sometimes also surf, but not in the same area. Advanced surfers will surf in the green wave area and beginners will surf in the white water area. The standards of surfers will depend on the level of practice, i.e., beginner, intermediate, advanced, and big wave rider, and on style, i.e., body board, skim, shortboard, longboard—for this reason, in general terms, some beaches are better for beginners and others for advanced surfers. Even so, as meteo-conditions are constantly changing, there is no general surf clue that can help the surfing community. For this reason, the present research has focused on developing those different needs identified from the survey profiling different kinds of surfers: beginners, intermediates, advanced, and big wave riders [9]. Relatedly, future research may explore the provision of an app with reactive programming for surfers that could help them to set preferences for meteo-oceanic variables.

Future research may also explore the needs of actual resort managers and/or developers by means of focus groups, adapting Font et al.'s [8] methodology to better re-design a climate service. The development of this kind of research will promote the maximization of the usage of surfing resources.

Research has explored the advances in climate services in multiple fields but determining a climate service for surfing destination management through downscaled wave data with a 100 m horizontal spatial resolution has not been done before. Further research may focus on developing the same/similar indicators but while also combining surfing forecasting with the downscaling method employed in the present research. This forecast data would help destination managers formulate better marketing plans and development. The next steps of the investigation can apply the computation of the same indicators with projection data considering the different climate scenarios to study how surfing resources will change in the future.

Author Contributions: Conceptualization, A.B.C., M.M., M.S.B. and E.A.; methodology, A.B.C., M.M., M.S.B. and E.A.; software, A.B.C., M.M. and M.S.B.; validation, A.B.C., M.M., M.S.B. and E.A.; formal analysis, A.B.C., M.M., M.S.B. and E.A.; investigation A.B.C., M.M., M.S.B. and E.A.; resources, A.B.C., M.M., M.S.B. and E.A.; data curation, A.B.C., M.M., M.S.B. and E.A.; writing—original draft preparation, A.B.C.; writing—review and editing, A.B.C., M.M., M.S.B. and E.A.; supervision, A.B.C., M.M., M.S.B. and E.A.; funding acquisition, A.B.C., M.M., M.S.B. and E.A. All authors have read and agreed to the published version of the manuscript.

Funding: This research is within the INDECIS project (INDECIS is part of ERA4CS, an ERA-NET initiated by JPI Climate and funded by FORMAS (SE), DLR (DE), BMWFW (AT), IFD (DK), MINECO (ES), and ANR (FR) with co-funding by the European Union Grant 690462). This article publication was possible with the support of the Secretaria d'Universitats i Recerca del Departament d'Empresa i Coneixement de la Generalitat de Catalunya, the European Union (UE), and the European Social Fund (ESF) (Doctoral Research Grant 2021FI_B2 00147—Formació personal investigador novell).

Institutional Review Board Statement: Not applicable.

Informed Consent Statement: Not applicable.

Data Availability Statement: The data presented in this study are available on request from the corresponding author.

Conflicts of Interest: The authors declare no conflict of interest.

References

1. World Meteorological Organization. *Implementation Plan of the Global Framework for Climate Services*; WMO: Geneva, Switzerland, 2014.
2. Hewitt, C.; Mason, S.; Walland, D. The global framework for climate services. *Nat. Clim. Change* **2012**, *2*, 831–832. [CrossRef]
3. Swart, R.; Celliers, L.; Collard, M.; Prats, A.G.; Huang-Lachmann, J.T.; Sempere, F.L.; Timmermans, W. Reframing climate services to support municipal and regional planning. *Clim. Serv.* **2021**, *22*, 100227. [CrossRef]
4. Mahon, R.; Petrie, J.A.; Trotman, A.; Eyzaguirre, J.; Burrowes, R.; Matthews, L.; Charles, A. Climate services for tourism: Insights from Caribbean Small Island Developing States. *Clim. Serv.* **2021**, *24*, 100262. [CrossRef]
5. André, K.; Järnberg, L.; Gerger Swartling, Å.; Berg, P.; Segersson, D.; Amorim, J.H.; Strömbäck, L. Assessing the Quality of Knowledge for Adaptation–Experiences from Co-designing Climate Services in Sweden. *Front. Clim.* **2021**, *3*, 11. [CrossRef]
6. Ibarra, A.M.S.; Hewitt, C.; Winarto, Y.T.; Walker, S.; Keener, V.W.; Bayala, J.; van den Hurk, B. Resilience through climate services. *One Earth* **2021**, *4*, 1050–1054. [CrossRef]
7. Demiroglu, O.C.; Saygili-Araci, F.S.; Pacal, A.; Hall, C.M.; Kurnaz, M.L. Future Holiday Climate Index (HCI) performance of urban and beach destinations in the Mediterranean. *Atmosphere* **2020**, *11*, 911. [CrossRef]

8. Font Barnet, A.; Boqué Ciurana, A.; Olano Pozo, J.X.; Russo, A.; Coscarelli, R.; Antronico, L.; Aguilar, E. Climate services for tourism: An applied methodology for user engagement and co-creation in European destinations. *Clim. Serv.* **2021**, *23*, 100249. [CrossRef]
9. Boqué Ciurana, A.; Aguilar, E. Which Meteorological and Climatological Information Is Requested for Better Surfing Experiences? A Survey-Based Analysis. *Atmosphere* **2021**, *12*, 293. [CrossRef]
10. De Freitas, C.R.; Scott, D.; McBoyle, G. A second generation climate index for tourism (CIT): Specification and verification. *Int. J. Biometeorol.* **2008**, *52*, 399–407. [CrossRef]
11. Scott, D.; Rutty, M.; Amelung, B.; Tang, M. An inter-comparison of the Holiday Climate Index (HCI) and the Tourism Climate Index (TCI) in Europe. *Atmosphere* **2016**, *7*, 80. [CrossRef]
12. Rutty, M.; Scott, D.; Matthews, L.; Burrowes, R.; Trotman, A.; Mahon, R.; Charles, A. An Inter-Comparison of the Holiday Climate Index (HCI:Beach) and the Tourism Climate Index (TCI) to Explain Canadian Tourism Arrivals to the Caribbean. *Atmosphere* **2020**, *11*, 412. [CrossRef]
13. Terrado, M.; Christel, I.; Bojovic, D.; Soret, A.; Doblas-Reyes, F.J. Climate change communication and user engagement: A tool to anticipate climate change. In *Handbook of Climate Change Communication*; Springer: Cham, Switzerland, 2018; Volume 3, pp. 285–302.
14. Terrado, M.; Calvo, L.; Urquiza, D.; Octenjak, S.; Nicodemou, A.; Bojovic, D.; Christel, I. Towards more effective visual-isations in climate services: Best practices and recommendations (No. EMS2021-355). In Proceedings of the Copernicus Meetings, Austria, 4 May 2021.
15. Simm, J.; Gouldby, B.; Lumbroso, D.; Matthewson, T. Effective Coastal Climate Services—An End-User Perspective for Resilient Infrastructure. *Front. Mar. Sci.* **2021**, 1135. [CrossRef]
16. WTO and UNEP. *Climate Change and Tourism—Responding to Global Challenges*; World Tourism Organization (WTO) and United Nations Environment Programme (UNEP); WTO: Madrid, Spain, 2008; p. 256.
17. Álvarez-Díaz, M.; Rosselló-Nadal, J. Forecasting British tourist arrivals in the Balearic Islands using meteorological variables. *Tour. Econ.* **2010**, *16*, 153–168. [CrossRef]
18. Rosselló-Nadal, J.; Riera-Font, A.; Cárdenas, V. The impact of weather variability on British outbound flows. *Clim. Change* **2011**, *105*, 281–292. [CrossRef]
19. Førland, E.J.; Jacobsen, J.K.S.; Denstadli, J.M.; Lohmann, M.; Hanssen-Bauer, I.; Hygen, H.O.; Tømmervik, H. Cool weather tourism under global warming: Comparing Arctic summer tourists' weather preferences with regional climate statistics and projections. *Tour. Manag.* **2013**, *36*, 567–579. [CrossRef]
20. Day, J.; Chin, N.; Sydnor, S.; Cherkauer, K. Weather, climate, and tourism performance: A quantitative analysis. *Tour. Manag. Perspect.* **2013**, *5*, 51–56. [CrossRef]
21. Falk, M. Impact of long-term weather on domestic and foreign winter tourism demand. *Int. J. Tour. Res.* **2013**, *15*, 1–17. [CrossRef]
22. Forster, J.; Schuhmann, P.W.; Lake, I.R.; Watkinson, A.R.; Gill, J.A. The influence of hurricane risk on tourist destination choice in the Caribbean. *Clim. Change* **2012**, *114*, 745–768. [CrossRef]
23. Hamzah, J.; Habibah, A.; Buang, A.; Jusoff, K.; Toriman, M.E.; Mohd Fuad, M.J.; Azima, A.M. Flood disaster, impacts and the tourism providers' responses: The Kota Tinggi experience. *Adv. Nat. Appl. Sci.* **2012**, *6*, 26–32.
24. Tsai, H.T.; Tseng, C.J.; Tzeng, S.Y.; Wu, T.J.; Day, J.D. The impacts of natural hazards on Taiwan's tourism industry. *Nat. Hazards* **2012**, *62*, 83–91. [CrossRef]
25. Arent, D.J.; Tol, R.S.J.; Faust, E.; Hella, J.P.; Kumar, S.; Strzepek, K.M.; Tóth, F.L.; Yan, D. Key economic sectors and services. In *Climate Change 2014: Impacts, Adaptation, and Vulnerability; Part A: Global and Sectoral Aspects*; Contribution of Working Group II to the Fifth Assessment Report of the Intergovernmental Panel on Climate Change; Field, C.B., Barros, V.R., Dokken, D.J., Mach, K.J., Mastrandrea, M.D., Bilir, T.E., Chatterjee, M., Ebi, K.L., Estrada, Y.O., Genova, R.C., et al., Eds.; Cambridge University Press: Cambridge, UK; New York, NY, USA, 2014; pp. 659–708.
26. Mieczkowski, Z. The tourism climatic index: A method of evaluating world climates for tourism. *Can. Geogr. Géographe Can.* **1985**, *29*, 220–233. [CrossRef]
27. Becken, S.; Hay, J.E. Tourism and climate change. In *Tourism and Climate Change*; Channel View Publications: Bristol, UK, 2007.
28. Adiguzel, F.; Bozdogan Sert, E.; Dinc, Y.; Cetin, M.; Gungor, S.; Yuka, P.; Vural, E. Determining the relationships between climatic elements and thermal comfort and tourism activities using the tourism climate index for urban planning: A case study of Izmir Province. *Theor. Appl. Climatol.* **2021**, *147*, 1105–1120. [CrossRef]
29. Adiguzel, F.; Bozdogan Sert, E.; Dinc, Y.; Cetin, M.; Gungor, S.; Yuka, P.; Vural, E.; Dogan, O.S.; Kaya, E.; Karakaya, K. Confort climático en la Argentina: Un recurso intangible para el turismo. *Cuad. Geográficos* **2021**, *60*, 52–72.
30. Scott, D.; Lemieux, C. Weather and climate information for tourism. *Procedia Environ. Sci.* **2010**, *1*, 146–183. [CrossRef]
31. Scott, D. Sustainable tourism and the grand challenge of climate change. *Sustainability* **2021**, *13*, 1966. [CrossRef]
32. Olano Pozo, J.X.; Boqué Ciurana, A.; Font Barnet, A.; Russo, A.; Saladié Borraz, Ò.; Anton-Clavé, S.; Aguilar, E. Co-developing climate services with local agents: The INDECIS Snow Tourism Index. In Proceedings of the EGU General Assembly Conference Abstracts, Virtual Event, 4–8 May 2020; p. 8926.
33. Rice, H.; Cohen, S.; Scott, D.; Steiger, R. Climate change risk in the Swedish ski industry. *Curr. Issues Tour.* **2021**, 1–16. [CrossRef]

34. Köberl, J.; François, H.; Cognard, J.; Carmagnola, C.; Prettenthaler, F.; Damm, A.; Morin, S. The demand side of climate services for real-time snow management in Alpine ski resorts: Some empirical insights and implications for climate services development. *Clim. Serv.* **2021**, *22*, 100238. [CrossRef]
35. Boqué Ciurana, A.; Aguilar, E. Expected distribution of surfing days in the Iberian Peninsula. *J. Mar. Sci. Eng.* **2020**, *8*, 599. [CrossRef]
36. Martin, S.A.; Assenov, I. The genesis of a new body of sport tourism literature: A systematic review of surf tourism research (1997–2011). *J. Sport Tour.* **2012**, *17*, 257–287. [CrossRef]
37. Amorim, R.C.; Rocha, A.; Oliveira, M.; Ribeiro, C. Efficient delivery of forecasts to a nautical sports mobile application with semantic data services. In Proceedings of theNinth International C* Conference on Computer Science & Software Engineering, Porto, Portugal, 20–22 July 2016; C3S2E '16. Association for Computing Machinery: New York, NY, USA, 2016; pp. 7–12.
38. Martínez Vázquez, R.M.; Milán García, J.; De Pablo Valenciano, J. Analysis and trends of global research on nautical, maritime and marine tourism. *J. Mar. Sci. Eng.* **2021**, *9*, 93. [CrossRef]
39. Portugal, A.C.; Campos, F.; Martins, F.; Melo, R. Understanding the relation between serious surfing, surfing profile, surf travel behaviour and destination attributes preferences. *Eur. J. Tour. Res.* **2017**, *16*, 57–73. [CrossRef]
40. Chen, N.; Liu, J.; Ba, Z.; Zhong, J.; Liu, X. The Construction and Research of Marine Tourism Management System Based on the Perspective of Industrial Integration. *J. Coast. Res.* **2020**, *112*, 132–135. [CrossRef]
41. Martin, S.A. The conservation of coastal surfing resources in Thailand: The Andaman Sea. In Proceedings of the International Conference on the Environment and Natural Resources (ICENR), Bangkok, Thailand, 5 November 2010; pp. 262–280.
42. Arroyo, M.; Levine, A.; Brenner, L.; Seingier, G.; Leyva, C.; Espejel, I. Indicators to measure pressure, state, impact and responses of surf breaks: The case of Bahía de Todos Santos World Surfing Reserve. *Ocean. Coast. Manag.* **2020**, *194*, 105252. [CrossRef]
43. Atkin, E.A.; Reineman, D.R.; Reiblich, J.; Revell, D.L. Applicability of management guidelines for surfing resources in California. *Shore Beach* **2020**, *88*, 53–64. [CrossRef]
44. Reineman, D.R.; Koenig, K.; Strong-Cvetich, N.; Kittinger, J.N. Conservation Opportunities Arise from the Co-Occurrence of Surfing and Key Biodiversity Areas. *Front. Mar. Sci.* **2021**, *8*, 253. [CrossRef]
45. Valencia, L.; Monterrubio, C.; Osorio-García, M. Social representations of surf tourism's impacts in Mexico. *Int. J. Tour. Policy* **2021**, *11*, 29–51. [CrossRef]
46. Fox, N.; Marshall, J.; Dankel, D.J. Ocean Literacy and Surfing: Understanding How Interactions in Coastal Ecosystems Inform Blue Space User's Awareness of the Ocean. *Int. J. Environ. Res. Public Health* **2021**, *18*, 5819. [CrossRef]
47. McGinnis, M.D.; Ostrom, E. Social-ecological system framework: Initial changes and continuing challenges. *Ecol. Soc.* **2014**, *19*, 30. [CrossRef]
48. Mindes, A.R. *The Perception of Hazards Among Surfers in Southern California*; California State University: Long Beach, CA, USA, 1997.
49. Brander, R.W. Rip currents. In *Coastal and Marine Hazards, Risks, and Disasters*; Elsevier: Amsterdam, The Netherlands, 2015; pp. 335–379.
50. Brander, R.W.; Bradstreet, A.; Sherker, S.; MacMahan, J. Responses of swimmers caught in rip currents: Perspectives on mitigating the global rip current hazard. *Int. J. Aquat. Res. Educ.* **2011**, *5*, 11. [CrossRef]
51. Dalrymple, R.A.; MacMahan, J.H.; Reniers, A.J.; Nelko, V. Rip currents. *Annu. Rev. Fluid Mech.* **2011**, *43*, 551–581. [CrossRef]
52. Attard, A.; Brander, R.W.; Shaw, W.S. Rescues conducted by surfers on Australian beaches. *Accid. Anal. Prev.* **2015**, *82*, 70–78. [CrossRef] [PubMed]
53. De Andrés, C.; Torrego, J.A.; Júnior, S.; Escamilla, V.; Ortiz, R.; Borgonovo, M.; del Estal, A. Valoración del Equilibrio con Y-balance Test en Surfistas de Competición y Población no Surfista en España. Available online: https://www.researchgate.net/profile/Alejandro-Martinez-22/publication/317175801_Valoracion_del_Equilibrio_con_Y-balance_Test_en_Surfistas_de_Competicion_y_Poblacion_no_Surfista_en_Espana/links/59297851458515e3d469c4af/Valoracion-del-Equilibrio-con-Y-bal (accessed on 9 June 2022).
54. Sariego López, I.; Moreno Melgarejo, A. El desarrollo turístico y territorial basado en el Surf: Ribamontán al Mar,"Surf a Toda Costa". *Estud. Turísticos* **2015**, *205*, 119–138.
55. Köppen, W.P. *Grundriss der Klimakunde*; Walter de Gruyter GmbH: Berlin, Germany, 1931.
56. Rafael, C. *Climate Atlas of the Cantabrian Region*; CMT CAS; Technical note; National Institute of Meteorology: Brasilia Brazil, 1992.
57. Hellín Medina, J. Análisis Climatológico del Mar Cantábrico y su Influencia en la Navegación; 2009. Available online: https://upcommons.upc.edu/bitstream/handle/2099.1/7451/An%E1lisis%20Climatol%F3gico%20del%20Mar%20Cant%E1brico%20y%20su%20influencia%20en%20la%20Navegaci%F3n.pdf?sequence=1 (accessed on 9 June 2022).
58. Camus, P.; Mendez, F.J.; Medina, R. A hybrid efficient method to downscale wave climate to coastal areas. *Coast. Eng.* **2011**, *58*, 851–862. [CrossRef]
59. Camus, P.; Mendez, F.J.; Medina, R.; Tomas, A.; Izaguirre, C. High resolution downscaled ocean waves (DOW) reanalysis in coastal areas. *Coast. Eng.* **2013**, *72*, 56–68. [CrossRef]
60. Reguero, B.G.; Menéndez, M.; Méndez, F.J.; Mínguez, R.; Losada, I.J. A Global Ocean Wave (GOW) calibrated reanalysis from 1948 onwards. *Coast. Eng.* **2012**, *65*, 38–55. [CrossRef]
61. Perez, J.; Menendez, M.; Losada, I.J. GOW2: A global wave hindcast for coastal applications. *Coast. Eng.* **2017**, *124*, 1–11. [CrossRef]

62. Saha, S.; Moorthi, S.; Pan, H.L.; Wu, X.; Wang, J.; Nadiga, S.; Tripp, P.; Kistler, R.; Woollen, J.; Goldberg, M.; et al. The NCEP Climate Forecast System Reanalysis. *Bull. Am. Meteorol. Soc.* **2010**, *91*, 1015–1058. [CrossRef]
63. ha, S.; Moorthi, S.; Wu, X.; Wang, J.; Nadiga, S.; Tripp, P.; Behringer, D.; Hou, Y.; Chuang, H.; Iredell, M.; et al. The NCEP Climate Forecast System Version 2. *J. Clim.* **2014**, *27*, 2185–2208.
64. Holthuijsen, L.H. *Waves in Oceanic and Coastal Waters*; Cambridge University press: Cambridge, UK, 2010.
65. Espejo, A.; Losada, I.J.; Méndez, F.J. Surfing wave climate variability. *Glob. Planet. Change* **2014**, *121*, 19–25. [CrossRef]
66. Yukawa, H.; Iino, M.; Fujiwara, T. Estimation and visualization of paddling effort for stand up paddle boarding with a geographical information system. *Procedia Eng.* **2015**, *112*, 552–555. [CrossRef]
67. Vermeersch, W.; Alcoforado, M.J. Wind as a resource for summer nautical recreation. Guincho beach study case. *Finisterra* **2013**, *48*, 95. [CrossRef]
68. Hutt, J.A.; Black, K.P.; Mead, S.T. Classification of surf breaks in relation to surfing skill. *J. Coast. Res.* **2001**, *29*, 66–81.
69. Koon, W.; Rowhani-Rahbar, A.; Quan, L. Do wave heights and water levels increase ocean lifeguard rescues? *Am. J. Emerg. Med.* **2018**, *36*, 1195–1201. [CrossRef]
70. Mazzone, W.F. *Development and Evaluation of a Swimmer's Rescue Suit*; Naval Submarine Base New London: Groton CN, USA, 1961.
71. Whitcomb, D. So Others May Live: Saving Lives, Defying Death with the Coast Guard's Rescue Swimmers. *Air Power Hist.* **2008**, *55*, 63–64.
72. Miloshis, M.; Stephenson, W.J. Rip current escape strategies: Lessons for swimmers and coastal rescue authorities. *Nat. Hazards* **2011**, *59*, 823–832. [CrossRef]
73. Scarfe, B.E.; Healy, T.R.; Rennie, H.G.; Mead, S.T. Sustainable management of surfing breaks—An overview. *Reef J.* **2009**, *1*, 44–73.
74. Tausía Hoyal, J. Spatial and temporal variability of surfing in Cantabria. Master 's Thesis, Universidad de Cantabria (UC), Santander, Spain, 2020.
75. Barlow, M.J.; Gresty, K.; Findlay, M.; Cooke, C.B.; Davidson, M.A. The effect of wave conditions and surfer ability on performance and the physiological response of recreational surfers. *J. Strength Cond. Res.* **2014**, *28*, 2946–2953. [CrossRef]
76. Peñas de Haro, P. La Geografía del Surf y el Bodyboard en Mallorca, Clima y Turismo Activo. PhD Thesis, Universitat de les Illes Balears, Palma de Mallorca, Spain, 2015.
77. Martín González, R.; Gil, A.M.L. Propuesta de diversificación de productos en destinos consolidados: El turismo de surf en la Costa del Sol Occidental. *Estud. Turísticos* **2014**, *199*, 63–88.
78. Short, A.D.; Hogan, C.L. Rip currents and beach hazards: Their impact on public safety and implications for coastal management. *J. Coast. Res.* **1994**, *14*, 197–209.
79. León, C.J.; Giannakis, E.; Zittis, G.; Serghides, D.; Lam-González, Y.E.; García, C. Tourists' Preferences for Adaptation Measures to Build Climate Resilience at Coastal Destinations. Evidence from Cyprus. *Tour. Plan. Dev.* **2021**, 1–27. [CrossRef]
80. Bradshaw, L. Surfing the Innovation Waves: Surf Tourism in Portugal. In *Tourism Innovation in Spain and Portugal*; Springer: Cham, Switzerland, 2021; pp. 149–166.
81. Borne, G. *Surfing and Sustainability*; Routledge: London, UK, 2018.
82. Kumar, U.; Werners, S.E.; Paparrizos, S.; Datta, D.K.; Ludwig, F. Co-producing climate information services with smallholder farmers in the Lower Bengal Delta: How forecast visualization and communication support farmers' decision-making. *Clim. Risk Manag.* **2021**, *33*, 100346. [CrossRef]
83. Spinrad, R.W. The new blue economy. In *Preparing a Workforce for the New Blue Economy*; Elsevier: Amsterdam, The Netherlands, 2021; pp. 87–111.
84. Mach, L.J. Surf Tourism in Uncertain Times: Resident Perspectives on the Sustainability Implications of COVID-19. *Societies* **2021**, *11*, 75. [CrossRef]
85. Mach, L.; Ponting, J. Establishing a pre-COVID-19 baseline for surf tourism: Trip expenditure and attitudes, behaviors and willingness to pay for sustainability. *Ann. Tour. Res. Empir. Insights* **2021**, *2*, 100011. [CrossRef]
86. Silva, G.; Correia, A.; Rachão, S.; Nunes, A.; Vieira, E.; Santos, S.; Fernandes, P.O. A Methodology for the Identification and Assessment of the Conditions for the Practice of Outdoor and Sport Tourism-Related Activities: The Case of Northern Portugal. *Sustainability* **2021**, *13*, 7343. [CrossRef]
87. Janowski, I.; Gardiner, S.; Kwek, A. Dimensions of adventure tourism. *Tour. Manag. Perspect.* **2021**, *37*, 100776. [CrossRef]

sustainability MDPI

Viewpoint

Beyond Climate Ready? A History of Seattle Public Utilities' Ongoing Evolution from Environmental and Climate Risk Management to Integrated Sustainability

Ann Grodnik-Nagle [1,*], Ashima Sukhdev [1], Jason Vogel [2] and Charles Herrick [3]

[1] Seattle Public Utilities, Seattle Municipal Tower, 700 5th Avenue, Seattle, WA 98104, USA
[2] Climate Impacts Group (CIG), University of Washington, Seattle, WA 98195, USA
[3] Washington Center, New York University, Washington, DC 20016, USA
* Correspondence: ann.grodnik-nagle@seattle.gov

Abstract: Seattle Public Utilities (SPU) is a municipal water supply, drainage, wastewater, and solid waste management utility in Seattle, Washington. This utility has explored the impacts of climate change and supported climate adaptation work since 1997. Faced with threats such as sea level rise, drought, wildfires, and extreme precipitation events, SPU has worked to "mainstream" climate science throughout its strategic planning, capital investments, management, operations, staffing, institutional culture, and more. This paper provides a descriptive, chronologically ordered account of how SPU's climate-change-related work has evolved to become an aspect of a broader social and environmental sustainability orientation, aimed at resilience against climate impacts, but also towards improving greenhouse gas emissions reduction, carbon sequestration, water and waste circularity, green infrastructure, ecosystem and species stewardship, green and blue workforce development, affordability, an intergenerational perspective, and environmental justice. We frame this transition as a movement from a core focus on risk management toward a proactive and integrated mode of sustainable operations. While SPU's journey has been enabled by a co-productive approach to climate services, we speculate on how this model can be broadened and diversified to help SPU pursue their goal of becoming a sustainable organization. It is our hope that this paper sparks reflection and discussion within the climate services community, amongst utilities, municipalities, and policy entrepreneurs that are interested in sustainability.

Keywords: sustainability; public utilities; resilience; climate change; climate adaptation; seattle; climate services; water utility; waste utility

Citation: Grodnik-Nagle, A.; Sukhdev, A.; Vogel, J.; Herrick, C. Beyond Climate Ready? A History of Seattle Public Utilities' Ongoing Evolution from Environmental and Climate Risk Management to Integrated Sustainability. *Sustainability* **2023**, *15*, 4977. https://doi.org/10.3390/su15064977

Academic Editors: Genovaitė Liobikienė and Mohammad Aslam Khan Khalil

Received: 26 January 2023
Revised: 28 February 2023
Accepted: 2 March 2023
Published: 10 March 2023

1. Introduction

Seattle Public Utilities (SPU) is a public utility serving the Seattle, Washington, metropolitan area and its surrounding communities. With service deliveries of water supply, drainage, wastewater, and solid waste services, it is important to understand that climate change first emerged as an unconnected issue in each of these service areas at different times and for different reasons. SPU first addressed climate change within water supply planning in 1997, following a series of extreme rainfall events and droughts between the mid-1980's and mid-1990's. Following that, SPU addressed climate change in its drainage and wastewater line of business in the wake of urban flooding and extreme rain events in 2006 and 2007. SPU's work in solid waste management has long been guided by environmental considerations, with a specific analysis conducted in the early 2000s around the potential of effective waste management systems to reduce greenhouse gas emissions, amongst other financial and environmental benefits.

While each of these stories could be told in isolation, the staff at SPU are increasingly seeing that these stories are not separate. This leads to a more complicated story, one in which it is helpful to understand the standpoint of SPU's leadership and staff, what they view as their future direction, and the kinds of assistance that they will need to realize their commitment to sustainability. The utility's early climate-change-related work is now one aspect of a broader effort to foster sustainable utility operations and management, a portfolio of activity, that, in addition to addressing climate impacts, includes greenhouse gas emissions reduction, carbon sequestration, water and waste circularity, green infrastructure, continuity of service during weather emergencies, ecosystem and species stewardship, green and blue workforce development, affordability, an intergenerational perspective, and environmental justice.

We believe that our story can be characterized as an evolution from a risk management viewpoint to a commitment to a broader and more inclusive set of factors, which are organized under the concept of sustainability. While SPU retains focused climate and environmental risk management strategies and protocols within its lines of business, it is striving toward a proactive, utility-wide approach to sustainability and adaptive management. This evolution includes a concerted effort to look beyond being "climate ready" or understanding climate impacts, and modifying utility strategic planning, capital investments, and operations and maintenance to account for the changing climate conditions. SPU's commitment to social and environmental sustainability and resilience is taking a more integrated, holistic approach by addressing the root causes of problems, often across multiple service areas, to realize conventional utility objectives such as affordability, service reliability, and service equity, while also pursuing new objectives such as racial equity, addressing displacement pressure, population growth, economic and environmental injustice, environmental stewardship, greenhouse gas reduction commitments, and the mitigating experiences of repeated climate impacts. The Director of Corporate Policy and Planning at SPU, Danielle Purnell, reflected that SPU's transformation is about a "restorative balance in our relationship with the planet and with people. It is a re-remembering that everything is connected, and we must work together within our means. Climate science provides clarity about some of the fundamental conditions requiring restoration if we are to ensure sustainability".

As emphasized in the final section of this paper, SPU is by no means "finished" with the difficult work of transforming our strategic orientation, capital investments, management, operations, staffing, and institutional culture, but we have moved far enough that the rough contours of our desired future state and required support can be discerned. We hope that SPU's evolution will provide grist for the climate services community to use for reflecting on its own forward-looking priorities, as well as a precedent that is useful for peer utilities, municipalities, and policy entrepreneurs that are interested in sustainability.

The remainder of this paper is organized into six sections. First, we describe our technical approach (Section 2), followed by a short background on SPU (Section 3). We then describe SPU's history with environmental and climate risk management (Section 4) and then expand into a summary description of SPU's efforts to take a more proactive and integrated approach to achieving sustainable operations across all aspects of our service portfolio (Section 5). The paper concludes with two discussions: the first explores SPU's need for and use of technical and scientific information that falls outside of the purview of traditional utility operational know-how (Section 6), followed by a discussion of the ongoing challenges and implications of this information for the climate services community (Section 7).

2. Technical Approach

This paper is an interpretive, mixed methods study of an American utility—Seattle Public Utilities. This paper is written to highlight SPU's historical evolution, with major sub-sections framed as narrative chronologies. The research integrates: (1) participant observation [1] with (2) outputs from several third-party assessments and evaluations; (3) a review of utility archival materials, including planning documents, technical memoranda, commissioned research reports and internal analyses, capital project application and approval documents, internal policy statements, and regulatory documentation (e.g., Consent

Decree); and (4) semi-structured, in-depth interviews with SPU staff working on different facets of sustainability, climate science, policy, and planning, between 1993 and the present.

(1) The authors of this article are participant observers to SPU's evolution. In total, two of the authors (Grodnik-Nagle and Sukhdev) lead the climate mitigation, adaptation, sustainability, and circular economy policy at SPU. The other two authors have worked in a consulting capacity for SPU or for entities that have drawn lessons from SPU's work for more than a decade (Vogel and Herrick).
(2) SPU has been the subject of several third-party assessments of climate resilience activity that have collectively documented the utility's efforts to address climate readiness over the course of about 18 years, since approximately 2005 [2–16]. The most comprehensive of these evaluations is a detailed 2016 case study, funded by the Kresge Foundation [12] and co-written by one of the authors of this essay [8]. This paper draws significantly from this collection of assessments in laying out the basic history, sequencing of events, and causal linkages of the SPU story. A total of two peer reviewed articles have emerged from this body of work [3,12].
(3) Archival materials were identified through participant–author familiarity with the utility (and predecessor organizations), its history, its standard operating procedures, and other formal and informal practices. Additional archival materials were identified through in-depth staff interviews. Other sources were identified through a limited review of the relevant technical literature and documentation produced by peer and partner organizations.
(4) To ensure authenticity, the author team developed a preliminary narrative of the SPU experience using the evaluation and archival sources above, and then solicited direct feedback from the professional staff involved in SPU's climate resilience and sustainability management, operations, and research, as far back as 1993. We conducted four semi-structured group interviews (1–4 interviewees per group) with a total of nine individuals. The interviewees included individuals involved in all three of SPU's lines of business, as well as management, leadership, and core technical staff who had focused on climate impacts and resilience over the past three decades. Once this manuscript was drafted, we asked each interviewee, as well as additional staff, to review the draft and provide critical feedback to ensure that our account was consistent with the lived experience of these key participants (see the list of participating SPU staff in Appendix A Table A1).

3. SPU Background

Established in 1997 by consolidating the Water Department with the Solid Waste, Drainage, and Wastewater Utilities Departments, Seattle Public Utilities provides drinking water to 1.5 million retail customers and 19 neighboring utility wholesale customers throughout the region, and provides drinking water, drainage, wastewater, and solid waste services directly to residents and businesses within the city of Seattle. SPU has organized these four essential services into three lines of business: water supply, drainage and wastewater, and solid waste (see Figure 1). SPU has about 1400 employees. The utility manages two mountain watersheds, the Cedar River watershed, and the South Fork Tolt River watershed. Its system includes almost 200 miles of water transmission pipelines, 1680 miles of water distribution main, 1400 miles of in-city sanitary and combined sewer mainlines, over 480 miles of drainage pipes, and two major garbage and recycling transfer stations that process an estimated 750,000 tons of garbage, recycling, and organic waste each year. SPU operates a fleet of 606 vehicles, including 73 construction vehicles, and contracts with providers for garbage, organics, and recycling collection. With operating revenues of over $1.4 billion per year, SPU is considered a relatively large U.S. utility, and is uncommon in its consolidated water supply, drainage, wastewater, and solid waste services, which allow it to have a broad purview over Seattle's resource management, environmental services, and pollution issues.

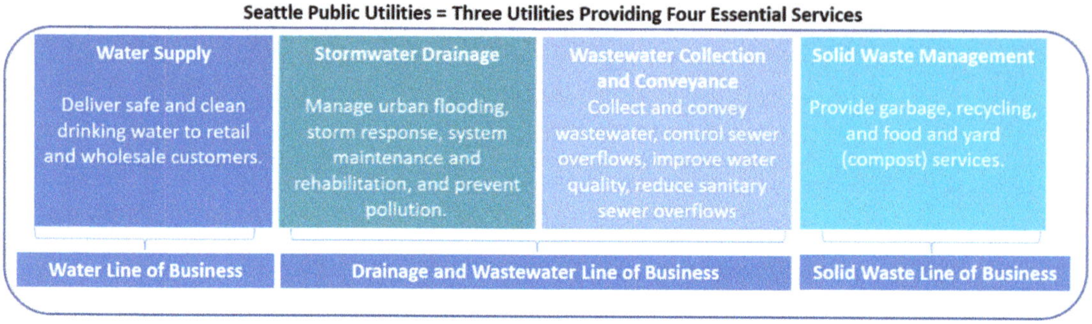

Figure 1. SPU operates three main lines of business, providing four essential services to Seattle residents and businesses.

In addition to customer-facing essential services, SPU provides the Seattle area with a diverse portfolio of critical services. While most SPU customers equate the utility's operations with traditional, high-visibility service streams, such as trash pick-up and the reliable delivery of safe, potable water, the reality is much more complicated, and includes operational priorities that are as diverse as managing wildland forests, salmon stewardship, urban tree planting programs, water and waste educational programs, pollution source control, graffiti removal, and recreational vehicle wastewater pump-out services.

The breadth of SPU's core and corollary services provides a backdrop for the variety and range of SPU's climate and sustainability risks and opportunities. The utility's interface with climate change, in terms of its impacts and contributions, is varied and multi-sectoral. Figures 2 and 3 provide an overview of SPU's understanding of climate change, and its relationship to SPU's management and operations.

Drought	SPU's two water supply reservoirs are vulnerable to drought conditions. Drought years with little snow stress the system's capacity to provide sufficient water for people and fish [17].
Sea level rise	SPU's drainage and wastewater infrastructure (including outfalls, wastewater pump stations, and tidally influenced CSO facilities) will be impacted by sea level rise. Puget Sound has risen by more than nine inches in Seattle over the past century. By 2100, sea level rise is projected to increase by another two to five feet [18,19].
Wildfire	Wildfires in the Cedar and Tolt river watersheds could impact the water quality and supply, as well as wildlife habitat. Warmer temperatures, droughts, and wind increase the flammability of forest fuels and increase fire intensity [20]. The forests on the west side of the Cascade Mountains are now more at risk of wildfire [21], including Seattle's two forested mountain watersheds.
Extreme rainstorms	Atmospheric river-driven rainstorms increased in intensity by 30% between 2003 and 2017 [22,23]. This poses capacity and water quality challenges for the drainage and wastewater system. With more frequent and intense rain events, it becomes more challenging to meet the state and federal regulations for combined sewer overflows (CSOs). Extreme rain events can also increase sewer backups, localized urban flooding, and landslides, which often have a greater impact on Seattle's vulnerable communities. In addition, extreme downpours can elevate the turbidity in SPU's water supply systems, creating challenges for water treatment.
Decreasing air quality	Air quality is expected to worsen due to increased heat waves and wildfire smoke. For the past five summers, Seattle has experienced periods of unsafe air quality due to wildfire events. This will negatively impact SPU employees, particularly our frontline operations and maintenance staff.
Increasing temperatures	The average annual air temperature in Seattle has increased by 1.3 °F since 1895, and is projected to be 5.5 °F warmer by 2050 [22,23]. This, combined with urban heat island effect, means acute heat risks to SPU employees, Seattle residents, animals, and ecosystems. Rising temperatures also increase the demand for water and the likelihood of water quality incidents, including bacterial outbreaks, algal blooms, and stress salmon recovery efforts.

Figure 2. Climate impacts on SPU's system [17–23].

Operational and supply chain GHG emissions	Citywide waste	Carbon sequestration
SPU's facilities, fleets, solid waste trucks, employee travel and commutes, and construction projects contribute GHG emissions. Fleet and facility electrification, energy efficiency and renewable energy investments have potential to reduce this contribution.	The product consumption and waste generated by Seattle's residents and businesses contributes to local and global GHG emissions. SPU's Waste Prevention and Diversion Programs can address the significant emissions associated with the global production and consumption of goods and food.	SPU's green infrastructure investments and composting services create nature-based, urban carbon sinks. Additionally, SPU's management of its 100,000 acres of watershed forest preserves a substantial carbon sink.

Figure 3. SPU's contributions to and opportunities to mitigate climate change.

4. SPU's History as a Risk Management Utility

Seattle Public Utilities has been working to characterize and address climate change and environmental stewardship issues since its establishment in 1997. We characterize SPU's management of environmental externalities and emergent climate risks in this earlier period as a risk management orientation. From a risk management viewpoint, distinct impacts (e.g., drought) are seen as threats to the stability of the existing and established utility services (e.g., the provision of reliable, affordable drinking water). In addition, environmental externalities (e.g., waste and pollution) are largely managed "downstream" (reactively), as opposed to "upstream" (preventatively). Such risks are typically addressed within a single line of business and are reactive within nature—two characteristics that are very different from the more integrated and adaptive sustainability standpoint are described in Section 5.

We focus our discussion in this section on the evolution of climate risk and environmental externality management at SPU. Notably, the risks that were associated with climate change were only one risk factor, among others, that were driving progress at SPU. Within water supply, other environmental externalities, such as salmon habitat protection, were critically important. Within drainage and wastewater, other environmental issues, such as improving surface water quality, were compelling and regulated. Within solid waste management, many environmental drivers (beyond greenhouse gas emissions) and financial considerations were taken into account when promoting recycling and composting programs for the city.

SPU's risk management work continues as an important effort under the current sustainability orientation described in Section 5. However, SPU is striving for risk management to apply across all of SPU's lines of business to treat the threats and opportunities in a more integrated and holistic fashion [24]. As such, the narrative of each line of business below does not have a clean break between the past and the present, nor is it intended to be comprehensive. Rather, each narrative provides background on the climate risk and environmental externality management activities at SPU that allows us to describe their evolution into a sustainability orientation in Section 5.

4.1. Water Supply

Following a series of extreme rainfall events and droughts between the mid-1980′s and mid-1990′s, SPU embarked upon an effort to characterize its climate-related exposures and risks, and to address its vulnerabilities. Early climate-related work was spurred on by the projection of a strong El Niño in 1997–1998, and an analysis of the city records to assess the historical water supply in El-Niño-like years [25]. SPU's first climate change study, in 2002, was carried out in partnership with the University of Washington Climate Impacts Group (CIG) to develop analysis techniques to help SPU's water supply planning staff and decision makers incorporate global climate change information into local long-range water supply planning processes. (Chinn, A., 1 September 2022, personal interview; [26]).

The concept of using climate services—relying on forward-looking projections of changing climate conditions—took root at SPU in their water line of business in the late 1990′s and early 2000′s [27–29]. This was before the Intergovernmental Panel on Climate Change and National Oceanic and Atmospheric Administration projection forecasting and the science behind the El Niño–Southern Oscillation (ENSO) patterns of climate variability were well established. As a result of these early efforts to understand hydrologic systems' changes and impacts and the implications of ENSO, the utility began an effort to integrate climate-related risks across all levels of its operations—an effort that continues to this day.

By the mid-2000's, SPU staff and leadership began to understand that climate change might affect SPU's ability to meet its water supply mission. However, there was a gap between the level and type of climate change information that water managers needed, and the level and type of the information that was being disseminated by the scientific community. Consequently, a considerable effort was put into building relationships to share this information about climate change and its impacts on water resources. Much of this activity was reflected in a two-year (2004–2006) study, jointly sponsored by the American Water Works Association Research Foundation and the National Center for Atmospheric Research, which included a case study that characterized SPU's efforts to use climate science to inform water supply planning [30,31].

SPU continued to partner with CIG, nearby water utilities in Everett and Tacoma, and stakeholders in King County, and in 2007, developed downscaled climatological data for the Tacoma, Seattle, and Everett water utilities. The utilities ran this climate information through their own system models to generate the 2009 water outlook, which provided a long-range view of the future water demand in this three-county region [7,32,33].

Shortly after a January 2007 Water Utility Climate Change Summit, attended by more than 200 water and wastewater utility executives, SPU worked with several other United States water utilities to form the Water Utility Climate Alliance (WUCA), a self-funded and collaborative effort to provide leadership on the climate change issues affecting the country's water supply agencies. This organization is now composed of 12 water providers nation-wide that supply water for more than 50 million people. The WUCA collaboration has funded its own research agenda to provide context-specific information on climate change and how to integrate that information into the utility decision making for SPU and its sister agencies. The WUCA launch built a bridge that is sustained to this day for the collaboration and information sharing between water utilities and the climate services community. WUCA-sponsored studies (see https://www.wucaonline.org/publications/ (accessed on 3 January 2023)) have provided important groundwork for SPU progress on integrating the climate into both planning and operations, as well as into capital project delivery.

In 2015, in partnership with WUCA and Oregon State University's Climate Impacts Research Consortium, SPU carried out a climate modeling effort, known as the Pilot Utility Modeling Applications (PUMA) Project. This effort, which created a set of 40 scenarios that were downscaled to several point locations in SPU's watersheds, fed into SPU's hydrology model and utility system model, in order to enable the utility to consider the future water supply under a range of conditions [8,11,12]. This work focused on SPU's system vulnerabilities instead of projecting reductions in supply, and shifted the focus from attempting to predict the future to considering adaptation measures to reduce any vulnerabilities. The results from this effort informed SPU's 2019 Water System Plan (see the Overview of SPU's Climate Change Approach in [34]).

SPU is currently working with CIG and King County to study the potential changes in the flood regime of major King County rivers due to climate change. Additionally, SPU is collaborating with scientists and west coast utility managers to improve the forecast tools and strategies for dealing with atmospheric rivers, which are anticipated to increase in frequency and intensity with climate change. Lastly, SPU is working to refine our supply and demand forecasting, and update our portfolio of options for improving the climate resilience of the water supply system.

SPU's Watershed Management Division, also within SPU's water line of business, has undertaken a climate-driven analysis and action beyond the issues of water supply, particularly in relation to wildfire risks, adaptive forest restoration, and watershed management [35]. The forest management plans for the Cedar River and South Fork Tolt

River Municipal Watersheds focus on forest restoration and climate adaptation in Seattle's two mountain-source watersheds. The watersheds group is also partnering with other organizations to pilot climate adaptive forest restoration, testing reforestation in the Tolt River watershed, with trees sourced from more southern regions where current climates are similar to the projected climates in western Washington later this century [36]. The current development of the Wildfire Risk Analysis focuses on planning for a climate-driven shift in fire regime and the potential impacts on drinking water quality and supply. SPU is collaborating with the US Forest Service and the US Geological Survey to collect ash samples and monitor the post-fire hydrology and water quality impacts from the 2022 Bolt Creek and Loch Katrine fires in the western Cascade Mountains, which will be used to improve the modeling analyses of the potential wildfire impacts on water supply, if a fire were to occur in one of SPU's supply watersheds.

4.2. Drainage and Wastewater

The integration of climate impacts into SPU's drainage and wastewater business started later than it did for water supply. The interest in climate change within SPU's drainage and wastewater planning began in the wake of severe urban flooding caused by extreme rain events in 2006 and 2007. These events triggered a significant investment in natural flood management strategies, including floodable open space in Madison Valley, and floodplain reconnection projects, such as Meadowbrook Pond.

The city experienced more extreme storms from 2012 through to 2017, and during that period, Combined Sewer Overflow (CSO) control projects were overflowing due to their insufficient control volume. The sizing decisions that had been made in 2009 based upon historic rainfall resulted in infrastructure constructed in 2012–2013 that had insufficient volume for the storms. Because of this, the utility's CSO program began considering climate change in its analysis of reduction strategies. This action first shows up in SPU's 2015 Long Term Control Plan (LTCP), which is the fifth CSO planning effort undertaken by SPU [37]. The LTCP was one volume of the city's comprehensive reduction strategy for CSOs and stormwater pollutants. The LTCP, which runs through to 2035, was driven by a regulatory obligation to reduce the CSOs in Seattle's water bodies and was developed under SPU's Consent Decree with the U.S. Environmental Protection Agency (EPA) and the U.S. Department of Justice.

The LTCP applied a 6% scaling factor to historic rainfall to mimic climate-perturbed rainfall projections, providing an "upper bound" for the control volume values [37]. Following the initial implementation of the Long-Term Control Plan in 2015, SPU's CSO program and systems maintenance staff continued to contend with extreme precipitation events and changing rainfall patterns. This challenge suggested that the sizing approach included in the 2015 plan did not amount to an "upper bound" for designing the infrastructure control volumes, which led to a renewed effort that focused on predicting and modeling future precipitation.

SPU staff updated the rainfall records used for establishing capital infrastructure sizing and included a climate-perturbed data set that used outputs from statistically downscaled global climate models [38]. Project modeling using the climate-perturbed rainfall data led to the utility's first major upsizing decision based on future precipitation projections. This upsizing happened for SPU's largest-ever infrastructure project: a 2.7-mile stormwater storage tunnel, known as the Ship Canal Water Quality Tunnel, which was designed to prevent 75 million gallons of stormwater and sewer pollution, annually, in Lake Union and the Ship Canal. The tunnel was upsized to accommodate future extreme precipitation and to provide more operational flexibility during highly localized extreme rainstorms. SPU chose to increase the planned 14′ diameter tunnel to 18′10″, increasing the project's storage volume from 16.1 million gallons to 29.6 million gallons [6].

SPU's work to develop climate-perturbed intensity–duration–frequency (IDF) curves led to additional climate risk management within their drainage and wastewater line of business. Sea level rise guidance was developed for drainage and wastewater capital projects in 2017 [39]. Shortly after the climate-perturbed IDF curves were developed, SPU's integrated drainage and wastewater planning effort, Shape Our Water, used these projections to assess the future capacity impacts upon SPU's drainage and wastewater conveyance systems. SPU's planning team also assessed the risks of inundation due to extreme storm events citywide, and again for creek watersheds specifically. Finally, they used sea level rise (SLR) projections to assess the SPU system risks due to SLR inundation [20,40,41]. These analyses, which embed climate uncertainty, inform the utility's 50-year drainage and wastewater system plan.

4.3. Solid Waste Management

In 1987, Seattle faced a waste management crisis. The city's last two remaining landfills, closed in 1983 and 1986, had been designated by EPA as Superfund sites and would cost more than $90 million to remediate and make environmentally safe. Seattle began hauling garbage to the King County landfill, which increased Seattle's garbage disposal costs and led to an 82 percent increase in solid waste customer rates. SPU's predecessor agency responded to public concerns and used the crisis to design and launch a waste reduction, compost, and recycling program.

Seattle's first solid waste plan was the 1989 Integrated Solid Waste Management Plan, On the Road to Recovery. In 1998, Seattle prepared its second Solid Waste Management Plan, On the Path to Sustainability. Seattle's 1998 Solid Waste Plan incorporated and began to operationalize the concepts of zero waste, waste prevention, sustainability, and product stewardship, that continue today to drive SPU's approach to solid waste management [42].

Climate change mitigation has become an increasingly important part of SPU's materials management story. The solid waste line of business' understanding of the climate benefits of waste prevention emerged early and has evolved over time. In 2004, as part of the Solid Waste Comprehensive Plan Amendment, SPU commissioned a consultant to estimate the environmental, human health, and economic benefits of recycling, including the impact of increased recycling (and the use of recycled materials in manufacturing), the reduced landfilling of organic materials, and waste prevention on greenhouse gas emissions [43].

The 2004 Solid Waste Comprehensive Plan Amendment included language indicating that "recycling programs could be an important element in the City's global warming solutions". The role that materials management could play in mitigating climate change was again recognized in the 2011 Solid Waste Comprehensive Plan Revision, acknowledging that "solid waste management as a cornerstone strategy in climate protection plans". When the city of Seattle's Climate Action Plan was released in 2013, it included actions around "Waste Reduction & Product Stewardship," and shared the results of a consumption-based emissions inventory conducted by King County. These early plans laid the groundwork for SPU's recent increased focus on "upstream" waste prevention and addressing of consumption-based emissions, as reflected in the 2022 Solid Waste Comprehensive Plan Update.

5. SPU's Broader Vision of Sustainability

The last decade of climate impacts experienced by Seattle—as well as the global context of environmental and public health inequities—have resulted in an accelerated effort to pursue a more holistic approach to social and environmental sustainability across SPU, particularly since 2017. Building on SPU's experience in climate and environmental risk management, as outlined in Section 4, there has been broad recognition that: (1) the urgency behind climate change impacts and global inequity requires a markedly different and accelerated approach toward action, (2) SPU will accomplish more on multiple fronts by striving for integrated, holistic utility services and adaptive management strategies across its operational service streams, and (3) this new approach must involve a focus "upstream" on solving the interconnected, systemic root causes of environmental externalities, as opposed to reacting to "downstream" risks or vulnerabilities.

In other words, SPU has begun to understand limitations of focusing predominantly on narrowly scoped problems and single-purpose solutions associated with the "risk management orientation," or looking at sustainability issues in a siloed manner. SPU began a process to pursue a mission of holistic environmental and social sustainability, where it engages in complex, inter-connected challenges with multi-benefit, multi-partner solutions. SPU sees this sustainability mission as both a social and environmental commitment. It includes managing the risks associated with climate impacts and other environmental externalities, but also includes commitments to environmental justice, greenhouse gas emissions reduction, carbon sequestration, water and waste circularity, green infrastructure, ecosystem and species stewardship, green and blue workforce development, and affordability.

This adaptive, integrated focus on sustainability can be seen today across four levels of organizational activity: (1) strategic planning, (2) capital investments and pilot innovation programs, (3) staffing, employee engagement, and culture, and (4) partnerships, collaboration, and alignment. SPU is working to establish an integrated sustainability approach in each of these areas. This section illustrates how the utility is prioritizing multiple benefits and contrasts the actions being taken within SPU today to its past efforts.

5.1. Strategic Planning

Sustainability principles are becoming institutionalized through SPU's major planning initiatives. These planning efforts define the sustainability priorities for the utility in the near- and long-term. We address the overarching SPU strategic business plan, which covers all three lines of business, and reflect on a selection of business plans in turn. The key observation here is that these strategic planning efforts are evolving from a risk management orientation towards a new sustainability orientation, which adopts a more integrated, holistic approach to social and environmental concerns, as well as climate resilience that addresses the root causes of problems, often across service areas and lines of business.

5.1.1. SPU 2021–2026 Strategic Business Plan

The development of SPU's 2021–2026 Strategic Business Plan was the utility-wide turning point that first brought its individual lines of businesses together to galvanize the utility's shared commitments to affordability, equity, and resilience [44]. While the turning points from the risk management approach to a proactive focus on sustainability for each service area were unique, and pre-dated 2021, the Strategic Business Plan acted as a culmination and acknowledgement of these shifts across the entire organization.

The plan provides a roadmap for the utility to meet the needs of customers and communities and establishes a rate path for six years. It also defines SPU's utility-wide vision as "Community Centered, One Water, Zero Waste," illustrating the multi-faceted goals to which SPU and the broader Seattle community aspire. It represents a shift from the prior business plan (which had been the first to mention climate change). In that plan, climate action was tied in a siloed way to expanding the implementation of green stormwater infrastructure and water supply system improvements [45]. The 2021–2026 plan goes beyond a risk management approach, with objectives being framed in terms of environmental justice, adaptation, and mitigation for water and waste. Its key commitments include adaptive management within water supply and stormwater management, sea level rise adaptation planning that supports anti-displacement goals, and a consumption-based emissions inventory to assess the impact that SPU's solid waste management and waste prevention programs have on community-wide emissions. The need to assure affordability, service reliability, and service equity across all three lines of business in the face of weather extremes and changing climatic conditions, population growth, economic and environmental injustice, and increasingly stringent regulations, is persistent throughout the plan.

5.1.2. Shape Our Water

Shape Our Water is SPU's integrated system plan to guide the utility's next 50 years of investments in its drainage and wastewater systems. Climate change vulnerability reduction is a key driver for Shape Our Water, along with population growth, affordability, and equity. Previous drainage or wastewater system planning efforts were conducted in relative isolation, and the primary foci included regulatory compliance, levels of services, and financial constraints. Shape Our Water represents a shift toward utility planning that focuses on the value of the investments that are made from both a community benefit and environmental benefit perspective.

Community visioning and technical analyses provide the foundation for this planning effort. The analysis stage of Shape Our Water occurred from 2018–2021 and was focused on identifying and prioritizing existing and future risks and opportunities citywide, including future climate change and growth impacts. The Shape Our Water team assessed a range of drainage and wastewater challenges, including flooding, sewer overflows, creek and shoreline health, water quality, and the sustainable operation and management of drainage and wastewater systems over time. For the first time, SPU assessed the citywide impacts of extreme storm events and sea level rise on drainage and wastewater systems, as well as on Seattle's communities. Finally, SPU assessed the community context through studies that highlighted Seattle's current racial inequities within health, wealth, and environmental quality.

The Community Vision for Shape Our Water was co-developed with community members, and hinges on an equitable, resilient, and community-centered infrastructure. SPU's community visioning process adopted a community-centered approach that illustrates the utility's evolution toward a sustainability-focused organization. From 2019–2021, the Shape Our Water team used pandemic-responsive engagement strategies to create a shared vision for SPU's infrastructure investment with the communities that SPU serves. The engagement strategies created space for community partners to share their enthusiasm, knowledge, lived experience, talent, and inspiration. The insights gained through this engagement process were distilled into the Community Vision for Shape Our Water [46].

This vision, together with the foundational analysis, will drive the next 50 years of drainage and wastewater infrastructure. In the next stage of Shape Our Water, SPU will brainstorm solutions for Seattle's future challenges. These solutions will build upon the ideas that arose during the visioning—from efficient resource use and reuse, like stormwater harvesting, to expanding the partnerships that support skill-building and job opportunities. The highest-value long-term and short-term solutions will be identified and included in the Shape Our Water plan. Throughout the next steps in planning, SPU will continue to create opportunities to co-create and evaluate future solutions with community-based organizations, other city departments, government agencies, and tribal governments.

5.1.3. 2022. Solid Waste Comprehensive Plan Update

The 2022 Solid Waste Comprehensive Plan Update, "Moving Upstream to Zero Waste", built on a foundation of leading policies in the solid waste management space, this time with a renewed and increased focus on waste prevention strategies that would advance SPU's Zero Waste goals and accelerate the transition to a circular economy. SPU is required to develop a comprehensive solid waste management plan and update it every six years. This plan provides a roadmap for how Seattle will manage and finance its solid waste services and facilities, and projects the system management needs over 20 years. The plan describes how the city handles, collects, processes, and disposes of Seattle's waste. It also describes current waste prevention, recycling, and composting programs, and SPU's progress towards its solid waste goals.

The 2022 Plan Update prioritizes waste prevention within solid waste management to eliminate waste and toxics, prevent pollution, conserve natural resources, and to reduce carbon emissions, while recognizing the role that SPU plays in addressing the community-wide emissions associated with the production and consumption of goods and food. This increased focus on preventing waste and toxins "upstream" is an effort to maximize the environmental and public health impacts of Seattle's solid waste and hazardous materials management system. The plan update represents a shift in SPU's approach, with an understanding that we must look upstream of SPU's traditional purview of "managing" waste, to partnering with others and preventing it in the first place. With this "upstream" focus, SPU is attempting to address the mitigation of climate change and environmental and public health pollution prevention through effective material management.

5.2. Capital Investments and Pilot Innovation Programs

SPU expends approximately $1.4 billion dollars a year on its operations, maintenance, and system expansion and upgrades. SPU's infrastructure and system improvement investments are designed with a multi-decadal service life. SPU is working to deliver capital projects that incorporate climate mitigation and adaptation opportunities, and that contribute to social and environmental sustainability for years to come through their design and procurement choices.

SPU's capital projects are required to demonstrate that climate change has been considered as an aspect of their design, construction, and planned operation. This began with the 2014 addition of climate change considerations into SPU's capital improvement program review process, known as Stage Gates [15,39]. Initially, the climate change component of Stage Gates was seen as cumbersome and difficult to implement due to a lack of guidance or actionable steps for the project teams. In some cases, it was criticized as more of a "box-checking" exercise than a critical review of the opportunities to reduce GHG emissions or to design flexible, climate-adaptive solutions. Since climate-perturbed rainfall data sets and localized SLR projections have become available, teams have been able to use these tools to evaluate the climate risks for various project alternatives. SPU continues to develop its approaches to further embed climate resilience, climate mitigation, and broader sustainability objectives into project Stage Gates and its decision making.

Over the last ten years, many of SPU's capital projects have taken on sustainability-focused metrics and outcomes, and the utility has begun funding pilot innovations that allow for the learning and testing of new approaches. This shift is due to several supporting factors:

1. Climate science has become mainstreamed, understood, and relatively more certain and regionally articulate, allowing utility leadership more clarity on the probability of future climate impacts, and therefore the risk (and cost) of inaction.
2. Contextual changes, such as an increased awareness of and emphasis upon racial, environmental, and public health inequities, the need to reduce displacement pressures, and consistent and negative lived experiences with climate impacts, have driven an urgency to take action.
3. An acknowledgment that the co-benefits of infrastructure (e.g., jobs, access to parks, and extreme heat mitigation) need to be considered when selecting a preferred option for addressing the primary system challenge.
4. A focus on "no regrets" investments has emerged, grounded in flexible practices that are informed by the best available science, where operational modes can be adaptively managed. This allows for the utility to capitalize on community input and innovations that we do not yet have.

Below, we describe a small sample of SPU's pilot innovation programs to illustrate the various ways that we are pursuing our sustainability vision, often across lines of business.

5.2.1. Green Stormwater Infrastructure for Climate Resilience and Community Wealth Building

SPU's green stormwater infrastructure (GSI) program is illustrative of the utility's actions towards proactive climate resilience and natural solutions. SPU began this work in 1999 with a green street retrofit project [47], and began working in close partnership with King County in 2010 to deliver the RainWise program. In 2013, SPU and the city of

Seattle's elected officials passed a policy to make GSI the preferred method of managing stormwater citywide, and set a reach target to manage 700 million gallons of stormwater runoff annually with GSI investments (such as bioretention, rain gardens, urban forestry, and pervious pavement), in order to improve water quality, manage flooding, reduce regulatory costs, build resilient infrastructure, and invest in nature-based, urban carbon sinks [48]. Thanks to an array of programs such as those highlighted below, Seattle appears to be on track to meet this 700 million gallons target by 2025 [49].

5.2.2. RainCity Partnerships

While SPU has long promoted and invested in GSI for its multiple benefits, the utility is now seeking ways to explicitly prioritize these benefits in Seattle's black, indigenous, and people of color (BIPOC) communities. RainCity is a 5-year, $15 million pilot program, focused on larger-scale GSI projects and riparian area restoration. The program includes required performance metrics for both water quality outcomes (e.g., the area of impervious surface managed and the area of riparian area restored) and community-identified benefits (e.g., the percentage of local hires, small business mentorship and subcontracting, 40% women- and minority-owned business contracting targets, and metrics for community-initiated, community-led projects). If the pilot proves successful, the utility is poised to invest $100 million over 20 years to generate stormwater management and wealth-building outcomes [50]. RainCity is new for the utility, not because of "what" project types it is delivering, but because of "how" it is being planned and operated. RainCity is SPU's first foray into utilizing a community-based public–private partnership, which has been promoted by the U.S. EPA and the Washington State Department of Commerce, as a contract mechanism that focuses on improving water quality, as well as a community's quality of life and opportunities [51]. Mami Hara, SPU's General Manager leading up to the launch of RainCity, underscored the intention of programs like this, "As we look to the future, we are intent upon demonstrating how job creation, workforce development, and community wealth building can fruitfully intersect with our missions of environmental enhancement and reliable, equitable service" [52].

5.2.3. Natural Drainage System (NDS) Partnering

SPU's 2016–2025 NDS Partnering Program is a multi-year capital improvement program that is focused on providing significant water quality improvements to Seattle's three major creek watersheds: Longfellow, Piper's, and Thornton Creeks, by managing roadway runoff. The program designs and constructs multi-block, roadside natural drainage systems—primarily vegetated bioretention systems that are located on the public right-of-way planting strip or shoulder—that filter and manage the stormwater runoff and improve neighborhoods with street trees, traffic calming, and, in some cases, new sidewalks or pedestrian walkways. To deliver holistic projects, SPU partners closely with the Seattle Department of Transportation (SDOT), the Seattle Office of Arts and Culture 1% for the Arts Program, and the King County Flood Control District, as well as with a wide variety of community-based organizations. Since its inception, this program has delivered projects in all three major creek watersheds and has integrated a range of public art and pedestrian infrastructure improvements [53,54].

5.2.4. South Park Water Quality Facility

This facility is a stormwater quality facility in Seattle's South Park neighborhood that will treat the stormwater from the surrounding industrial roads so that it is clean before it is pumped into the adjacent Duwamish River. This is a second example of SPU's efforts to embrace community-led infrastructure planning, and is described in more detail in Section 5.4, as an example of philanthropic and community-based organization partnerships. The facility is part of an effort to provide equitable development and environmental justice for a historically underserved, overburdened community, as prioritized in the City of Seattle's Duwamish Valley Action Plan [55].

Like the Natural Drainage Systems Partnering and RainCity partnerships, the goals of this facility include conventional stormwater quality improvement metrics, but also community benefit metrics that are defined by the adjacent residential and business communities, and which will tie to community wealth building and sea level rise adaptation. It is an example of the utility aspiring to use its water quality investments as anchor investments, which are designed to anchor additional community investments to benefit current businesses and residents. While stormwater quality improvement is a priority in this neighborhood, so is community-owned space, affordable housing, and local career pathways for youth. SPU is working to partner with public and private entities in the development of this water quality project, so that the final outcomes of the SPU's investment will span beyond just stormwater quality improvement.

5.2.5. On-Site Non-Potable Water Reuse

Because of its relatively abundant water supply, Seattle has been slow to embrace water reuse. SPU is now laying the groundwork to support voluntary action by the private sector to advance a more widespread adoption of on-site non-potable water reuse systems, enabling the utility to recapture, clean, and reuse water within the footprint of one or more buildings. In 2021, the Washington State Legislature passed a bill requiring the Department of Health to develop statewide rules for the use of on-site non-potable water reuse systems, and SPU is preparing to work with other public sector agencies and private sector partners to advance this work, once the rules are finalized. SPU is also an active member of the U.S. Water Alliance National Blue Ribbon Commission for Onsite Non Potable Water Systems. The Commission develops tools based on the best management practices and current science to support the advancement of on-site non-potable water systems. This initiative marks the beginning of a broader conversation on water reuse at SPU, which will connect the work of all the water-relevant lines of businesses and view drinking water supply, drainage, and wastewater as part of an integrated system.

5.2.6. Promoting Multi-Benefits of Composting

Composting organic materials, such as yard and food waste, recycles them into a beneficial soil amendment and imitates the natural processes of decay and regeneration. However, when organic materials such as food and yard waste are landfilled, they produce large amounts of methane as they decompose in this anaerobic environment. Composting organic materials avoids these potent greenhouse gas emissions, and the finished composted organic material is a critical tool for sustainability because of its many environmental benefits. Compost supports the restoration of soil health, stormwater management through improved infiltration, biofiltration, erosion control, water conservation, and soil carbon sequestration. Moreover, compost supports healthy plant growth in urban landscapes and agricultural sites alike. To gain these broader environmental benefits, it is critical to ensure that the compost is good quality, free of harmful chemical and physical contaminants, and widely used. SPU requires residents and businesses to participate in organics recycling programs (it is illegal to place food and yard waste in the garbage in Seattle), and SPU creates programs to encourage the use of compost. SPU works collaboratively with King County and other agencies across Washington state to develop compost markets, including the expansion from landscaping practices into agriculture (encouraging the compost created from Seattle's organic waste to be used in regional agriculture). In recognition of the interconnected nature of this work, SPU has created a Landscape and Organics Resource Conservation Planner and Program Lead position, which works for both its solid waste and drainage and wastewater lines of business. This position is a hub for sustainability work, "connected with the water conservation team, the urban forestry staff, and Green Stormwater Infrastructure staff" [Kurtz, K., 1 September 2022 personal interview]. The initiative is also a part of SPU's broader efforts to create nature-based carbon sinks, alongside green infrastructure investments, the forest management of SPU's 100,000 acres of watershed, and investments into urban forestry programs such as Trees for Seattle.

5.2.7. Sustainable Energy Management

SPU has created a Sustainable Energy Management Program to coordinate SPU's greenhouse gas reduction and energy management efforts across all three lines of business. This program aims to manage utility-wide energy use and associated greenhouse gas (GHG) emissions throughout its operations, contracting, construction projects, and service delivery [56]. The program has three goals: (1) achieve carbon neutrality by 2030, (2) encourage energy efficiency and awareness, and (3) generate renewable energy. In pursuit of these goals, SPU is conducting an operational greenhouse gas inventory and a supply chain greenhouse gas inventory, has developed building and fleet electrification strategies, and is taking part in energy efficiency programs. SPU is also exploring ways to generate its own renewable energy, using sources within the utility's existing infrastructure through pilot projects. The North Transfer Station, for example, was the first SPU facility to install solar panels in 2016, with the potential to generate enough electricity to power up to 130 homes. SPU is also exploring the installation of its first in-line hydropower generation station at the Lake Forest Park Reservoir, which would take advantage of the excess pressure in our water distribution network to generate as much as 700,000 kilowatt hours of electricity annually. These carbon-free sources of electricity can not only help to offset the operating costs within the utility's facilities, but could also provide a pathway to help offset some of the most carbon-intensive electricity in our emissions profile. SPU plans to build upon the results of these pilot projects to prioritize new renewable energy generation opportunities throughout the utility's infrastructure.

5.3. Staffing, Employee Engagement, and Climate-Aware Culture

SPU is building on its long history of staff-led climate initiatives, and as sustainability emerges as a guiding vision for the utility, that work is being highlighted and celebrated in a more prominent way. Andrew Lee, SPU's General Manager and Chief Executive Officer, serves as the environmental justice chair for the National Association of Clean Water Agencies, and has made a "holistic approach a priority, out of necessity, because challenges and impacts are coming at us so fast" [Lee, A., 30 August 2022 personal interview]. However, this commitment to sustainability has deep support from SPU staff, as the organizational culture shifts toward the connectivity, coordination, and orchestration of efforts, with a focus on intergenerational planning. We include two examples of this staff culture shift below.

5.3.1. All Utility Staff Are Climate Practitioners

In 2006, SPU hired its first climate program manager and established a Climate Resiliency Group to help the utility to understand its exposure and sensitivity to climate change, and to build its capacity to adapt. When this group was formalized, it was widely seen as a separate enterprise from the daily decision making of the utility's strategic planning, capital investments, operations, and maintenance,. While this group still exists and is co-led by a climate adaptation policy lead, alongside a climate mitigation and circular economy policy lead, it is no longer seen as separate or as an add-on: the group's focus is to embed climate science and sustainability into strategic planning, capital investments, operations, and maintenance. As climate science has been mainstreamed, the mantle of "climate staff" has spread beyond this team to include staff from all lines of business and all branches. Reflecting this trend, SPU's Climate Community of Practice has emerged as an internal force for climate-related work. Acting as a locus of climate-related activity, this group of nearly 100 staff gather quarterly to learn, share information, and build collaborative partnerships [8,50,57]. This community of practice does not focus solely on the impacts of climate change, but on the broader set of issues entrained in our sustainability-oriented utility.

5.3.2. Frontline Staff Are Precipitation First Responders

Climate planning has historically been the work of desk-bound, science- and policy-focused staff. However, water utility crew staff, including those who perform system maintenance and operations in the field, are experiencing climate impacts firsthand, and are SPU's "precipitation first responders". These crew members have relevant, experiential knowledge about precipitation risks that is often not communicated and integrated into the utility's strategic planning and implementation activities. SPU partnered with the University of Minnesota to survey 115 frontline staff in the drainage and wastewater and water lines of business about their experience with rain, and their thoughts on priority adaptation investments for the utility. These frontline staff experience climate change impacts on a daily basis, and anticipate the need to take actions around communication, infrastructure/facilities, equipment, and workforce capacity. Throughout this survey, and throughout additional related initiatives that are focused on building a better connection between frontline staff and leadership, SPU is learning that intra-utility communication and worker engagement is a critical strategy for mainstreaming adaptation and sustainable operations [58].

5.4. Partnerships, Collaboration, and Alignment

Sustainability is showing up in the networks and collaborations that the utility prioritizes and invests in. SPU benefits from and builds upon collaboration with scientific-, peer-, and community-based partners throughout all three lines of business.

5.4.1. Philanthropic and Community Organization Partnerships

In 2018, SPU was awarded a $200,000 Connect Capital grant from the Center for Community Investment (CCI). This grant brought value to the utility beyond financial support: it seeded an effort to leverage the SPU's drainage and wastewater investments in Seattle's South Park neighborhood to drive the planning and investment in sea level rise adaptation and anti-displacement policies [59]. South Park is a majority people of color community in South Seattle's Duwamish Valley that has a documented average life expectancy of thirteen years less than other less diverse, wealthier neighborhoods in Seattle. It suffers from poor air quality due to nearby highways and freight traffic, chronic flooding, and a dearth of green space. It is also the area in Seattle that is most vulnerable to sea level rise, due to its low elevation, flat topography, and adjacency to the tidally influenced Duwamish River [60]. The CCI grant work ultimately led to a subsequent climate cities equity grant from the Robert Wood Johnson Foundation for the city of Seattle to develop a resilience district in the Duwamish Valley to implement these adaptation and anti-displacement goals [61,62].

In 2018, SPU was at the beginning of a three-project suite of drainage and wastewater investments in the neighborhood, including road improvements and conveyance, a pump station, and a water quality facility. This grant-funded effort to leverage these projects for a broader community benefit illustrates SPU's transition to a sustainability focus, which is due to how the development strategy has evolved: the original climate-focused emphasis upon drainage infrastructure, as described in the Stults et al., 2016 case study, was in elevating the South Park pump station to ensure that it would continue to function alongside the rising seas in the adjacent Duwamish River. As SPU's sustainability vision evolved, the focus of the project shifted from asset protection to using the investment as an anchor to address community-identified challenges such as displacement pressure and future sea-level-rise-related flooding.

These grant-funded efforts have fostered a collaboration between SPU, the Duwamish River Community Coalition (DRCC), and the Seattle Foundation. This collaboration is remarkable because DRCC and SPU sit on opposing sides of the Lower Duwamish Waterway Superfund Cleanup, where the city of Seattle is a liable party and DRCC is a community advocacy and technical advisory group. It is intended to provide a platform for the long-term sea level rise adaptation strategies that will ultimately be integrated into the design for the water quality facility, while also addressing the long-standing needs of the local community.

5.4.2. Tribal Partnerships

SPU has worked with indigenous peoples on salmon recovery, the preservation and repatriation of cultural resources, sediment cleanups, land access for cultural practices, and permanent artworks for the Ship Canal Water Quality Project. Water to support fisheries is key to maintaining indigenous communities, as is their access to protected natural lands for the hunting and gathering of food and medicine to sustain their cultural practices and community health. For example, the Muckleshoot Indian Tribe (MIT) has access to the Cedar River Municipal Watershed under their reserved treaty rights to hunt and gather. SPU continues to work with the MIT on fisheries and forest management to ensure that these resources are available, and plans to work with other local and regional tribes on similar sustainable management challenges in the future.

5.4.3. Collaborating with Private Sector Partners for Waste Prevention

SPU is engaging in public–private partnerships to encourage waste prevention in the areas of food and packaging, recognizing the critical role of the private sector in the development of a circular economy. As a signatory of the Pacific Coast Food Waste Commitment, SPU is collaborating with grocery retailers and manufacturers in an effort to reduce the food that goes to the garbage across the west coast by 50% by 2030. In addition, SPU is partnering with businesses and nonprofits to improve how edible, unsold food gets diverted from organics or garbage streams and donated to those who need it in the Seattle area. SPU has also formed a public–private partnership (Reuse Seattle) to create a standardized, city-wide reusable food and beverage container system. This system was piloted in 2022 in over 10 participating entertainment venues, including the Woodland Park Zoo, Paramount Theatre, and The Showbox. The goal is to make food and beverage container reuse scalable and affordable for customers, businesses, and the city. Greenhouse gas emissions reduction, solid waste diversion, and economic development are among the drivers for these programs.

5.4.4. Impact Investment in Waste and Water

In 2021, SPU launched an impact investment pilot program, Seeds of Resilience, to invest in and to incubate the water- and waste-related businesses that advance the community resiliency, circular economy, and green job opportunities for underrepresented communities. This program directs $600,000 annually into private sector endeavors that help SPU to achieve its waste and water management goals. In addition to helping advance SPU's mission of better managing waste and water, these investments grow Seattle's green economy, deliver environmental benefits, and expand equity and opportunity. With mixed funding from all three lines of business, the program can invest in projects and activities that cross the traditional utility service silos, addressing waste and water issues in an integrated manner. One of the first projects funded by Seeds of Resilience is aimed at increasing the access to and demand for water cisterns on residential properties, by finding ways to make captured rainwater more easily usable inside the home, in order to lower drinking water bills [63].

5.4.5. Intersectional Peer Networks

SPU staff are heavily engaged in a number of peer networks, including the Water Utility Climate Alliance (WUCA), the US Water Alliance, the National Association of Clean Water Agencies (NACWA), the Evergreen Chapter of the Solid Waste Association of North America, C40 Cities, Water Environment Foundation, and the West Coast Climate & Materials Management Forum. SPU has benefitted from modeling scenario planning projects and other thought leadership collaborations with these organizations, and is now working with them to integrate One Water and Zero Waste principles, community leadership and engagement, and equity into planning and operations.

SPU's founding membership in WUCA has been productive, and WUCA's trajectory has mirrored the utility's evolving focus on sustainability. WUCA was formed in 2007 to provide leadership and collaboration on the climate change issues affecting the country's water agencies. SPU's membership bolstered the utility's early efforts to mainstream climate science and downscale the climate models for western Washington applicability. Today, SPU is building out its environmental justice efforts as a part of WUCA's water equity workgroup and is learning from peer utilities about how best to plan for sea level rise and inventory, and how to reduce operational and supply chain greenhouse gas emissions. The US Water Alliance and NACWA's Environmental Justice committee are also valued partners for the utility.

6. SPU, Sustainable Operations, and New Questions for the Climate Services Community

As illustrated above, SPU is evolving from a risk management mode of operation toward a sustainability orientation. This evolution includes looking beyond being "climate ready" and modifying utility strategic planning, capital investments, operations, and maintenance to account for the changing climate conditions. SPU's commitment to social and environmental sustainability is taking a more integrated, holistic approach by addressing the root causes of problems, often across business lines, to realize conventional utility objectives such as affordability, service reliability, and service equity, while also pursuing new objectives such as racial equity, anti-displacement policies, environmental justice and stewardship, and climate mitigation commitments.

Given this juncture in planning and operations, we wonder if the climate services enterprise can also evolve into a more technically diversified, value-driven, and integrated realm of activity, something that more closely matches the operational commitments of SPU and other utilities. Below, we illustrate three examples, out of a much larger universe of newly relevant questions, of how this is playing out today, and outline the new types of questions being posed:

(1) Analysis of financing and affordability challenges: While the climate crisis compels near-term action, there are difficult questions that remain less than fully answered. For instance, it is not clear who will ultimately be called upon to pay for climate-resilient investments and incremental add-ons to absorb or buffer future climate impacts and protect SPU's core service delivery operations. Additionally, more broadly, who should be paying for the efforts to create a city that is more climate resilient? Should all ratepayers bear an equal burden of these costs, or should individual carbon footprints be used to prorate cost allocations? Will costs be borne by today's customers or future generations? Should this be government funded, and if so, at what level? Or should the private sector be playing a role as well? If it is entirely left for SPU to pay for climate resiliency, it is important to bear in mind that SPU rates are already unaffordable for an unacceptable number of customers, and keeping essential services affordable is a key concern for the utility. Can we make incremental "no regrets" investments that can be expanded in future decades to spread that cost out? Are there ways for climate service providers to engage in this sort of value-driven discourse and analysis?

(2) Development of strategies, programs, and support mechanisms to build community resilience and wealth alongside climate-resilient infrastructure: Infrastructure is only part of the solution to preparing communities for climate impacts. SPU is striving to drive policy, standards, and job creation opportunities to support incumbent communities, particularly low-income communities and communities of color, so that they can continue to thrive in place instead of being displaced as a result of public infrastructure improvements. The potential for wealth-building "green jobs" and blue/green workforce development in the water supply, drainage and wastewater, and solid waste arenas is significant [52,64–66]. Can climate services be designed to help us identify the existing spatial relationships between climate vulnerability and systemic racism? Or, more positively, how can the climate services community

work with utilities and municipalities to develop scenarios, models, or other tools that reflect this fundamental commitment to communities-in-place?

(3) Holistic and standardized approaches to accounting for greenhouse gas emissions and sinks: Like many utilities and companies, SPU has begun to track its operational greenhouse gas emissions [56]. Additionally, while SPU has adopted a protocol for tracking these emissions, significant uncertainties remain with respect to the emission tracking methodologies. Additionally, there is no correlative protocol for tracking carbon sinks. This is important because SPU's carbon story is broader than the emissions inventory that we currently maintain. How can we accurately measure, track, and account for the carbon sinks that we maintain in our watershed forests, in our urban forests and vegetation, and in our soils? In addition, SPU is uniquely positioned as a solid waste management service provider to have a significant impact on the community-wide GHG emissions related to the production and consumption of goods and food. How can this broader carbon impacts story be tied into our existing climate and emissions story and tracking?

7. Moving from Climate Resilience to Sustainable Operations: Ongoing Challenges and Observations

Although this paper has been framed as a series of promising developments, the authors are under no illusion that SPU's transformation to sustainable operations is a foregone conclusion. As conceptualized by a range of professional experts and academic researchers, meaningful sustainability will be disruptive of current practices and pathways, and can be expected to radically alter incumbent technological regimes, institutional structures, and organizational culture [67–71]. At any scale and in any context, sustainability is a wicked problem fraught with challenges [72,73]. The challenges facing SPU include, but are likely not limited to, the following.

Going forward, SPU will continue to focus on building momentum at all levels of the organization around sustainability, resilience, and climate preparedness. This will include supporting and educating executives, management, and field staff about the need to make decisions that are robust under current and future climates. Overlaying a sustainability orientation on top of departments that were—and to some extent, remain—driven by a traditional mindset of linear problem solving is an organizational and cultural challenge, but also a personnel and staffing issue. It is clear that staff can experience frustration because of a lack of definition and a sense of occupational scope creep [71,74,75]. For example, SPU still struggles with competing priorities in its project delivery, as extensive and meaningful community engagement can prolong and complicate the project scope, schedule, and budget. In addition, having to assess the sustainability considerations of new projects (e.g., the greenhouse gas emissions associated with a capital investment) requires a new skillset, and this work may often be seen as a trade-off for expedient project delivery or budget limitations. SPU remains challenged by the need to help staff effectively address the uncertainties associated with climate change and other projections of future conditions. Staff feel pressured "to take the median or to take one of the scenarios we are using and base all decisions on that scenario" [13]. In the case of capital improvement funding, project managers want to know what range of temperature, precipitation, or sea level rise they are expected to plan for [13]. Because of this, SPU continues to work on techniques to help its staff become more comfortable with this uncertainty and to be able to make informed judgements regarding which future projections to privilege in their planning exercises. According to Paul Fleming, SPU's former Climate Resiliency Group manager, the goal remains "to understand and embrace uncertainty so that you can make informed decisions that are robust under multiple futures" [8].

Another set of challenges to the achievement of sustainable operations can arise due to organizational structure. In the context of a large water and waste system, structural demarcations can act to impede the recognition and deployment of cross-disciplinary, integrated solutions to environmental-, resilience-, and sustainability-related problems.

So-called "siloing" within agencies or among departments can frustrate even concerted top-down efforts to impose change upon an organization [71,76,77]. As an organization, SPU is not—and likely will never be—a monolith. Champions of sustainability within SPU recognize that changes will accrue slowly and that the issues that arise among and between branches and divisions must continue to be navigated with care.

It is undeniable that tensions exist—and will continue to exist—between the future goals for sustainable operations and service affordability in the present day. The rising cost of service delivery is a vexing challenge faced by utility and municipal leadership. With basic service provision already too expensive for some residents and customers, the question of how to finance new, sustainability-related practices and technologies is critical. Clearly, economic downturns and utility-scale financial issues could disrupt the achievement of SPU's sustainability initiatives.

Another potential chokepoint in SPU's transition toward sustainable operations involves the development of rigorous yet practically applicable metrics to help evaluate the utility efforts to reduce vulnerability and increase resiliency and sustainability. According to James Rufo-Hill, a former SPU meteorologist and climate science advisor, SPU needs to improve its efforts to document and monitor the effectiveness of its operations, decision making, and planning processes [8]. Without such metrics, SPU will be less able to provide robust analyses that demonstrate whether and how its efforts have increased sustainability and reduced the utility's vulnerability to climate change. The development and operational implementation of benchmarks and metrics is important to the long-term viability of SPU sustainability initiatives.

As demonstrated in this essay, the enterprise of climate services has positioned SPU to be better prepared for and more resilient against the present and future impacts of extreme weather, climate variability, and climate change. In other words, climate services were—and are—crucial to SPU's evolution as a risk management utility. As illustrated above, SPU still has a need for scientific and informational expertise beyond the capabilities of its current utility staff to make progress as a sustainability-oriented utility. It stands to reason that climate services can and will continue to play a vital role and help organizations like SPU, as they work to move beyond the goal of climate resilience to pursue the broader objective of social and environmental sustainability. Climate resilience and sustainability share important characteristics. They both require:

- That utilities become aware of and are competent in nontraditional areas of science and technical understanding, such as the science of climate change and the impacts of redlining on community wealth and opportunity;
- That utilities, stakeholders, and technical specialists must co-produce models, decision aids, scenarios, data sets, plans, and other boundary objects;
- That utility staff and their partners understand and appreciate that their interactions and outputs will necessarily involve a mix of factual materials and public values.

However, the SPU experience also suggests that climate resilience and sustainability differ in ways that may necessitate changes in emphasis, or perhaps even basic alterations to the co-productive model.

As articulated by Dilling and Lemos [78], the co-production of knowledge refers to the contribution of multiple knowledge sources and capacities from different stakeholders, spanning the science–society interface with the goal of jointly creating knowledge and information to inform decision making. In general terms, the quest for sustainable modes of operation entails a multi-generational perspective and integrates economic vitality, social equity, and environmental stewardship. Studies of sustainability are necessarily multi-disciplinary, cross-sectoral, and inter-organizational. As recognized by Cvitanovic et al. [79], the practitioners implementing sustainability programs "do not necessarily consider scientific information to be more important than other knowledge..." [80]. They recognize that sustainability initiatives involve a mix of scientific characterization and projection, technological and engineering applications, professional standards and expectations, and clearly articulated commitments to value-based objectives. Furthermore, sustainability seems to be place-based and circumstantially specific [72] and is sometimes characterized as a societal process of learning and creation.

Based on this, we observe the following:

- As recognized elsewhere in this Special Issue [81], a hard and fast distinction between the knowledge producers and knowledge users is neither helpful nor realistic in the context of sustainability. Sustainability is always a composite of values, knowledge, natural conditions, technological capabilities and constraints, and stakeholder life experience. Within such a milieu, there is no one who is simply and purely a knowledge user or a knowledge producer.
- The co-productive enterprise requires scientific and technical inputs across a far broader range of knowledge and competencies than can be provided by the "traditional" disciplines of climate science, i.e., climate model projections. Additionally, for this reason, climate service providers need to expand their networks and prepare to interact with numerous other disciplines.
- Climate literacy in the water sector is defined as "water managers' knowledge of the climate system and the impact of climate variability on the availability of water relative to annual operating decisions and long-term plans" [82]. While it is inarguable that responsible managers and decision makers need to be aware of the factors and conditions that can influence utility operations, it is not clear to us that a concept like "climate literacy" is helpful in the water or waste sectors, especially if one shifts their point of reference from climate change to sustainability. As the examples in Section 5 illustrate, sustainability can be site- or situation-specific, making it nearly impossible to stipulate in advance how much of any given knowledge domain will be necessary to inform a particular effort to pursue sustainable operations. SPU sustainability initiatives involve perhaps dozens of disciplines or topical domains, including the physiological factors that influence the population-level dynamics of endangered species; the socio-cultural determinants for equitable and generational planning; the hydro-geological variables that affect watershed functions; the principles of sustainable landscape design that emphasize native species; and the financial forecasting and modeling capabilities that can help to actualize concepts such as intergenerational planning.
- As discussed above, the efforts to achieve lasting organizational sustainability need to accommodate an increased operational comfort with stubborn uncertainties, possibly through the adoption of robust decision-making practices. The exercise of professional judgement is therefore critical to the development of pragmatic, usable, relevant, and acceptable outcomes. As described by Donald Schon in his seminal work The Reflective Practitioner: How Professionals Think in Action, decision making under conditions of uncertainty involves a tacit skill set called "reflection in action", or a willingness and ability to reframe problems and adjust the means to ends (or ends to means) in real time. As demonstrated by Lempert and others in this volume [83], climate service providers may need to adopt an elastic, on-the-fly mode of interactive support.

Climate change represents a clear challenge to efforts to forge a sustainable future. In our view, the effort to develop and apply climate services is—or at least ought to be—part of the larger enterprise of sustainable development.

Author Contributions: Conceptualization, A.G.-N., A.S., J.V. and C.H.; methodology, A.G.-N., A.S., J.V. and C.H.; investigation, A.G.-N., A.S., J.V. and C.H.; writing—original draft preparation, A.G.-N., A.S., J.V. and C.H.; writing—review and editing, A.G.-N., A.S., J.V. and C.H. All authors have read and agreed to the published version of the manuscript.

Funding: This research did not receive any specific grants from funding agencies in the public, commercial, or not-for-profit sectors. Partial funding for this work was provided through a budget proviso from the Washington State legislature that supports the Climate Impacts Group "to conduct data modeling and provide technical assistance on climate impact analysis to Washington communities, businesses, and governments".

Institutional Review Board Statement: Not applicable.

Informed Consent Statement: Not applicable.

Data Availability Statement: Not applicable.

Conflicts of Interest: Authors Ann Grodnik-Nagle and Ashima Sukhdev are both employed by Seattle Public Utilities.

Appendix A

Table A1. List of seattle public utilities staff who contributed to this article.

Interviewees	
Name	**Title**
Chinn, Alan	Water Resources Planning Unit Supervisor
Edgerly, John	Water Resources Hydrologist
Garcia, Elizabeth	Water Resources Planner
Kennedy, Katie	Waste Diversion Planner
Kurtz, Kate	Landscape and Organic Resource Conservation Planner
Lee, Andrew	General Manager
Purnell, Danielle	Director of Corporate Policy and Planning
Schwenger, Stephanie	Solid & Hazardous Waste Lead Planner
Webster, Leslie	Drainage and Wastewater Planning Manager
Additional Reviewers	
Name	**Title**
Emerson, Pam	Green Stormwater Infrastructure Planner
Gersonde, Rolf	Forest Ecologist
LaBarge, Amy	Watershed Management Director
Hilton, Chris	Risk and Resilience Advisor
Marre, Ben	Drainage and Wastewater Planning & Program Management Director
Munger, Julia	Watersheds Natural Resources Manager
Tackett, Tracy	Drainage and Wastewater Capital Projects Manager

Appendix B. SPU Staff Interview Questions—30 August and 1 September 2022

Appendix B.1. Background

As a contribution to a special issue of the journal Sustainability, SPU staff are preparing an interpretive history of the Utility's approach to climate change adaptation over the last two decades. A draft of this manuscript is attached for your review. It is our observation that SPU has evolved from an early emphasis on managing externalities and risks as they emerged and in a siloed manner and shifted toward a broader focus across all aspects of climate adaptation and mitigation that is part of a more integrated, strategic, and proactive focus on sustainability across operational service streams. We seek your critical reaction to this perspective. Please review the draft manuscript and consider the questions below prior to our scheduled discussion.

Appendix B.2. SPU Staff Interview Questions

Appendix B.2.1. SPU's Purpose

- How has SPU's purpose or statement of values evolved over time? Do you see a difference between what SPU was trying to accomplish in the 2000s versus what it is trying to accomplish in the 2020s?
- Would you describe SPU's evolution as systemic or incremental? Do you think it is accurate to say that we have fundamentally transformed how we operate?

Appendix B.2.2. SPU's early climate/sustainability work

- Describe the utility's climate risk management work in the 1990s, 2000s.
- How and when did the utility's earliest climate work begin? What do you remember about the early climate work—1990s and 2000s?
- How connected were the various planning/stewardship efforts to climate in 1995–2015? Was there a throughline of sustainability or climate resilience? Or were they being developed more independently?

Appendix B.2.3. SPU & climate services

- In our effort to address climate impacts, SPU has sought external, expert advice. We have worked with private consultants, university institutes, community-based organizations and industry networks. Please reflect critically on this process. What aspects went well, what could be improved? What lessons have we learned that could be applied as we seek increasingly sustainable operations?
- How could the climate services community better comprehensively address SPU's needs? Please identify the leading challenges facing the utility today.
- What is the relationship between SPU's mission and vision and our work to address climate impacts? Are they meaningfully integrated?

Appendix B.3. Notes on Methodology and Interview Protocol

To ensure authenticity, the author team developed a preliminary narrative of the SPU experience drawing on participatory involvement, evaluation reports, and archival sources, and then solicited direct feedback from professional staff involved in SPU's climate resilience and sustainability management, operations, and research as far back as 1993. We conducted four semi structured group interviews (1–4 interviewees per group) with a total of nine individuals. Interviewees include individuals involved in all three of SPUs lines of business as well as management and leadership, and core technical staff who have focused on climate impacts and resilience over the past three decades. Although guided by this questionnaire document, interview subjects were permitted to linger and focus on topics of particular interest. Once this manuscript was drafted, interviewees and additional staff were encouraged to review the draft and provide critical feedback to ensure that our account is consistent with the lived experience of key participants

References

1. Iacono, J.; Brown, A.; Holtham, C. Research methods—A case example of participant observation. *Electron. J. Bus. Res. Methods* **2009**, *7*, 39–46.
2. Barsugli, J.; Anderson, C.; Smith, J.; Vogel, J. Options for Improving Climate Modeling to Assist Water Utility Planning for Climate Change. Prepared for Water Utility Climate Alliance. 2009. Available online: http://www.wucaonline.org/assets/pdf/pubs-whitepaper-120909.pdf (accessed on 14 July 2022).
3. Barsugli, J.; Vogel, J.; Kaatz, L.; Smith, J.; Waage, M.; Anderson, C. Two faces of uncertainty: Climate science and water utility planning methods. *J. Water Resour. Plan. Manag.* **2012**, *138*, 395–398. [CrossRef]
4. Ferguson, D.B.; Rice, J.; Woodhouse, C.A. Linking Environmental Research and Practice: Lessons from the Integration of Climate Science and Water Management in the Western United States. Prepared for Climate Assessment for the Southwest, University of Arizona. 2014. Available online: https://www.climas.arizona.edu/sites/climas.arizona.edu/files/pdflink-res-prac-2014-final.pdf (accessed on 20 August 2022).

5. Inha, L.; Hukka, J. Policies enabling resilience in Seattle's water services. *Eur. J. Creat. Pract. Cities Landsc.* **2009**, *2*, 1. Available online: https://cpcl.unibo.it/article/view/8688/10130 (accessed on 20 August 2022).
6. Hart, J.; May, C.; Mak, M.; Ramirez Lopez, D.; Badet, Y.; Cohn, A.; Rockwell, J.; Anbessie, T. *Scaling and Application of Climate Projections to Stormwater and Wastewater Resilience Planning*; Pathways Climate Institute for Water Utility Climate Alliance: Las Vegas, NV, USA, 2022. Available online: https://www.wucaonline.org/assets/pdf/stormwater-wastewater-report-2022.pdf (accessed on 3 January 2023).
7. Vano, J.; Voisin, N.; Cuo, L.; Hamlet, A.; Elsner, M.; Palmer, R.; Lettenmaier, D. Climate change impacts on water management in the Puget Sound region, Washington State, USA. *Clim. Change* **2010**, *102*, 261–286. [CrossRef]
8. Stults, M.; Carney, K.; Vogel, J. Mainstreaming Climate Change into Internal Planning and Decision Making: Seattle Washington. In *Climate Adaptation: The State of Practice in U.S. Communities*; Vogel, J., Carney, K.M., Smith, J.B., Herrick, C., Stults, M., O'Grady, M., St. Juliana, A., Hosterman, H., Giangola, L., Eds.; The Kresge Foundation: Troy, MI, USA, 2016. Available online: http://kresge.org/climate-adaptation (accessed on 3 January 2023).
9. U.S. EPA. *Climate Change Vulnerability Assessments: A Review of Water Utility Practices*; EPA/800/R-10/001; U.S. Environmental Protection Agency: Washington, DC, USA, 2010.
10. U.S. EPA. *Climate Change Vulnerability Assessments: Four Case Studies of Water Utility Practices*; EPA/600/R-10/077F; U.S. Environmental Protection Agency: Washington, DC, USA, 2011. Available online: http://cfpub.epa.gov/ncea/global/recordisplay.cfm?deid=233808 (accessed on 20 January 2023).
11. Vogel, J.; Carney, K.; Smith, J.; Herrick, C.; Stults, M.; O'Grady, M.; Juliana, A.S.; Hosterman, H.; Giangola, L. *Climate Adaptation: The State of Practice in U.S. Communities*; The Kresge Foundation: Troy, MI, USA, 2016. Available online: http://kresge.org/climate-adaptation (accessed on 15 January 2023).
12. Vogel, J.; McNie, E.; Behar, D. Co-producing actionable science for water utilities. *Clim. Serv.* **2016**, *2–3*, 30–40. [CrossRef]
13. Vogel, J.; Smith, J.; O'Grady, M.; Fleming, P.; Heyn, K.; Adams, A.; Pierson, D.; Brooks, K.; Behar, D. Actionable Science in Practice: Co-Producing Climate Change Information for Water Utility Vulnerability Assessments. Final Report of the Piloting Utility Modeling Applications (PUMA) Project. Prepared for the Water Utility Climate Alliance. 2015. Available online: https://www.wucaonline.org/assets/pdf/pubs-puma-white-paper-20150427.pdf (accessed on 15 January 2023).
14. Means, E.; Laugier, M.; Daw, J.; Kaatz, L.; Waage, M. *Decision Support Planning Methods: Incorporating Climate Change Uncertainties into Water Planning*; Water Utility Climate Alliance: Las Vegas, NV, USA, 2010.
15. Raucher, K.; Raucher, R. A Case Study Examination of How Climate Change is Shifting Water Utility Planning. Prepared for the Water Utility Climate Alliance, American Water Works Association, Water Research Foundation, and American Metropolitan Water Agencies. 2015. Available online: https://www.wucaonline.org/assets/pdf/pubs-uncertainty.pdf (accessed on 15 September 2022).
16. U.S. E.P.A. *Expanding the Benefits of Seattle's Green Stormwater Infrastructure*; U.S. Environmental Protection Agency: Washington, DC, USA, 2018. Available online: https://www.epa.gov/sites/default/files/2017-03/documents/seattle_technical_assistance_010517_combined_508.pdf (accessed on 15 February 2023).
17. Seattle Public Utilities. *Overview of SPU's Climate Change Approach for 2019 Water System Plan*; Seattle Public Utilities: Seattle, WA, USA, 2019.
18. Roop, H.A.; Mauger, G.; Morgan, H.; Snover, A.; Krosby, M. *Shifting Snowlines and Shorelines: The Intergovernmental Panel on Climate Change's Special Report on the Ocean and Cryosphere and Implications for Washington State*; Climate Impacts Group, University of Washington: Seattle, WA, USA, 2020. [CrossRef]
19. Seattle Public Utilities. *Drainage System Analysis. Flooding Topic Area: Sea Level Rise Analysis*; Seattle Public Utilities: Seattle, WA, USA, 2020.
20. Littell, J.; Oneil, E.; McKenzie, D.; Hicke, J.; Lutz, J.; Norheim, R.; Elsner, M. Forest ecosystems, disturbance, and climatic change in Washington State, USA. *Clim. Change* **2010**, *102*, 129–158. [CrossRef]
21. Morgan, H.; Bagley, A.; McGill, L.; Raymond, C. Managing Washington Wildfire Risk in a Changing Climate. Workshop Summary Report Prepared by the Northwest Climate Adaptation Science Center and the Climate Impacts Group, University of Washington, Seattle. 2019. Available online: https://nwcasc.uw.edu/wp-content/uploads/sites/23/2019/04/Managing-Western-Washington-Wildfire-Risk-in-a-Changing-Climate-1.pdf (accessed on 10 December 2022).
22. U.S. Global Change Research Program. *Climate Science Special Report: Fourth National Climate Assessment*; U.S. Global Change Research Program: Washington, DC, USA, 2017; Volume I, Chapter 6, 7 and 9.
23. Intergovernmental Panel on Climate Change. Working Group 1, Summary for Policymakers. 2013. Available online: http://www.climatechange2013.org/images/uploads/WGIAR5-SPM_Approved27Sep2013.pdf (accessed on 10 December 2022).
24. Seattle Public Utilities. SPU's Risk and Resiliency Strategic Plan, 2019 Final Report. 2019. Available online: https://www.seattle.gov/documents/Departments/SPU/AboutUs/SPU_SBP_AppendixF_RiskandResilience_CouncilSubmittal.pdf (accessed on 10 September 2022).
25. Seattle Public Utilities. *Seattle Public Utilities Water System Plan 2001 Update*; Seattle Public Utilities: Seattle, WA, USA, 2001.
26. City of Seattle and University of Washington. *Memorandum of Agreement: Developing Analysis Techniques to Incorporate Climate Change Information into Seattle's Long Range Water Supply Planning (DA2001-09)*; City of Seattle: Seattle, WA, USA, 2001.
27. Seattle Public Utilities. *The Water Supply Outlook in the Face of El Nino*; Seattle Public Utilities to the Seattle City Council Utilities and Environmental Management Committee: Seattle, WA, USA, 1997.

28. Seattle Public Utilities. *Seattle's Water Supply & El Nino*; Seattle Public Utilities Website: Seattle, WA, USA, 1998.
29. Chinn, A.; Schneider, G. Managing Seattle's Water Supply in the Face of Hydrologic Uncertainty (or, Want to Bet on the Weather?). In Proceedings of the AWWA Conference, Dallas, TX, USA, 24 June 1998.
30. City of Seattle. *Press Release: New Climate Change Report: Seattle's Water Planning Among Nation's Best National Center for Atmospheric Research Calls City a "Pioneer" in Climate Change Planning*; City of Seattle: Seattle, WA, USA, 2005.
31. Miller, K.; Yates, D. Climate Change and Water Resources: A Primer for Municipal Water Providers. *AWWA Research Foundation*. 2006. Available online: https://ral.ucar.edu/sites/default/files/public/product-tool/Primer_2006_cr_project_2973_Climate_Change_and_Water_Resources.pdf (accessed on 10 September 2022).
32. Miller, J. Impact of climate change on water supplies of Everett, Seattle, and Tacoma. In Proceedings of the AWWA-PNWS Annual Conference, Vancouver, WA, USA, 8–10 May 2008. Available online: http://www.pnws-awwa.org/Page.asp?NavID=299 (accessed on 10 September 2022).
33. Seattle Public Utilities. Seattle Public Utilities Water System Plan 2007 Update. 2007. Available online: https://www.seattle.gov/documents/Departments/SPU/Documents/Archive/2007WaterSystemPlan.pdf (accessed on 10 September 2022).
34. Seattle Public Utilities. Seattle Public Utilities Water System Plan 2019 Update. 2019. Available online: https://www.seattle.gov/utilities/about/plans/water/water-system-plan (accessed on 10 September 2022).
35. Gustin, R.; Hanan, E.J.; Adam, J.C.; Ren, J.; Garcia, E.; LaBarge, A.; Liu, M.; Kolden, C.; Abatzoglou, J.T. Is forest management a safeguard against a climate change-altered wildfire regime in the City of Seattle's largest source watershed? In Proceedings of the American Geophysical Union, Fall Meeting 2019, abstract #GC11F-1167, San Francisco, CA, USA, 9–13 December 2019.
36. Hintz, D. Stossel Creek: A Forest Planted for the Future. Mountains to Sound Greenway. 2020. Available online: https://mtsgreenway.org/blog/stossel-creek/ (accessed on 10 November 2022).
37. Seattle Public Utilities. Long-Term Control Plan, Volume 2. 2015. Available online: https://www.seattle.gov/documents/Departments/SPU/Documents/DraftLongTermControlPlan.pdf (accessed on 10 November 2022).
38. Rath, J.; Roy Sujoy Butcher, J. Intensity Duration Frequency Curves and Trends for the City of Seattle. In *Technical Memorandum*; Tetra Tech Inc.: Pasadena, CA, USA, 2017.
39. Seattle Public Utilities. *Sea Level Rise Guidance for Drainage & Wastewater Capital Projects, DWW-110*; Seattle Public Utilities: Seattle, WA, USA, 2017.
40. Seattle Public Utilities. *Drainage System Analysis. Flooding Topic Area: Extreme Storms Analysis*; Seattle Public Utilities: Seattle, WA, USA, 2020.
41. Seattle Public Utilities. *Drainage System Analysis. Flooding Topic Area: Creeks Analysis*; Seattle Public Utilities: Seattle, WA, USA, 2020.
42. Seattle Public Utilities. Draft 2022 Solid Waste Plan Update: Moving Upstream to Zero Waste. 2022. Available online: https://www.seattle.gov/utilities/about/plans/solid-waste/2022-plan-update (accessed on 10 November 2022).
43. Morris, J.; Bagby, J. Economic Analysis of New Waste Prevention and Recycling Collection Programs Projected to Increase Recycling from 40% to 60%. 2004. Available online: https://www.seattle.gov/Documents/Departments/SPU/Documents/2004SolidWastePlanAmendment.pdf (accessed on 10 November 2022).
44. Seattle Public Utilities. 2021–2026 Strategic Business Plan. 2021. Available online: https://www.seattle.gov/documents/Departments/SPU/AboutUs/SBP_2021-2026.pdf (accessed on 3 August 2022).
45. Seattle Public Utilities. 2015–2020 Strategic Business Plan. 2015. Available online: https://www.seattle.gov/documents/Departments/SPU/AboutUs/SPU-StrategicBusinessPlan-2015.pdf (accessed on 3 August 2022).
46. Seattle Public Utilities. Shape Our Water Community Vision: Creating a Water Resilient Future in Seattle. 2021. Available online: https://static1.squarespace.com/static/5efcd020ca4aa07f7ef991ef/t/6179b9dab9e13b0d63cf18b3/1635367391392/Shape%2BOur%2BWater%2BCommunity%2BVision%2BOct+2021.pdf (accessed on 17 August 2022).
47. Seattle Public Utilities. Street Edge Alternatives Project. Available online: https://www.seattle.gov/utilities/neighborhood-projects/street-edge-alternatives (accessed on 10 November 2022).
48. City of Seattle. Executive Order 2013-01: Citywide Green Stormwater Infrastructure Goal & Implementation Strategy. Available online: https://www.seattle.gov/Documents/Departments/OSE/GSI-exec-order.pdf (accessed on 5 December 2022).
49. Seattle Public Utilities. King County, The Road to 700 Gallons: A Natural Approach to Stormwater Management. 2020. Available online: https://www.seattle.gov/documents/Departments/SPU/Documents/GSI-ProgressReport2020.pdf (accessed on 5 December 2022).
50. Seattle Public Utilities. Request for Proposals, Contract 20-187-S, RainCity Partnerships. 2021. Available online: https://www.risc.solutions/wp-content/uploads/2022/05/Seattle-RainCity-Partnership-RFP-July-2021.pdf (accessed on 27 January 2023).
51. U.S. EPA. *Community-Based Public-Private Partnerships and Alternative Market-Based Tools for Integrated Green Stormwater Infrastructure*; 2015. Available online: https://www.epa.gov/sites/default/files/2015-12/documents/gi_cb_p3_guide_epa_r3_final_042115_508.pdf (accessed on 27 January 2023).
52. Bozuwa, J. Building Resiliency Through Green Infrastructure: A Community Wealth Building Approach. (Foreword by Mami Hara, Seattle Public Utilities). The Democracy Collaborative. 2019. Available online: https://democracycollaborative.org/learn/publication/building-resiliency-through-green-infrastructure-community-wealth-building (accessed on 14 August 2022).

53. Seattle Public Utilities. Natural Drainage Systems Overview and Projects. 2007. Available online: https://www.waterboards.ca.gov/rwqcb2/water_issues/programs/stormwater/muni/nrdc/natural%20drainage%20projects.pdf (accessed on 1 February 2023).
54. Seattle Public Utilities. Seattle's Natural Drainage and Complete Streets. In Proceedings of the National Association of City Transportation Officials Designing Cities Conference, Seattle, WA, USA, 24–26 October 2016. Available online: https://nacto.org/wp-content/uploads/2016/07/Seattles-Natural-Drainage-Systems-and-Complete-Streets.pdf (accessed on 1 February 2023).
55. City of Seattle. Duwamish Valley Action Plan. 2018. Available online: https://greenspace.seattle.gov/wp-content/uploads/2018/06/DuwamishValleyActionPlan_June2018.pdf (accessed on 12 February 2023).
56. Seattle Public Utilities. *Greenhouse Gas Emissions and Energy Usage Report 2019–2020*; Seattle Public Utilities: Seattle, WA, USA, 2023.
57. Kaatz, L.; Vano, J.A.; Sullivan, A.; Brooks, K.; Mahmoud, M.I.; Asefa, T.; Cohn, A. *Leading Practices in Climate Adaptation*; Water Utility Climate Alliance: Clearwater, FL, USA, 2021. Available online: https://www.wucaonline.org/assets/pdf/WUCA-leading-practices-report-2021.pdf (accessed on 10 November 2022).
58. Gonzales, K.; Roop, H.; Grodnik-Nagle, A.; Purnell, D.; Rozance, M.; Rack, M. Mainstreaming Adaptation to Extreme Precipitation at a West Coast Water Utility by Engaging Precipitation First-Responders. In Proceedings of the National Adaptation Forum, Baltimore, MA, USA, 25–27 October 2022.
59. Center for Community Investment. Connect Capital: Seattle, WA. 2018. Available online: https://centerforcommunityinvestment.org/connect-capital-seattle-wa (accessed on 10 August 2022).
60. U.S. Army Corps of Engineers. *Preliminary Flood Risk Management Study for the Duwamish River at South Park*; U.S. Army Corps of Engineers: Seattle, WA, USA, 2017.
61. City of Seattle. *City to Create a 'Resilience District' with Award from Robert Wood Johnson Foundation*; Greenspace Blog, Office of Sustainability & the Environment: Seattle, WA, USA, 2020. Available online: https://greenspace.seattle.gov/2020/12/city-to-create-a-resilience-district-with-award-from-robert-wood-johnson-foundation/#sthash.9f0RrMLh.MRhS0bA6.dpbs (accessed on 10 December 2022).
62. Zehner, E. Resilience District Concept Gathers Momentum in Seattle. Lincoln Institute of Land Policy. 2021. Available online: https://www.lincolninst.edu/publications/articles/2021-04-climate-health-equity-resilience-district-concept-gathers-momentum-in-seattle (accessed on 7 September 2022).
63. Seattle Public Utilities. SPU 'Seeds of Resilience' Impact Investment Concept, Prepared for SPU's Customer Review Panel. 2020. Available online: https://www.seattle.gov/documents/Departments/SPU/AboutUs/SeedsOfResilienceImpact_InvestmentProgram.pdf (accessed on 7 September 2022).
64. Smith, C. Wastewater Has the Best Green Jobs Workers Don't Know about. Governing. 2021. Available online: https://www.governing.com/work/wastewater-has-the-best-green-jobs-workers-dont-know-about.html (accessed on 7 September 2022).
65. Kane, J.; Tomer, A. *Renewing the Water Workforce: Improving Water Infrastructure and Creating a Pipeline to Opportunity*; Brookings Institute: New York, NY, USA, 2018. Available online: https://www.brookings.edu/research/water-workforce/ (accessed on 7 September 2022).
66. Innovation Network for Communities; Earth Economics; The Democracy Collaborative. "How Can Cities Use Investments in Climate Resilience to Generate Economic Benefits for Low-Income Neighborhoods?" Supported by Seattle Public Utilities and Summit Foundation. 2020. Available online: http://lifeaftercarbon.net/wp-content/uploads/2020/03/Climate-Resilience-and-Low-Income-Neighborhoods-Final.pdf (accessed on 15 August 2022).
67. Wolfram, M.; Borgstrom, S.; Farrelly, M. Urban transformative capacity: From concept to practice. *Ambio* **2019**, *48*, 437–448. [CrossRef] [PubMed]
68. Frantzeskaki, N.; Holscher, K.; Bach, M.; Avelino, F. (Eds.) Co-creating sustainable urban futures. A primer on applying transition management in cities. In *Future City 11*; Springer: Cham, Switzerland, 2018.
69. Romero-Lanko, P.; Bulkeley, H.; Pelling, M.; Burch, S.; Gordon, D.J.; Gupta, J.; Johnson, C.; Kurian, P. Urban transformative potential in a changing climate. *Nat. Clim. Change* **2018**, *8*, 754–756. [CrossRef]
70. Newton, P.; Meyer, D.; Glackin, S. Becoming urban: Exploring the transformative capacity for a suburban-to-urban transition in Australia's low-density cities. *Sustainability* **2017**, *9*, 1718. [CrossRef]
71. Herrick, C.; Pratt, J.; Surbaugh, H.; Grumbels, B.; Loken, L.; Abhold, K. *Changing Organizational Culture to Promote Sustainable Water Operations: A Guidebook for Water Utility Sustainability Champions*; Water Research Foundation: Denver, CO, USA, 2013.
72. Herrick, C.; Pratt, J. Sustainability in the water sector: Enabling lasting change through leadership and cultural transformation. *Nat. Cult.* **2012**, *7*, 285–313. [CrossRef]
73. Grimm, N.; Faeth, S.; Golubiewski, N.; Redman, C.; Wu, J.; Bai, X.; Briggs, J. Global change and the ecology of cities. *Science* **2008**, *317*, 756–760. [CrossRef] [PubMed]
74. Wilby, R.; Vaughan, K. Hallmarks of organizations that are adapting to climate change. *Water Environ. J.* **2010**, *25*, 271–281. [CrossRef]
75. Brown, R. Impediments to integrated urban stormwater management: The need for institutional reform. *Environ. Manag.* **2005**, *36*, 455–468. [CrossRef] [PubMed]
76. Funfgeld, H. Institutional challenges to climate risk management in cities. *Curr. Opin. Environ. Sustain.* **2010**, *2*, 156–160. [CrossRef]
77. Shrivastava, P.; Hart, S. Creating sustainable corporations. *Bus. Strategy Environ.* **1995**, *4*, 154–165. [CrossRef]

78. Dilling, L.; Lemos, M.C. Creating usable science: Opportunities and constraints for climate knowledge use and their implications for science policy. *Glob. Clim. Change* **2009**, *21*, 680–689. [CrossRef]
79. Cvitanovic, C.; Marshall, N.; Wilson, S.; Dobbs, K.; Hobday, A. Perceptions of Australian marine protected area managers regarding the role, importance, and achievability of adaptation for managing risks of climate change. *Ecol. Soc.* **2014**, *19*, 33. [CrossRef]
80. Meadow, A.; Ferguson, D.; Guido, Z.; Horangic, A. Moving toward the deliberate co-production of climate science knowledge. *Weather. Clim. Soc.* **2015**, 179–191. [CrossRef]
81. Herrick, C.; Vogel, J. Climate adaptation at the local scale: Using federal climate adaptation policy regimes to enhance climate services. *Sustainability* **2020**, *14*, 8135. [CrossRef]
82. Niepold, F.; McCaffery, M. *Essential Principles of Climate Literacy*; NOAA Climate Program Office: Silver Spring, MD, USA, 2008. Available online: http://climateliteracynow.org/files/Climate_Literacy_K-12.pdf (accessed on 15 August 2022).
83. Lempert, R.; Busch, L.; Brown, R.; Patton, A.; Turner, S.; Schmidt, J.; Young, T. Community-level 2018, participatory co-design for landslide warning with implications for climate services. *Sustainability* **2023**, *15*, 4294. [CrossRef]

sustainability

MDPI

Article

Slowing Down Climate Services: Climate Change as a Matter of Concern

Werner Krauß

Artec Sustainability Research Center, University of Bremen, Enrique-Schmidt-Straße 7, 28334 Bremen, Germany; wkrauss@uni-bremen.de

Abstract: This article addresses the appropriate place for and design of climate services drawing upon a case study of three different forms of climate service delivery in a coastal landscape in Northern Germany. Each of these forms addresses different audiences and provides different types of knowledge about climate change and a different orientation toward policy support. The three-part case study includes a regional, a municipal and a social climate service. Drawing upon this comparative, case-based research, I develop the idea of 'slowing down climate services', based on the 'slow science manifesto' introduced by the science philosopher Isabelle Stengers, by postnormal science and by political ecology as suggested by Bruno Latour. How does climate change become a matter of concern? Slowing down climate services means following the social life of scientific facts, engaging with the public and exploring ways to improve democratic and place-based decision making. I argue that there is an urgent need to overcome the big science orientation of climate services and to add what Stengers calls 'public intelligence', the integration of a sense of place and of the social, cultural, political and other performative aspects of climate change in specific landscapes.

Keywords: slow science; postnormal science; political ecology; matters of concern

Citation: Krauß, W. Slowing Down Climate Services: Climate Change as a Matter of Concern. *Sustainability* **2023**, *15*, 6458. https://doi.org/10.3390/su15086458

Academic Editors: Susan Clayton, Charles Herrick, Jason Vogel and Glen Anderson

Received: 15 November 2022
Revised: 3 April 2023
Accepted: 5 April 2023
Published: 11 April 2023

1. Introduction

Climate services provide climate data and information on global, national, regional and local scales, and many areas are already well-served. In regions such as Northern Germany, for example, there is a dense infrastructure of climate services. Stakeholders, decision makers, the media and ordinary citizens have access to science-based information about changes in climate [1]. However, data derived from models or empirical observation alone do not provide solutions for complex climate-related problems or roadmaps for the decarbonization of societies. There is an increasing demand for active engagement with stakeholders and decision makers in communities, for taking into account the complexities and uncertainties of climate change and for including a wider range of voices and actors. Research programs such as Horizon Europe, the Joint Program Initiative (JPI) and especially the European Research Areas for Climate Services (ERA4CS) encourage research on policy support and the co-development of climate services. This article results from my participation as a social anthropologist in two of these European projects, 'Co-development of place-based climate services for action' (CoCliServ 2017–2021) (http://cocliserv.cearc.fr/ (accessed on 4 April 2023)) and a recently started project about the standardization of climate services and policy support, 'climateurope2' (2022–2027) (https://climateurope2.eu/ (accessed on 4 April 2023)). This change in perspective, from the production of data for climate services to participant observation—the main anthropological method—is the starting point of this article. For whom is climate change a matter of concern, how does climate change come to matter [2], and what does this mean for the practice of climate services?

The shift from providing scientific evidence of climate change and climate risks, which was the main task of climate science for a long time, to actually dealing with climate-related problems on the ground includes an important epistemological aspect. It is not data and information alone that is needed but knowledge about the land, the people and their heritage and the political ecology of climate-related problems [3]. In this article, I present elements for a more inclusive and participatory approach, based on longtime ethnographic research in the coastal landscapes of Northern Germany. The theoretical umbrella for my approach is 'slow science', a concept promoted by the science philosopher Isabelle Stengers [4], with its constituent parts based on premises from political ecology and postnormal science. Slow science asks for a more reflective and deliberative approach to scientific research, to engage in dialogue with the public and to develop a sense for the geo-social constitution of the respective locations. Climate change always happens somewhere, in some place, and there is a difference between results gained from a model or a set of data and the actual changes caused by climate in the real world [5].

Postnormal science is a concept introduced by Silvio Funtowicz and Jeremy Ravetz [6], and it indicates situations and cases that cannot be solved by 'normal' science. Municipalities, stakeholders or decision makers have to take action despite uncertain knowledge, financial risks and value conflicts. Both concepts, slow science and postnormal science, rest on the assumption that climate is not only a scientific fact but a matter of concern. This distinction goes back to Bruno Latour's [7,8] conception of political ecology. He challenged the separation of science on the one side and society, law or politics on the other, and he shifted attention to the assemblies of actors that are involved in any given issue where climate data come to matter. This shift from climate as a matter of fact—which played an important role in providing evidence of human impact on climate change and was challenged by climate skeptics [7]—to climate as a matter of concern implies a shift from model and information to practice and collaboration. In this article, I argue for slowing down climate services and closely reconsidering, analyzing and also imagining new forms of provider (climate services) and user (all kinds of public actors) encounters in order to develop new and more effective forms of climate protection.

As an empirical basis for my argument, I discuss three different forms of climate service delivery in a coastal landscape in Northern Germany. Each addresses different audiences, provides different forms of knowledge about climate change and offers different forms of policy support. The three-part case study includes (1) a regional setting in which a science-based climate service provides climate data and information for Northern Germany; (2) a 'social climate service' [9], consisting of an emergent, loosely organized group of citizens, which aims to bring climate-related problems into the public sphere, to put pressure on politics and to promote climate-friendly practices; and (3) a municipal setting in which government managers develop place-based climate and energy plans to mitigate greenhouse gas emissions and to adapt to climate change impacts. Each of these different forms of climate service application has its own history, which in turn informs its approach to climate change. At a very basic level, the purpose of this article is to step aside and reflect upon climate services as a social practice.

Climate Services as an Object of Research: Background and Methods

During my extended fieldwork in Northern Germany, I both observed and initiated various encounters between different climate service providers and the public [10]. In this first part of the article, I will provide some background concerning my research, followed by epistemological reflections on the concepts of slow science, postnormal science and political ecology. In the main parts, I will provide descriptions and vignettes of these encounters and discuss them accordingly. Maybe more than figures and tables, anecdotal depictions are a method to keep the heterogeneity, complexity and messiness of real-life situations alive.

My starting point in the research about the co-development of climate services was my interest in 'narratives of change', in the role that climate and weather and its changes play in all kinds of past and present narratives, including those of climate services [11]. In this context, I practiced the ethnographic method of participant observation in the strict sense of the term: I both observed and initiated communication and interaction between climate services and the public. In current anthropology, participant observation has turned into the co-design of research and para-anthropology, which is a more precise terminology. Both co-design and para-anthropology entail active collaboration with local partners for a shared matter of concern and are methods that have evolved as a result of the increasing integration of anthropology in inter- and transdisciplinary research projects [12]. In the first part of the CoCliServ project, I worked with the 'North German Coastal and Climate Office', a regional climate service from the Helmholtz Center Geesthacht and partner in the CoCliServ project. Together, we organized a workshop with local and regional decision makers, stakeholders and concerned citizens. In the second part of the project, I engaged with local climate activists, with whom I co-developed a citizen's initiative. During this time, I also followed the successive implementation of municipal climate protection managers in this area, an initiative by the German government. Altogether, the coastal landscape of Northern Germany provides a dense and diverse network of climate services. It reflects the European and German efforts to initiate a transformation of society toward decarbonization. However, recent surveys of this process show that there is no plausibility that Germany will reach its ambitious climate goals with current measures [13,14]. At the same time, these surveys highlight the role of social movements and public pressure on politics for a successful transition toward decarbonization. As a consequence, climate services have to expand their scope and situate themselves in a highly politicized and complex environment, and they have to follow the communal life of facts [2], too, in order to make an impact.

Global climate discourse, from the IPCC to climate services and everyday talk, is science-based, managerial and technology-oriented [15]. For climate services, this raises many difficult questions. Climate services are based on the premise that appropriately configured data and information about climate will enable better political decisions. However, commentaries in Nature or Science concede that there is no undisturbed transfer of knowledge [16]. This is especially true when the knowledge base is uncertain, stakes are high, morals are included and action is urgent—which is the situation for postnormal science as defined by Funtowicz and Ravetz [6]. Climate services and decision makers alike have to deal with these 'postnormal' situations, and there is a need for experimenting with different forms of interaction with local or regional actors. It is a long way from providing data and information to the co-development of climate services for action. One of the main preconditions is to understand climate science and climate services as part of an assemblage of actors concerned with climate change. In a regional setting such as the coastal landscapes of Northern Germany, climate services have to deal with administrations, municipalities, dike and sluice organizations, various forms of stakeholders, farmers and NGOs, and most of them are climate-literate [10]. In order to make a meaningful and especially place-based contribution, climate services have to go beyond the mere purveying of data derived from models and observation. This challenge demands a certain level of introspection and self-reflection, as well as a sense of place, the geo-social formation of the coastal landscapes and the actual political constellations [17].

The science philosopher Isabelle Stengers discusses in her book 'Another science is possible: A manifesto for slow science' [4] the example of genetically modified organisms (GMOs). On the one hand, genetic modification was praised as the solution for poverty and hunger in the world, while on the other hand, there was significant protest of field scientists, local farmers and environmental groups. Stengers illustrates how the laboratory conditions under which scientific knowledge is produced have little to do with 'those situations we are confronted with as citizens' ([4], p. 2f). As a consequence, she argues that political ecology has 'to put the sciences into politics, but without reducing them to politics. This requires fully developing, around each issue, the primordial question: who can talk of what, be the spokesperson of what, represent what, object in the name of what?' ([4], p. 148).

Stengers suggests slowing down science, engaging in a debate with the 'public intelligence' ([4], p. 14) and integrating additional aspects of reality, instead of only opting for science-based technological solutions for all problems. Conway [18] defines the concept of public intelligence as a collective phenomenon, 'an intelligently distributed and contested arrangement of roles, defined by the agents themselves'. This can also be understood as a call for climate services to engage with a wider range of users, defined as those segments of society that understand climate change as a matter of public concern. In order to do so, climate services have to leave the comfort zone of science and confront climate change as a political object. This implies the virtue of self-reflection, of their own role in the field and in society.

At the same time, climate service providers are professionalizing and becoming competitors in a contested knowledge market. Either they have to professionally compete for third-party funds, or they are private consultant agencies developing a new market segment in cooperation with insurance companies and other interested parties. This tendency is also reflected in science-based climate discourse and its terminology. In official documents from the IPCC to regional climate services, neo-liberal concepts, such as innovation, markets, growth, providers, users, decision makers, stakeholders and others, are commonplace [19]. Everything is pressed into the narrow imagery of a market terminology with the result that climate services become agents for market expansion. In many climate service programs, including the IPCC, there is little focus on the need for climate protection, for care and well-being, or emphasis on the catastrophic situation we are in [19]. The reduction of climate to numbers and statistics does not help; numbers have agency, too, they frame the perception of climate, and in doing so, the solutions are designed accordingly [20]. The political, economic and social causes of climate change are turned into technical problems which supposedly will be solved by engineering, by technical solutions for adaptation and mitigation. The call to 'follow the science', popularized by Fridays for Future, has a flipside, which is the strong belief in the 'general authority' [18] of science. Stengers [4] argues sarcastically that the public is supposed to trust in science, 'but they have to know how to wait, and understand that scientists owe it to themselves to remain deaf to any noisy or anxious demands', and that people should not 'be urged to get involved in questions they are not, in any case, capable of understanding'.

In the context of European governance strategies, climate services are understood as competitors in an emerging knowledge market, and there is hardly any mention of climate protection or care for the environment. The European Green Deal is easily depicted as a Janus-faced strategy that either serves to develop new—climate-friendly—markets or serves to change the system of growth and depletion of natural resources. In between are the municipalities, landscapes and nations that have to make decisions about how to proceed into the future. For them, there is more at stake than only markets and statistics; landscapes are also life-worlds, where people interact with geo-social conditions which were formed over a long time and materialize in everyday activities, the sense of place, identities and customary laws [21,22]. This is where social anthropology and interdisciplinary landscape studies intersect with the concept of slow science.

The term 'slow science' has roots in the 'slow food' movement, which originated in Italy as a form of regional protest against the standardization of food [23]. The slow-food movement emphasized regional identity and the singularity of the geo-history of European landscapes and their people. For slow-food protagonists, the acquisition and preparation of food is a matter of concern. This provides support for Stengers' argument that 'matters of concern' have to be taken as seriously as 'matters of fact'. Slow science means here not only the quality control of the scientific process but also the integration of what Stengers ([4], p. 14) calls 'public intelligence'. Scientific facts have a social life, they travel through landscapes and households, and they change global and municipal policies as well as individual ways of life. In reference to the slow-food movement, Stengers asks for 'connoisseurs of science' who act as 'agents of resistance against a scientific knowledge that pretends it has general authority; they partake in the production of what Donna Haraway calls 'situated knowledges'' ([4], p. 19). This is where the concept of slowing down climate services comes in: climate services are privileged to coproduce these situated knowledges and to instigate climate action, as demanded in the European calls for the co-development of climate services for action. In my participation in two of these European projects, I put this call for action into practice, as an anthropological observer, as a participant and as an instigator of climate service action.

2. Regional Climate Service

In the ERA4CS project about 'the co-development of place-based climate services for action', I worked together with one of our project partners, the North German Coastal and Climate Office. This climate service is the outreach of the Helmholtz Center of Geesthacht (now Hereon), and its service includes the Hamburg area and Northern Germany. It serves as a contact point between climate science and the public, and one of its main tasks is to provide an overview on regional climate change in Northern Germany, serving interested citizens, scientists, educators, economic actors, public authorities, media, civil society organizations and political officials [24,25].

For several months, we practiced a division of labor; while the Climate Office documented existing climate services in the coastal area, I interviewed mayors, administrators, representatives of NGOs, farmers and other stakeholders. We realized early that there was a qualitative difference between the information provided by climate services and the narratives that I collected in the field. One example involved the extreme weather events during our research period in this area, namely an unusually warm, dark and wet winter, followed by a drought in the summer. For the climate services in this area, this was a matter of statistical interest (sometimes followed by general warnings), while for my interlocutors, these changes occurred as an aspect of their daily 'weatherworld' ([26], p. 120). During the winter, the fields were inundated by water, and the farmers had difficulties applying manure, while in the summer, many had to sell cattle because of a lack of feed, which became exceptionally expensive in the international market [10].

Finally, we organized a public workshop in the coastal village of Dangast and invited about 30 persons, chosen from our field sites. In the invitation, we asked for 'the regional effects of climate change and for possible answers, from the world climate council, the IPCC, to the municipal council, from climate research to local knowledge' (translation by the author). As our goal, we wanted to start a conversation among diverse members of the public and asked what it takes to make the region climate-friendly and fit for the future.

The organization of the workshop reflected our division of labor: the first part was organized by the Climate Office and the second part by me, an anthropologist from the university. The two parts could not have been more different: in the first half, the moderator of the Climate Office presented data about global and regional climate change, followed by an intensive Q&A section. In the second part, I organized a discussion among our guests. While the first part was orderly and quiet, the second part was like a marketplace, loud and chatty with people moving around in groups. Like a mirror ball, these different settings reflected our different backgrounds and origins and also the different conclusions we drew from this event.

2.1. A Regional Climate Service in Action

When I arrived at the venue maybe an hour before the workshop started, the climate service team was already there. They had prepared the room like a theatre: in front of the chairs for the audience, there was a lectern, framed by two standing posters. The photos on the posters displayed a dramatic scene, with a research vessel cutting through the waves of a stormy sea. In the center of the poster there was the logo of the Helmholtz Research Center and the motto *'erkennen, verstehen, handeln'* (identify, understand, act). On a table next to the desk, there were two computers for public use, introduced as 'web tools' where people could navigate climate scenarios. Everyone entering the room could not help but be aware of the scientific authority of Helmholtz research, of big science.

The workshop started with a presentation by the moderator of the climate service. She outlined how climate changed globally, followed by an outlook into the future. The forecast effects of emission reduction and mitigation were presented in the form of model outputs based on IPCC findings and data, showing the different scenarios from worst- to best-case. The data for the regional development, which were mostly derived from empirical research, attracted the greatest attention from the audience. There has been a 0.8 degree rise in temperature since 1961, the beginning of apple blossom starts two to three weeks earlier (while late frost remains unchanged), and there are now more heat days (from 1–2 to 4–5). Rainfall in the summer remains fluctuating, while winter becomes warmer and wetter, sea level rise (20 cm) conforms with global tendencies, with an increased threat of more and higher storm surges. As a result, the moderator stated that the North German coastline is one of the hot spots of climate change in Germany.

The presentation lasted maybe 30 min, and it left the audience in a state of shock. The confrontation of the global scenario with data from the region made climate change real, in an uncomfortable way. Climate change is not somewhere out there, but in front of our own doorsteps, as one discussant stated. In the following discussion, people expressed concern with discrepancies in climate discourse and politics. As one of the first questions, the moderator was asked how she perceives the discrepancy between existing knowledge about climate change and the less-than-adequate political action. Another aspect was the discrepancy between knowledge about mitigation technology and its lack of implementation. There was also a discussion about the gap between knowledge about climate change and actual consumer behavior, a discrepancy that was attributed to the powerful marketing strategies of the industry. In the meandering discussion, the issue was raised of which strategy might be best for the farmers to bring the livestock through a drought, organic or conventional farming. The moderator put great efforts to maintain a neutral position, as with other controversial issues that were raised. She insisted on the neutrality of climate services, as she summarized in her report:

'Thus, a place based climate service for action implies that the role of science needs to be neutral. Rather than supporting a particular favoured action, science can support decision-making processes by analysing how certain decisions, compared to others, may initiate specific changes and impact developments' ([25], p. 53).

For the climate service, the job was done, followed by the anthropological part. The second part of the workshop was dedicated to the question of what it takes to achieve a climate-friendly regional future. One of the main goals was to bring together different actors who normally do not meet or engage in discussions, such as a mayor and a member of an NGO, a priest and a farmer or a student and a coastal manager. We carefully organized an incremental discussion with random groups of four, who in the end, presented their conclusions to the plenary. After the strict discipline of a Q&A discussion in the first part, here people discussed freely their ideas in various constellations.

As a result, there were dozens of cards with keywords on the whiteboard, which we tried to group into themes such as mobility, energy, water management, consumption, habitation, land use and agriculture. The response to the exercise was lively and appreciative. In conversations at the end, people pointed out that they learned a lot during the afternoon and that they hoped for a follow-up workshop to discuss things in more detail. The workshop was considered a first step, and we promised to send feedback and to organize a second step, which unfortunately never happened, at least in this constellation and due to the different understandings of the public role of a climate service.

2.2. *Discussion: Regional Climate Services*

In the evaluation of events, we came to different conclusions. For the North German Coastal and Climate Office, the purpose of the discussion was to identify knowledge gaps and information needs, while I looked for possibilities to expand the range of climate change issues and to motivate the audience for further collaborations. We had different understandings of 'matters of concern', and thus we followed different epistemologies. In the first part, the Climate Office confirmed the separation between scientific facts on the one side and local matters of concern on the other, in a hierarchical way. The result was the identification of further knowledge needs and information gaps, which now can be added to the already existing corpus of place-based climate knowledge. From this perspective, the workshop served as a venue for performing the general authority of climate science. The slogan of 'identify, understand, act', displayed on the posters, left the public in a position where they have little choice but to trust the science and wait until it presents the solution, as Stengers puts it. The public is left in a state of helplessness: either they confirm the power of science, or they are viewed as biased, ideological and non-objective. The main mantra of the climate service is that science and politics have to remain neatly separated: once and again, the moderator of the Climate Office stated that it is up to politics to decide what to do. Both the public and politics are left in a double bind: if they make a decision, they are at risk that the decision will easily be dismissed as 'non-scientific' or not up to scientific standards; if they ask science what to do, they are told that it is up to political decision makers. Due to our different interpretations of the event, our cooperation ended here. We did not agree on a common evaluation for the audience, nor did we manage to stage a follow-up event. There were other obstacles, too: the partners from the Climate Office did not even share the recordings of the event with me, due to new data policies of the Helmholtz Center. They had already enough material for their project deliverable and moved on, while I stayed for the rest of the project in the region. This kind of sad story unfortunately is common in projects and is a tribute to the permanent pressure in a research landscape, where the duration of projects is short and dependency on third-party funding is high. However, serendipity is an important feature in life and in anthropological fieldwork [27] and so did not disappoint me. The experimental co-development of climate services, together with 'the public', found its continuation in another place, a few months later.

3. Social Climate Services

In November 2019, I organized a follow-up workshop in the neighboring district of Ammerland. The initiative came from an environmental activist, who had already participated in the previous workshop in Dangast, and she wanted to carry the format of public participation to her district. The workshop was held in November 2018, and this time, 'the co-development of place-based climate services for action' was successful. As a result of the workshop, we founded a citizen's initiative, the 'Klimamarkt Ammerland' (climate market). It is a loosely organized group of concerned citizens, whose goal is to bring climate change into the public sphere, put pressure on politics and help turn Ammerland into a climate-friendly district.

The term 'social climate services' was coined by Bremer et al. [9] in their article 'Recognizing the social functions of climate services in Bergen, Norway', where they propose 'a field of 'social climate services' that configures relationships between scientists and social actors, built on technologies of humility, for enriching the ongoing culturally and politically charged debates and practices around climatic change in informal institutional settings' ([9], p. 1).

I use the term social climate service for the purpose of this article, even though the members of the Klimamarkt would never do so. In everyday German, the term 'service' is associated with institutions such as banks, administrations or public transport—institutions where citizens are turned into 'consumers' or 'clients'. The Klimamarkt identifies as a public forum, a network, a multiplier or perhaps even—though we did not discuss that yet—as diplomats between science, the public and the landscape of Ammerland. The Klimamarkt has no preconceived agenda, and there are hardly any strategic debates. In my terms, as the participant anthropologist, we organize public performances of climate change as a matter of concern. For press releases, brochures or public communication, we use, in variations, the following text module:

'The Climate Market Ammerland was launched at the end of 2019 to express concern about climate change in Ammerland. Ideas for a climate-friendly Ammerland from the areas of energy, mobility, food and agriculture, land use, water, construction and renovation and health were collected by concerned citizens. Together, projects are now being initiated that will help avoid CO_2 and prepare the communities in Ammerland for the effects of climate change. The Klimamarkt Ammerland is independent, autonomous and open to everyone. It is a public forum to initiate debates and initiatives in order to make a difference. The Klimamarkt wants to help shape a sustainable Ammerland.' (https://klimamarkt-ammerland.de/ (accessed on 4 April 2023)).

Currently, the Klimamarkt consists of a core group of seven people, most members are engaged in environmental organizations and other forms of public life. We fit maybe in the category of public intelligence as suggested by Stengers [4]: we are able to read and understand scientific assessments, some have a long career in environmental activism, and some of us are specialized in certain areas such as biology or, in my case, in environmental anthropology, and we are widely connected, in Ammerland and in climate research, from Scientists4Future to local activism of all kinds.

3.1. A Social Climate Service in Action

From the first workshop in 2019 until today, the Klimamarkt developed a series of activities. While the terminology of climate services is borrowed from the neoliberal market economy, the term Klimamarkt reminds of a farmers' market, a public place where different people meet and exchange the produce of the soil as well as news and gossip. The idea of the Klimamarkt is to bring climate change into the marketplace, or into the public sphere, as a matter of concern. Our trademark is the organization of climate markets in the literal sense of the term.

3.1.1. Climate Market #1

The structure of our first 'Klimamarkt' was similar to the workshop in Dangast which I described in Section 2. We publicly invited people to the workshop, and about sixty of them attended. There was one main difference in the staging of the workshop: there was no official climate service involved. The climate data set about global and regional developments is publicly available, so we presented it ourselves as an introduction. For the public discussion, we posed the question 'How does a climate friendly Ammerland look like in 2030?' We organized this conversation along seven main topics: energy, water, land use, health, mobility, food and construction. We staged whiteboards and encouraged the audience to walk around and fill in cards and pin them down. At each of the stands, there were lively discussions. In the end, the seven stands presented the results to the public. Many issues were addressed, from sustainable household management to the circulation, distribution and consumption of regional products, from communal gardening and the conception of heat islands as protection from heat waves to the rewetting of the moors and the fight against a new Autobahn which is supposed to cross the local moorlands. Despite the concrete measures and proposed actions, the event was an exercise in reclaiming the public sphere and facilitating civic engagement. The Klimamarkt brought the climate problem into the public sphere, in the marketplace where opinions, no matter how qualified, can be expressed. Climate as a matter of public concern goes far beyond the reduction of climate to its physical and chemical composition; it is an exercise in democratizing climate.

3.1.2. Climate Market #2

The second climate market was staged as a result of collective confusion and emotional upheaval. In the summer of 2020, forest fires in California, Southern Europe and Germany made the news, COVID-19 brought public life to a halt, and the new IPCC report was published with dire projections for the future. Many people were concerned about this culmination of bad news, and so were we. How to deal with such a situation? We spontaneously organized another public Klimamarkt. Maybe 20 or more people met in an old barn; many of them engaged in care activities in schools and public institutions, others were environmental activists and concerned citizens, there were a couple of mothers with recent-born babies, and the district administrator made her appearance, too. We did not necessarily discuss possible solutions or activities. Instead, we tried to situate ourselves in a world where climate change was not merely a statistical construct but a lived reality. Instead of using words for a final statement, we formed a human exclamation mark and sent the photo to a regional newspaper, where it was published. Climate change as a public concern includes emotions, and the Klimamarkt served as a place where these feelings could be responsibly expressed, before people returned again to their families and to their jobs in the kindergarten, the school or the administration. It is not only individual well-being that matters; climate as a matter of concern means taking care of people and the environment. The Klimamarkt served to embody this sense of place and care, a sense that informs this kind of social climate services for action.

3.1.3. Climate Market #3

The title of the third climate market was 'Climate (protection) needs to be down-to-earth' (Klimaschutz braucht Bodenhaftung). We invited twenty different private or communal initiatives to exhibit their work in an old railway station which now serves as a community center. Among the presenters were local bee-keepers, a bicycle organization, environmental educators, a repair shop, a one-world shop, collectives presenting their herb garden or their communal gardening, an initiative of retired citizens who drove electric busses to maintain public transportation in the countryside, a clothes swap initiative, regenerative energy collectives, municipal climate managers and others. There was a café and folk music, and the public enjoyed strolling through this emergent venue of climate-friendly alternatives on a rainy Sunday afternoon. The district administrator took the stage and delivered a welcome speech, in which she outlined her program to turn Ammerland into a climate-friendly district. Our goal was to present activities that represented sustainable and climate-friendly forms of living, consuming and producing. It was a playful and friendly atmosphere that invited people to think about climate change without being indoctrinated or educated. As a motto served a quote from the anthropologist Margret Mead that it only takes a handful of people to change the world. In retrospect, this was again an epistemological exercise in alternative market metaphors; metaphors of care, exchange, circulation and sustainability were performed and subtly replaced the usual neoliberal terminology of innovation, management, stakeholder, participation or growth. From an anthropological perspective, we performed what it means to be 'down to earth' in the sense of Bruno Latour's manifesto for 'Politics in the new climate regime' [28]. Recently, soil has become a prominent feature in research about the effects of climate change and life in the Anthropocene. In Ammerland, there is a surprising variety of lifestyles and activities that practice a down-to-earth mentality in the literal sense of the term. Our third climate market brought these human resources into the public sphere with the intention to create networks and cross-connections for sustainable and climate-friendly forms of land use.

3.1.4. Mixed Activities

Before the national elections in 2021 and the elections in Lower Saxony in 2022, we organized public debates with the political candidates, we made a workshop about water management and webinars about agriculture, which all were well attended. We organized an art competition 'Dem Ammerland ein Gesicht geben' (Give the Ammerland a face), and currently we have a call for a writing contest, 'Klimageschichten aus dem Ammerland' (Climate stories from the Ammerland). These activities make climate real, bring it into everyday life and give climate a face, a history and make it part of our life. Most of all, we have managed to become a household name in the area and an address for networking. The non-partisan nature of the Klimamarkt, its support by the head of the district and municipal climate managers, the public events and the networking activities make it indeed a social climate service.

3.2. Discussion: Social Climate Services

Following the work of Bremer et al. [9] about citizen science, I dubbed the Klimamarkt as a social climate service. The Klimamarkt discusses the coastal landscape in terms of categories such as energy, water, land use, etc., and it adds a sense of place which is not covered by the algorithms of top-down climate software tools. Besides its networking activities, the Klimamarkt emphasizes the performative foundations of our climate interactions and creates its own vernacular climate narrative. In the opening remarks of the first Klimamarkt, we argued that climate is more than statistics and highlighted the importance of bringing climate into democracy. The second Klimamarkt was dedicated to climate change as a reality that has no precedence and is emotionally frightening. The third Klimamarkt was literally place-based and down-to-earth. The exhibition of alternative forms of life and care expanded the scope of climate service activities into the public and social sphere. These performative acts provide an additional meaning to Stengers' 'public intelligence', understood as an emergent, community-based effort with situated knowledges at disposal. A citizen's initiative such as the Klimamarkt serves well as an occasional mediator, as a networker, diplomat and producer of new forms of climate knowledge and, importantly, as a productive counterpart to institutionalized climate services, such as municipal climate managers in the next example. It is also important to keep in mind that the Klimamarkt is not an established institution or NGO. The website is only updated occasionally, activities depend on the individual time schedules and ideas of its members, and they are voluntary and spontaneous. From an informed everyday perspective, the activists of the Klimamarkt address many aspects of the long-term and also immediate effects of climate change which are not captured in official discourse. The goal is not only to include climate change into existing political and administrative structures but also to challenge the current forms of decision making in order to make democracy fit for climate change.

4. Municipal Climate Managers

The implementation of a municipal climate (protection) manager program is a recent development in the national climate service market and in Northern Germany. The main task of a climate manager is to produce an integrated climate protection plan in order to translate international and national climate goals into municipal practice. A governmental program, the National Climate Initiative (https://www.klimaschutz.de/de/ (accessed on 4 April 2023)), promotes municipal climate protection projects, such as plans for emission reduction, renewable energies or the production of communal energy and climate balances. The program is quite successful; by the end of 2019, 3650 municipalities had employed a climate manager [29]. The focus on this new form of climate service goes hand-in-hand with my current and ongoing participation in the Horizon project about the standardization of climate services. What can actually be standardized, and where are the limits of standardization? In the following, I will present a first description of this kind of service.

During my research, I interviewed several climate managers and followed their activities in two municipalities. There is a shift in perspective from the model of regional climate service to this form of municipal planning activity. It is a shift from the provision of climate data and information to the actual design of municipal planning and politics. Municipal climate managers introduce a new matter of concern, climate, into the municipal agenda, and they do so in the form of a standardized procedure.

The implementation of the program is highly contested in many places. For example, I followed discussions in the municipality of Varel, in the district of Friesland, where the majority of the local council vehemently argued against 'another administration', another 'bureaucratic nuisance' or a 'green paper tiger' expected to cost a lot of money in the long run. Others considered climate managers as agents of the 'green ideological agenda', and the mayor proudly argued that their municipality already does a lot for climate protection and is not in need of special advisory services. NGOs, concerned citizens and the local Agenda 20 group campaigned for several years for the implementation of a climate manager, until finally a young graduate from a nearby university was hired. In other municipalities and districts, the process went more smoothly; climate managers were welcomed and served as a sign that climate change is a matter of municipal concern.

In my interviews, several climate managers complained about how difficult it is to find their place in the hierarchy of the municipal administration. As one recently installed manager puts it, 'When you are fresh out of university, you first have to build a reputation. For a start, they gave me a small room under the rooftop, which I had to share with the nature conservation representative'. Communal politics more often than not are based on established networks between an administration, personal and/or party affiliations and representatives of various public interests, and it can be difficult to establish new issues on the political agenda. Climate managers have to start from scratch in a political and administrative environment with a long tradition and well-established hierarchies. In any case, statistics and general opinion suggest that the implementation of climate managers is a valuable program.

4.1. Standardization of Municipal Climate Services

In 2019, the municipal council of Edewecht in the district of Ammerland decided to hire a climate manager. Between January 2020 and June 2022, this climate manager produced an integrated climate protection plan for the municipality, involving the participation of local actors and the following criteria:

> 'The climate protection concept serves the municipality of Edewecht as a strategic basis for decision-making and a planning aid for its climate protection activities. With the help of the climate protection concept and the climate protection management, climate protection is sustainably anchored in the municipality as a cross-sectional task. These (tasks) include the analysis of the climate protection situation, the calculation of a municipal energy and greenhouse balance according to the territorial principle, the determination of potentials for the generation and utilization of renewable energies and energy efficiency, the calculation of a climate protection scenario until 2050, the development of strategies to increase the climate protection potentials and the derivation of prioritized fields of action'. [30], p. VIII (Translation by the author)

The production of the climate plan is supported and guided by a software program called 'Der Klimaschutzplaner' [31], which offers guidance in climate monitoring. This software is standardized and certificated by BISKO, a systematic communal greenhouse gas emission balancing tool provided by the Federal Environment Agency. It offers guidance regarding energy use in the municipality, provides optional paths and enables comparability with other municipalities in Germany.

The six main fields of action covered in the final report of the climate protection plan are (1) construction, sanitization and heat transition, (2) renewable energies, (3) mobility, (4) education, advisory and participation, (5) climate adaptation and emission sinks and (6) a climate neutral administration. The final plan provides detailed insight into the infrastructure of the municipality, identifies the main sources of emissions and discusses potential sinks and other details. To create and track accountability, there are in-built controlling mechanisms and incentives such as the 'Edewecht climate bonus' for the transition toward climate-friendly housing.

There were several workshops with local actors, including young people. Due to the pandemic, the workshops were mostly online. Additionally, there were online tools such as a map of the municipality, where the public could write comments, share impressions and make suggestions. An online portal provides access to resources for use by individual households, for spatial planning, communal energy saving and so on. The mayor and especially green politicians and activists considered the completion of the plan a great success.

4.2. Discussion: Municipal Climate Managers

Municipal climate services narrow the scope of climate services and intervene in municipal politics. They do so on the basis of a standardized process, which is accountable and calculable and enables various forms of control. The fragility of these plans is obvious: the implementation of the suggested measures to reduce the climate impact is not binding, even after the approval of the municipal council. At least, this is my latest information.

It seems to be indeed a question of standards and liabilities, what is binding and what is optional. This means that each of the suggested measures has to be debated, leaving it unclear what might be lost along the way and what is actually implemented. In terms of making use of public intelligence, it is a double-edged tool; because the framework is based on scientific calculations, the goals are reduced to technical and economic possibilities within the jurisdiction. Many potentially relevant factors fall outside their scope, such as major transport systems, economic processes and the management of territory, as well as infrastructure and buildings that are in private hands. Thus, the climate protection plan is reduced to a narrow window of opportunity for municipal politics. There is public participation, but it is channeled and constrained by the conditions of the standardized procedure and its technical framing. However, within this framework, climate indeed becomes part of the political agenda, the administration, spatial planning and communal life.

In the course of the current EU project 'climateurope2', I will bring together municipal climate managers, the Klimamarkt, selected stakeholders, administrators and local politicians in order to discuss what actually is and should be standardized, what is not and should not be standardized and what it takes to assure trust, transparency and the implementation of climate proposals and projects. To establish those encounters, it takes long-term research, and it takes time. Most climate services work in the rhythm of short-term projects, and it is another serendipity that this research can be stretched over two project terms.

5. Conclusions: Slowing Down Climate Services

In Northern Germany, like in many other regions in Europe, many people are what I call climate-literate. Climate services are needed to downscale global data and to provide long-term empirical observations about local changes in temperature, sea level or the frequency of extreme weather events. There is a dense infrastructure, and many administrations have their own experts and routines to gain relevant climate data. In the public sphere, scientific facts about climate change are framed as a matter of public concern and, as such, gain their own social life. The close relation of climate services to climate science makes them prone to call the public to order, to return to the science of climate. In this article, I argue that it is time to change direction and engage with the manifold performances of climate change in public life.

Slowing down climate services means situating climate science within the complex reality of the geo-social landscapes where climate change actually happens. It means overcoming the reduction of climate to statistics, of service to information and of solutions to technology. It implies a willingness to move beyond simple co-production exercises and the immersion of climate data, information and tools within the performative arenas of local life-worlds. Ideally, slow climate services take their time to explore the many contours of climate change in the everyday world, to link the insights from the laboratory with the experiences in the field, the statistics with the sense of place. Last but not least, slow climate services carefully situate themselves in the respective landscape instead of viewing it from above; place-based climate protection is based on partial knowledges, and science is one among other actors whose concerns are as meaningful as scientific facts. Slow sciences slowly follow the social life of climate data and how they come to matter in public life.

In this article, I portrayed three different forms of climate services that are active in the same area. Each type of climate service provides different contours: there is the statistical view of regional climate services, the applied statistics and planning of local climate services and finally the encompassing view obtained through social climate services. Slowing down climate services means wandering from one perspective to the other, back and forth, in order to get a more complete picture of what climate services can and should do. We should keep in mind that current climate politics and climate services are far from decarbonizing society within the goals set by the Paris Treaty, and thus it takes the reflection of current practices and courage to test new forms of climate activism, inside and outside of established institutions.

Funding: This project has received funding from the German Ministry for Education and Research BMBF, grant agreement FK 01LS1701B (ERA4CS project Co-development of place-based climate services for action CoCliServ) and from the European Union's Horizon Europe research and innovation program under grant agreement No. 101056933 (project Climateurope2).

Institutional Review Board Statement: Not applicable.

Informed Consent Statement: Not applicable.

Data Availability Statement: Not applicable.

Conflicts of Interest: The author declares no conflict of interest.

References

1. Gerkensmeier, B.; Meinke, I.; Breton, F. Assessment of Climate Service Components for Each Case Study Site. Project Report. 2018. Available online: https://cocliserv.cearc.fr/sites/cocliserv.cearc.fr/files/resultats/CoCliServ_D3.1.pdf (accessed on 4 April 2023).
2. Callison, C. *How Climate Change Comes to Matter: The Communal Life of Facts*; Duke University Press: Durham, NC, USA, 2014.
3. Krauß, W.; von Storch, H. Post-normal practices between science and local knowledge. *Nat. Cult.* **2012**, *7*, 213–230. [CrossRef]
4. Stengers, I. *Another Science Is Possible: A Manifesto for Slow Science*; e-book ed.; Polity Press: Cambridge, MA, USA, 2018.
5. Herrick, C. Self and place constructs in climate change vulnerability assessments: Gaps and recommendations. *Sustainability* **2021**, *13*, 2990. [CrossRef]
6. Funtowicz, S.O.; Ravetz, J.R. Science for the post-normal age. *Futures* **1993**, *25*, 739–755. [CrossRef]
7. Latour, B. Why has critique run out of steam? From matters of fact to matters of concern. *Crit. Inq.* **2003**, *30*, 15–20. [CrossRef]
8. Latour, B. *Reassembling the Social: An Introduction to Actor-Network-Theory*; Oxford University Press: Oxford, UK, 2005.
9. Bremer, S.; Bremer, A.; Iversen, L.; Soares, M.B.; van der Sluijs, J. Recognising the social functions of climate services in Bergen, Norway. *Clim. Serv.* **2022**, *27*, 100305. [CrossRef]
10. Krauß, W. Narratives of change and the co-development of place-based climate services for action. *Clim. Risk Manag.* **2020**, *28*, 100217. [CrossRef]
11. Krauß, W.; Bremer, S. Introduction: The role of place-based narratives of change in climate risk governance. *Clim. Risk Manag.* **2020**, *28*, 100221. [CrossRef]
12. Rabinow, P.; Marcus, G.E.; Faubion, J.D.; Rees, T. *Designs for an Anthropology of the Contemporary*; Duke University Press: Durham, NC, USA, 2008.
13. Stammer, D.; Engels, A.; Marotzke, J.; Gresse, E.; Hedemann, C.; Petzold, J. *Hamburg Climate Futures Outlook 2021: Assessing the Plausibility of Deep Carbonization by 2050*; CLICCS: Hamburg, Germany, 2021.
14. Engels, A.; Marotzke, J.; Gresse, E.G.; López-Rivera, A.; Pagnone, A.; Wilkens, J. (Eds.) *Hamburg Climate Futures Outlook 2023: The Plausibility of a 1.5 Degree Limit to Global Warming—Social Drivers and Physical Processes*; CLICCS: Hamburg, Germany, 2023. [CrossRef]
15. Oels, A. Rendering climate change governable: From biopower to advanced liberal government? *J. Environ. Policy Plan.* **2005**, *7*, 185–207. [CrossRef]
16. Herrick, C. Why objective science does not resolve environmental policy issues. *Environ. Sci. Policy Sustain. Dev.* **2022**, *64*, 19–28. [CrossRef]
17. Krauß, W. Klimawandel und die Ethnologie von Landschaften. In *Schlüsselwerke der Sozialwissenschaften*; Youssef, I., Rödder, S., Eds.; Transkript: Bielefeld, Germany, 2022; pp. 373–378.
18. Conway, P. *Another Science Is Possible by Isabelle Stengers*; Book Review; Polity Press: Cambridge, MA, USA, 2018. Available online: https://www.societyandspace.org/articles/another-science-is-possible-by-isabelle-stengers (accessed on 4 April 2023).
19. Ghosh, A. *The Great Derangement*; Penguin Books: Gurgaon, India, 2016.
20. Mennicken, A.; Salais, R. (Eds.) The new politics of numbers: An introduction. In *The New Politics of Numbers: Utopia, Evidence and Democracy*; Palgrave Macmilan: Cham, Switzerland, 2022; pp. 1–44. [CrossRef]
21. Olwig, K.R. *The Meanings of Landscape: Essays on Place, Space, Environment and Justice*; Routledge: London, UK; New York, NY, USA, 2019.
22. Krauß, W. Postenvironmental landscapes in the Anthropocene. In *The Routledge Companion to Landscape Studies*, 2nd ed.; Peter, H., Thompson, I., Waterton, E., Eds.; Routledge: London, UK; New York, NY, USA, 2019; pp. 62–73.
23. Petrini, C. *Slow Food: The Case for Taste*; Columbia University Press: New York, NY, USA, 2001. [CrossRef]
24. von Storch, H.; Meinke, I.; Stehr, N.; Ratter, B.; Krauss, W.; Pielke, R.A., Jr.; Grundmann, R.; Reckermann, M.; Weisse, R. Regional Climate Services illustrated with experiences from Northern Germany. *Z. Umweltpolit. Umweltr.* **2011**, *34*, 1–15.
25. Meinke, I.; Baztan, J.; da Cunha, C.; Jensen, C.; Wardekker, A. Lessons Learned from the Use of Various Formats for Disseminating and Communicating Climate Knowledge. 2020. Available online: http://cocliserv.cearc.fr/sites/cocliserv.cearc.fr/files/resultats/CoCliServ_D3.2.pdf (accessed on 10 April 2022).
26. Ingold, T. *Being Alive: Essays on Movement, Knowledge and Description*; Routledge: London, UK; New York, NY, USA, 2011.
27. Marcus, G.E. *Ethnography through Thick and Thin*; Princeton University Press: Princeton, NJ, USA, 1998.
28. Latour, B. *Down to Earth: Politics in the New Climatic Regime*; Polity: Cambridge, MA, USA; Oxford, UK, 2018.

29. Klimaschutz Niedersachsen (No Date) Klimaschutz- und Energieagentur Niedersachsen. Available online: https://www.klimaschutz-niedersachsen.de/ (accessed on 4 April 2023).
30. Ross, S. Gemeinde Edewecht Integriertes Klimaschutzkonzept. 2022. Available online: https://edewecht.de/wp-content/uploads/2022/07/Klimaschutzkonzept_Edewecht_final_2022-1.pdf (accessed on 4 April 2023).
31. Der Klimaschutzplaner (No Date). Available online: https://www.klimaschutz-planer.de/ (accessed on 4 April 2023).

sustainability

MDPI

Article

Climate Adaptation at the Local Scale: Using Federal Climate Adaptation Policy Regimes to Enhance Climate Services

Charles Herrick [1,*] and Jason Vogel [2]

[1] Washington Center, New York University, Washington, DC 20016, USA
[2] Climate Impacts Group (CIG), University of Washington, Seattle, WA 98195, USA; jmvogel@uw.edu
* Correspondence: cherrickdc@gmail.com

Abstract: This paper is an interpretive reanalysis of 17 in-depth case studies of community-based climate adaptation sponsored by the Kresge Foundation between 2014–2016. Drawing from the political science and international relations literature, we use the policy regime construct to characterize U.S. federal policies and programs that drive and enable climate adaptation at the local scale. While the regime construct has been used to evaluate the international governance of climate change mitigation, it has not been used in the context of climate adaptation. We find that numerous federal policies are used by localities to pursue adaptation objectives. We find that local adaptation initiatives based on federal policy tend to be non-prescriptive, are situational in their application, utilize common policy tools, and adopt a de-centered mode of governance. While a truly sustainable and resilient society may entail fundamental "transformation", we suggest that such a paradigm shift might be constructively cultivated through the blueprint laid out in the 17 case studies examined here—using existing know-how and tools. Based on our analysis and characterization of a federal climate adaption policy regime, we propose that the enterprise of climate services may need to move beyond existing models of co-production to embrace an 'apprenticeship' model, immersing technical information providers in the milieu of policy and governance in order that they might learn to recognize factors that influence the applicability, usefulness, and uptake of climate products and services.

Keywords: climate change; climate services; adaptation; policy regime; governance; local; co-production

Citation: Herrick, C.; Vogel, J. Climate Adaptation at the Local Scale: Using Federal Climate Adaptation Policy Regimes to Enhance Climate Services. *Sustainability* **2022**, *14*, 8135. https://doi.org/10.3390/su14138135

Academic Editor: Samuel Asumadu-Sarkodie

Received: 27 May 2022
Accepted: 30 June 2022
Published: 4 July 2022

1. Introduction and Background

Climate change has become one of the most pressing issues of our time. Due to the global nature of greenhouse gas emissions, political attention and policy efforts related to climate change have principally focused on the need for international and national scale policy interventions. But the impacts of climate variability, extreme events, and climate change are often local in nature—affected by accidents of microclimate, geography, development, demographics, and governance. Because of this, climate adaptation scholars and professionals have come to recognize the fundamental necessity and validity of local scale efforts to adapt to climate change.

In the United States, localities have taken proactive steps to address climate change, both in terms of mitigation and adaptation to emerging and potential impacts [1–3]. Local governments exercise extensive authority over a wide range of day-to-day decisions that influence greenhouse gas emissions as well as climate impacts. This realm of municipal decision making utilizes a broad array of policy, fiscal, and administrative tools to implement and maintain programs and projects to reduce vulnerability to climate variability, extreme events, and climate change. A meaningful proportion of local scale adaptation activities have been crafted to fall under the purview of federal laws, programs, and policy. Because so much progress reducing vulnerability and building resilience occurs at this local scale, it is important to track, characterize, and as appropriate integrate this experience into our evolving map of how decisions about climate impacts and adaptation are made and how to best support and accelerate such decision making.

Most calls for government action on climate impacts and adaptation have followed the climate mitigation policy playbook by focusing on a "top-down" model, for example, enacting Presidential Executive Orders or proposing omnibus legislation at the national level. However, while climate mitigation activities entrain only a few sectors (i.e., energy production, transportation, buildings), climate impacts play out across many more sectors with a much wider range of legal, administrative, and management rules and norms. Because of this breadth of relevance and potential impacts, a wide range of federal policies either drive and/or enable adaptive initiatives at the local scale. With so much adaptation and resilience activity currently happening at the local scale, we submit that better understanding the scope and nature of federal policies that currently facilitate this ongoing activity should be foundational to proposals for additional federal level adaptation policy, and perhaps should lead us to contemplate whether working at the national level is the most direct path to support and accelerate resilience at the community level. The key, we think, is to identify policy and governance arrangements with potential to enhance and expand local uptake of resilience measures, and then use those lessons to help guide the development and deployment of climate services.

2. Technical Approach

This paper is an interpretive re-analysis of 17 in-depth case studies of community-based climate adaptation sponsored by the Kresge Foundation [3]. The authors were senior members of the research team that designed and conducted the overall project, including the case studies, a literature review, thought leader interviews, and a cross case analysis. The original case studies were developed by five core researchers under the supervision of two senior researchers subject to the guidance of a 16-member project advisory committee. As described in the Kresge case study report, the project developed a systematic process for empirical inquiry into the cases dubbed its 'research protocol' (https://kresge.org/sites/default/files/library/climate-adaptation-the-state-of-practice-in-us-communities-full-report.pdf, accessed on 25 June 2022, see page 14). This protocol guided the core researchers across all primary data collection and case study development steps, including literature review and desk-top research, site visits, in-person interviews, draft case study development, follow-up email or telephone interviews, review by the senior researchers, and review by the interviewees in each case. The methodology is extensively covered in a chapter of Applied Policy Research [4]. The case studies systematically identified and characterized factors that shaped policy actions that tangibly reduced vulnerability in each community.

It is important to emphasize that this project was selective rather than comprehensive or randomized. The research team "select [ed] only cases with distinct outcomes that already have resulted or are likely to result in tangible reductions of vulnerability to climate variability, extreme events, or climate change" [3], p. 15. Summarized in Table 1, the 17 cases were winnowed from a universe of 93 candidate localities identified through literature review and thought leader interviews. The outcomes identified in these cases included community-level practices that reduced exposure, reduced sensitivity, or enhanced adaptive capacity [5]. The locales differ, however, in terms of size, demographics, economic base, political orientation, geography, and types of relevant climate impact variables.

Table 1. Case Study Summaries, Associated Federal Policy, and Partners in Governance.

Case Study Community	Description of Profiled Action	Associated Federal Policy	Partner Organizations
Avalon, New Jersey	Comprehensive Shoreline Protection Strategy: Developed several physical shoreline barriers, acquired undeveloped land, limited development, and created and maintained shorefront dunes.	U.S. Army CoE beach nourishment and subsidized construction and maintenance of protective seawall and breakwater; FEMA National Flood Insurance Program (NFIP) ratings drive development ordinances	Borough government agencies, federal agencies
Baltimore, Maryland	Integration of Climate Change Adaptation into a FEMA All-Hazard Mitigation Plan: Added a climate risk and vulnerability assessment into an existing Disaster Preparedness Project and Plan.	Required update of FEMA All-Hazard Mitigation Plan using FEMA "Hazus-Multi-Hazards" tool and NOAA data	City government agencies, federal agencies, citizen advisory commissions, public-private resiliency hubs
Boston, Massachusetts	Climate Change and Resiliency Checklist: Mandate that climate change be considered in city approval process for large new developments and renovation projects.	Process informed using National Climate Assessment scenarios; compliance with FEMA floodplain mapping stipulations necessary for permitting of new development	City government agencies, federal agencies, public-private partnership
Chula Vista, California	Cool Roofs Ordinance and Shade Tree Policy: Implemented a shade tree policy and cool roofs ordinance to address raising temperatures in the San Diego region.	Recipient of 2014 EPA Climate Leadership Award	City government agencies, local foundation, state government agencies, public utility, collaborative association of local governments
Cleveland, Ohio	Neighborhood Action Toolkit and Associated Fund: Augmented existing neighborhood revitalization initiative to help vulnerable neighborhoods increase adaptive capacity and anticipate a climate-altered future.	Financial support through HHS Community Economic Development Grant	City government agencies, community development corporation, citizen advisory committee, private enterprises designated as 'community assets', federal agencies, collaborative fund
El Paso County, Texas	Inland Desalination Facility: Due in part to projected climate change, the County planned and developed a desalination plant to manage stormwater runoff and augment water supplies.	Public-private partnership to develop the desalination facility involved Department of Defense though siting on Fort Bliss. Design of the facility was enabled in part by USGS technical support. Actions driven by EPA Stormwater Master Planning and need to maintain compliance with Clean Water Act regulations. Siting on the grounds of a federal facility necessitated assessment under the National Environmental Policy Act.	Federal agencies, county government departments, state government agencies, city government
Flagstaff, Arizona	Watershed Protection Project: Passed USD 10 million bond to reduce catastrophic fire risk on nearby U.S. Forest Service lands.	Project conducted in accordance with USDA Forest Service Forest Management Plans, and subject to the National Environmental Policy Act	Federal government agencies, state government agencies, city government, Native American tribal nation

Table 1. *Cont.*

Case Study Community	Description of Profiled Action	Associated Federal Policy	Partner Organizations
Fort Collins, Colorado	Water Demand Management: Revised Water Supply and Demand Management policy to better prepare for severe drought and reduce water use through conservation and increased storage measures.	Expansion of reservoir storage capacity requires permit from Corps of Engineers.	Public utility, city government, federal agencies
Grand Rapids, Michigan	Vital Streets and Sidewalks Spending Guidelines: Developed guidelines and implemented tax support to improve stormwater management with green infrastructure.	Initiative subject to EPA stormwater management regulations	City government, state government agencies, public utility, citizen commission, regional council of governments, federal agencies
Miami-Dade County, Florida	Integrating Climate Adaptation into Comprehensive Development Master Plan: Plan updates require county departments to consider potential climate change impacts for approval of capital improvement projects.	Initiative subject to EPA stormwater management regulations; partnership with USGS to develop hydrological modeling tools unique to Miami-Dade circumstances	County government agencies, federal agencies, county advisory task force, regional council of county governments
Mobile County, Alabama	Oyster Reef Restoration: Federal grant monies provided support for a public-private effort to restore oyster reefs to provide protection against storm surge and raising seas.	American Recovery and Reinvestment Act grant issued through NOAA	Federal agencies, county government agencies, non-profit funding, local chapter of national NGO
Norfolk, Virginia	Coastal Resilience Strategy: Flood and coastal zone ordinance revised to require that new structures have at least 3-foot freeboard.	Initiative pursued to reduce FEMA NFIP ratings; financial support sought through Disaster Resilience grants.	City government; federal agencies, local NGOs, civic leagues, citizen commission
Oakland, California	Climate Action Coalition: Formation of a diverse coalition that developed a strategy to address sea level rise through a social justice lens.	Financial support through Department of Housing and Urban Development grant programs	City government, federal agencies, local NGOs and foundations, community-based organizations, public utility
Seattle, Washington	Mainstreaming Climate Change into Internal Planning and Decision Making: Public utility requires climate change impacts to be considered in strategic planning, division-level planning, capital investment evaluation, and day-to-day operations.	Worked with a NOAA-funded research consortium to develop climate-impacted water supply scenarios; utility initiatives subject to EPA drinking water and stormwater management regulations	Public utility, city government, utility customer panel, federal agencies
Southwest Crown, Montana	Forest Restoration: Forest thinning and prescribed fires used to reduce risk of catastrophic fire.	USDA Forest Service Forest Management Plans, National Environmental Policy Act	Public-private partnership, federal agencies, town governments, state government, Native American tribes, county governments, citizen councils, collaborative organization

Table 1. *Cont.*

Case Study Community	Description of Profiled Action	Associated Federal Policy	Partner Organizations
Spartanburg, South Carolina	Mainstreaming Climate Change: Initiative to integrate climate change consideration into utility operations, management practices, program delivery, and culture.	Initiative embedded within FEMA All Hazard Plan update, subject to EPA wastewater regulations. Utility partnered with EPA to develop a resilience options evaluation tool (CREAT); and enhanced public outreach through EPA "WaterSense" program. Future climate conditions were assessed in part through review of U.S. GCRP reports	Public utility, county government, city government, federal agencies
Tulsa, Oklahoma	Acquisition and Relocation: Program to acquire repeatedly flooded properties and convert into parks and other public uses.	Project was framed to exceed FEMA NFIP strictures and partially funded through a FEMA Project Impact grant.	County government, city government, community-based organizations, local non-profit organizations

For this reanalysis, the case studies and their supporting information were culled by means of formal rubric to characterize and assess (1) the role—if any—played by federal law, policy, or programs; (2) the organizations and agencies involved in formulation and implementation; and (3) the manner and degree to which technical scientific data, methods, and information were utilized in planning, project design, and decision making. We follow the lead of local actors as illustrated in the Kresge case studies by addressing 'climate impacts' broadly and inclusive of expected climate variability (e.g., drought years), extreme events (e.g., severe floods), as well as climate change (e.g., sea level rise).

This interpretive reanalysis begins by building off the empirical foundation of the 17 Kresge case studies to address two basic research questions: (1) do federal policies influence climate adaptation at the local scale; and (2) how do federal climate adaptation policy regimes drive and enable local adaptation initiatives? In this paper, we address these questions through a four-step process. First, we draw from political science and international relations literature to characterize federal influence on local scale climate adaption activities through application of the concept of 'policy regimes'. Second, utilizing the policy regime construct, we provide a summary of the federal adaptation policy regime as it played out in each of the 17 Kresge case studies to illustrate a federal climate adaptation policy regime. This analysis is augmented through a literature review to summarize federal adaption policies that did not appear to impact the direction or nature of adaptation initiatives in the case study communities, but which help to illustrate the broader relevance of defining federal climate adaptation policy regimes in different policy or management contexts. Third, we provide a descriptive summary of specific federal policies and the role they played in the Kresge case communities. Fourth, we draw upon and integrate across the first three steps to explore implications for the ongoing enterprise of climate services. In the final section we outline several limitations in our analytical approach and articulate potential areas for further research.

3. Policy Regimes Defined

Rarely is an environmental, resilience, or sustainability issue simple enough to address with a single policy. Multi-dimensional, or wicked problems [6] such as urban renewal, health insurance reform, and global climate change tend to be addressed through multiple policies using a variety of tools, something that political scientists and scholars of international relations refer to as 'policy regimes'. Policy regimes are constructs that depict the mix of institutional mechanisms that make up the governing arrangements addressing a particular problem [7–12]. A regime may be comprised of multiple laws, rules, and administrative actions that together specify the contours of governance with respect to an issue or topic. It is important to emphasize that the impact of a regime may or may not be wholly consistent with the stated policy positions of a given administration or agency. Perhaps countering or tangential to political rhetoric, platform statements, or even the titles and captions of legislative acts, the regime construct provides a way to characterize and evaluate the on-the-ground, situational, and empirical impact of federal policies. Academic analysis of regimes typically considers factors such as stated or inferred objectives, strategic focus, scope of mandate, prescribed policy tools, implementation preferences (e.g., legalistic, corporatist, market-based), implementing agencies, and institutional actors. Policy scientists use the concept of regimes to analyze current dynamics, but also to help illuminate historical trends and changes in governance.

Policy regimes vary significantly from issue-to-issue. They can be centralized and tightly linked or loosely coordinated and disjointed across multiple agencies, statutory authorities, and/or jurisdictions. For example, while U.S. monetary policy is the centralized province of the Federal Reserve Board, U.S. health care policies are scattered among as many as 15 departments, bureaus, institutes, and agencies [13]. Regimes can be comprehensive or piecemeal, addressing most or only limited aspects of a problem. They are sometimes characterized as a continuum: "At one end are fully integrated institutions that impose regulation through comprehensive, hierarchical rules. At the other extreme are fragmented collections of institutions with no identifiable core and weak or nonexistent linkages between regime elements" [14], p. 4. In some cases, the term 'regime complex' is used to describe loosely coupled arrangements located somewhere in the middle of the continuum.

Policy regimes can be the intentional product of policy design or accretions that form over a span of disassociated activity. As described by Thelen, Mahoney, and others, historical and circumstantial change within regimes can be characterized in terms of various processes, including exhaustion, replacement, layering, and drift [15,16]. 'Exhaustion' is a condition under which a policy is no longer effective and requires change. 'Replacement' is a state of affairs in which most elements of an institutional arrangement are replaced. 'Layering' occurs when new elements are added to the extant complex without abandonment or material alteration of incumbent policies. Finally, 'drift' is a situation in which some aspects of a policy mix are maintained even as major aspects of the overall policy environment shift [9].

While the regime construct has been used to evaluate the international governance of climate change mitigation [14,17], we know of no other efforts to characterize national-scale policy clusters as they pertain to climate adaptation at the local scale. In the next section, we apply the regime lens to assess the role played by federal law and policy with respect to climate adaption efforts at the local scale.

4. Do Federal Policies Influence Climate Adaptation at the Local Scale?

It is common to hear that the United States lacks a meaningful or 'transformational' climate adaptation policy [18–20]. There is no law, program, or suite of governmental activities that can be described as an omnibus vehicle that defines a nation-wide strategy or prescribes an enforceable state-by-state or sector-by-sector implementation approach. There is no bureau or agency with a mission charter centered on adaptation to climatic impacts. Nor is there even an executive proclamation in support of preferred tools of adaptation.

However, if reviewed through a regime lens, it can be demonstrated that the U.S. has already implemented and maintains a wide range of public policy that bears in a substantive way upon local efforts to adapt to climate impacts. Informed by review of the Kresge case studies, Table 2 provides a summary of 26 federal policies that have been utilized in local efforts to plan for and adapt to climate impacts. While none of the vehicles in Table 2 include specific textual enunciation of climate adaptation as a policy objective in the title or caption of its statutory charter, all have been implemented, operationalized, and/or enforced in a manner that drives on-the-ground, empirical consideration of projected climate impacts. It is this use-based symmetry of application that—in our estimation—qualifies the cluster of programs and policies listed in Table 2 as a pragmatic, de facto federal climate adaptation policy regime.

Table 2. Components of the Federal Adaptation Policy Regime Associated with Local Adaptation Initiatives Profiled in Kresge Case Studies.

Policy or Programmatic Vehicle (Statutory Authorization, Year of Enactment)	Implementing Agency	Degree of Focus on Climate Change Adaptation	Influence on Community-Level Governance
Stormwater master planning (Clean Water Act, National Pollutant Discharge Elimination System, 1972)	U.S. Environmental Protection Agency (EPA)	One of many factors included in guidance materials	Serves as a driving influence
Stormwater discharge permitting and regulations (Clean Water Act, National Pollutant Discharge Elimination System, 1972)	EPA	One of many factors that may be considered in evaluating a permit application	Serves as a driving influence
State Revolving Fund (SRF) financial assistance for drinking water and POTW development (Safe Drinking Water Act of 1974)	EPA	One of many factors that may be considered in making SRF Capitalization Grant awards	Serves to enable
Climate leadership award	EPA	Promotion and outreach initiative focused on climate change mitigation and adaptation	Serves to enable
Technical Assistance to Water Utilities	EPA	Climate change adaptation is one of many factors addressed through Agency research activities	Serves to enable
'Watersense' program	EPA	Climate change is one of many factors considered in this outreach program	Serves to enable
All hazard mitigation plans (Disaster Mitigation Act of 2000)	Department of Homeland Security, Federal Emergency Management Agency (FEMA)	One of many factors that may be assessed in plan approval	Serves as a driving influence
Floodplain designations, mapping, and flood insurance ratings (Biggert-Waters Flood Insurance Act of 2012; National Flood Insurance Act of 1968)	FEMA	One of many factors that may be considered under a designation	Serves as a driving influence
Hazard Mitigation Assistance Grants (National Flood Insurance Act of 1968 and Department of Homeland Security (Annual) Appropriations Act)	FEMA	One of many factors that may be considered as a basis for award	Serves to enable
Atmospheric research	Department of Commerce, National Oceanic and Atmospheric Administration (NOAA)	Explicit and stipulated, development of climate-related data, information, and tools is central to mission	Serves to enable
National Weather Service	NOAA	Explicit and stipulated, development of climate-related data, information, and tools is central to mission	Serves to enable
Coastal resilience grants (Coastal Zone Management Act, 1972)	NOAA	One of many factors that may be considered as a basis for award	Serves to enable

Table 2. *Cont.*

Policy or Programmatic Vehicle (Statutory Authorization, Year of Enactment)	Implementing Agency	Degree of Focus on Climate Change Adaptation	Influence on Community-Level Governance
Coastal zone planning (Coastal Zone Management Act, 1972)	NOAA, Department of the Interior Bureau of Ocean Energy Management	One of many factors that may be assessed during agency and stakeholder reviews of plan	Serves both as driving and/or enabling influence
Reservoir siting approval and permitting (Rivers and Harbors Act of 1899; Clean Water Act of 1972)	United States Army, Corps of Engineers (CoE)	One of many factors that may be considered in a permitting decision	Serves as a driving influence
Dredge and fill permitting (Clean Water Act of 1972)	CoE	One of many factors that may be considered in a permitting decision	Serves as a driving influence
Operations and facility support	CoE	One of many factors that may be relevant to operations, planning, and budget implementation	Serves to enable
Integrated forest management plans (National Forest Management Act of 1976)	United States Department of Agriculture, Forest Service	Revised at least every 15 years, NEPA Guidance requires that Forest Management Plans address climate change impacts	Serves as a driving influence
Resiliency grants (Stafford Disaster Relief and Emergency Assistance Act of 1988)	Department of Housing and Urban Development	Eligible to communities that declared disasters in 2011–2013, grant resources are available to address a variety of threats, including climate change	Serves to enable
Water supply and hydrologic research	U.S. Geological Survey (USGS)	Agency charter authorizes water supply, hydrologic, and related areas of research, all of which may be subject to climate change impacts. Climate services stipulated under Agency workplans for USGCRP and related programs.	Serves to enable
Federal Environmental Impact Assessment Process (National Environmental Policy Act of 1968)	Council on Environmental Quality (CEQ)	Federal actions and programs require assessment of environmental impacts, including actions that might be subject to climate change impacts	Serves as a driving influence
Guidance on Greenhouse Gases and Climate Change (National Environmental Policy Act, 1968)	CEQ (and implementing agencies)	NEPA guidance directs agencies to include climate change impacts in their Environmental Impact Assessments. Focus of the guidance is on GHG emissions reduction, adaptation not central but applicability to land use decisions makes climate services relevant	Serves as a driving influence
U.S. Global Change Research Program (Global Change Research Act of 1990)	Thirteen agencies, coordinated by The White House Office of Science and Technology Policy	Program conducts and sponsors a wide range of climate-related research, with periodic reports that address "Impacts, Risks, and Adaptation". Program has a clear mandate to develop "decision aids"	Enables vulnerability actions and informs development of climate services

Table 2. *Cont.*

Policy or Programmatic Vehicle (Statutory Authorization, Year of Enactment)	Implementing Agency	Degree of Focus on Climate Change Adaptation	Influence on Community-Level Governance
Community Economic Development Grant Program	Department of Health and Human Services	Not stipulated, but may be considered as basis for award and subsequent reporting	Serves to enable
Agency Annual Appropriation Acts	All agencies	Agencies are sometimes directed and funded to conduct adaptation—or resiliency—related actions	May enable actions for municipalities partnering or otherwise interacting with federal entities or facilities
Endangered Species Act	All agencies	As applicable, Endangered Species Management Plans may be required to address impacts due to anticipated climate change	May serve to drive or constrain adaptation initiatives

In some cases, implementation of laws or programs with a putative, textualist focus on other topics are used to drive climate adaptation initiatives at the local scale. For instance, EPA stormwater discharge permit applications and renewals may be denied if a utility or municipality fails to demonstrate adequate consideration of potential operational deficits due to precipitation changes or the changing likelihood of extreme storm events. Similarly, economic development plans may be scrapped or reformulated due to excessive flood insurance premiums if a municipality allows unchecked siting in federally designated floodplains. In other cases, federal statutes and associated programs provide resources that are used to enable local adaptation initiatives. Federal enablement can come in the form of funding, technical assistance, or promotional support. And in nearly all cases, federal policy is used as an authoritative mechanism through which to structure deliberation and translate sentiment for change into revisions in the structure and/or processes of local governance.

Per their statutory language or statements of programmatic charter, these vehicles were intended to address policy issues other than climate adaptation, such as water pollution abatement, provision of safe drinking water, construction of publicly owned water treatment infrastructure, usage restrictions and protections over public lands, preservation of endangered species, stewardship of National Forests, wise use of coastal zones, and disaster mitigation or recovery. However, agency implementation of these policies has come to allow their routine application in the service of climate adaptation, with formalization occurring not as articles of black letter law, but rather by means of administrative and operational mechanisms such as guidance documents or guidebooks, memoranda of interpretation or implementation, annual workplans from agencies to the Office of Management and Budget (OMB), technical specifications documents, docket notations, circulars, and a wide range of informational materials on agency websites [10,21].

With respect to change over time, the federal climate adaptation policy regimes outlined in this paper cannot be characterized in terms of 'exhaustion' or 'replacement'. The policies through which local scale adaption is being addressed did not take the place of earlier mechanisms due to recognition of a need to adapt to climatic impacts. The cluster of policies outlined in Table 2 was not comprehensively designed but emerged over a period of approximately five decades due to many political actions, initiated by different administrations, at different times, and in response to different issues. Most of the policies summarized in Table 2 are environmental protection, resource management, or disaster response vehicles that have come to be utilized in the context of local scale climate adaptation. In other words, the federal influence on local climate adaptation has resulted from 'layering' and 'drift', with new conditions being addressed through existing agencies, policy vehicles, and programmatic activities.

We emphasize that the federal climate adaptation policy regime outlined through reanalysis of the Kresge case studies is a contingent, situational, and incidental 'snap-shot' of the potentially applicable policy domain. The literature review indicates other federal policies with potential to drive and/or enable climate adaptation at the local scale [1,2]. For example, conservation provisions within the five-year 'Farm Bill' have been used by Conservation Districts to support adaptation programs in rural counties [22]. And the Superfund provision of the Comprehensive Environmental Response, Compensation and Liability Act (CERCLA) has been used to reduce the exposure and vulnerability of urban toxic waste sites [23]. Owing to its origins within a single series of case studies, it can only be said that this analysis characterizes a federal climate adaptation policy regime, but not the federal climate adaptation policy regime. We encourage further exploration into the practical boundaries of federal climate adaptation policy regimes. Because climate adaptation is an issue that crosses many policy domains at the local scale, it is likely that it will prove useful to describe multiple regimes that affect particular domains, e.g., a federal water resources adaptation policy regime, a federal land management adaptation policy regime, a federal infrastructure adaptation policy regime, etc. Provisos and scope limitations aside, the Kresge case study reanalysis indicates a pervasive, ongoing, and formative role being played by federal policies with respect to the incidence and nature of local adaptation policies and programs.

It is important to note that the complex of policies summarized in Table 2 differs markedly from recent literature-based inventories of U.S. federal adaptation policy. Reviews conducted by [18,20] include only policy vehicles with an explicit, textually enunciated focus on climate adaptation, sometimes even more narrowly defined as climate change to the exclusion of climate variability and extreme events. These inventories treat an Obama-era Executive Order [24] as the administrative and strategic center of the federal climate adaptation mission. Revoked under the Trump Administration but largely restored by President Biden [25], this order establishes a high-level adaptation planning and coordination task group and directs all federal agencies to develop mission-specific adaptation plans. These Executive Orders spawned a variety of interagency working groups and resulted in a number of sector- and geography-specific planning activities and technical support programs. While it is clearly possible that the Biden EO and associated planning activities will—at some point—alter the fundamental contours of U.S. climate adaptation policy, we emphasize that this set of activities was not in play and did not impact adaptation initiatives undertaken by the Kresge case localities.

5. How Do Federal Climate Adaptation Policy Regimes Drive and Enable Local Adaptation Initiatives?

As summarized in the third column of Table 1, all local adaptation initiatives profiled in the 17 Kresge case studies were influenced by federal policy. In some cases, this influence was relatively incidental, certainly not a necessary factor in the local decision process. Examples of such incidental influence include reference to a federal study or information exchange through a voluntary program. However, the bulk of the cases involved a substantive federal role, often through application of multiple policies. For example,

- Avalon, New Jersey developed a Comprehensive Shoreline Protection Strategy driven partially by the need to reduce flood insurance premiums due to a high rating under the Federal Emergency Management Agency (FEMA) National Flood Insurance Program (NFIP). The Avalon comprehensive strategy also included beach nourishment and creation of a system of breakwaters and protective dunes, enabled in part by planning, technical assistance, and subsidization by the U.S. Army Corps of Engineers. Maintenance of the artificial dune system is conducted in compliance with a Corps of Engineers 'dune template'. Avalon's actions were the subject of high-profile recognition in the form of a Hurricane Mitigation Award, sponsored in part by the National

Oceanic and Atmospheric Administration (NOAA). But in an interesting twist, city officials note that dune maintenance activities are constrained due to the presence of a federally listed endangered species (piping plover), subject to protection by the U.S. Fish and Wildlife Services under the Endangered Species Act.

- In Spartanburg, South Carolina, a major initiative was driven by a requirement to update a FEMA All-Hazards Plan, and to do so in a manner that maintained compliance with U.S. Environmental Protection Agency (EPA) wastewater regulations. This two-pronged planning process was enabled through use of EPA-developed analytical and decision-support tools and was promoted to local stakeholders through an EPA outreach program called WaterSense. The planning process was enabled through application of U.S. Global Change Research Program future scenarios.
- In the Southwest Crown region of Montana, a coalition of local governments and stakeholders conducted a major forest restoration initiative that included forest thinning and prescribed burns to reduce the risk of catastrophic wildfire. This activity was conducted pursuant to a periodic U.S. Forest Service Forest Management Plan update process and subject to Council on Environmental Quality guidance and applicable provisions of the National Environmental Policy Act.
- The El Paso, Texas Water Utility worked in partnership with the U.S. Department of Defense to design and construct a desalinization facility on the grounds of Fort Bliss. The initiative was driven in part by EPA Stormwater Master Planning requirements and the need to maintain compliance with Clean Water Act and Safe Drinking Water Act regulations. Since the facility is sited on a federal property, design activities were overseen by the Corps of Engineers and enabled by the U.S. Geological Survey through development of specialized hydrological modeling capabilities. Planning and development activities were subject to the National Environmental Policy Act.

As described above, most of the federal policies that affected the Kresge case study communities were originally enacted to address environmental protection, resource management, or disaster response concerns. Nevertheless, these federal laws and policies constitute a legal framework that circumscribes appropriate or feasible local policy activity as well as providing technical and fiscal resources beyond what any individual locality could muster on its own.

While the federal climate adaptation policy regime identified and characterized through the Kresge case studies exhibits a clear influence on local adaptation initiatives, it would be wrong to say that these efforts to increase adaptive capacity and reduce vulnerability were top-down impositions by federal agencies or authorities. In response to the question 'How are communities implementing adaptation actions?' the Kresge project illustrates a variety of localized strategies, including community-level leadership, consciously building community support, tailoring discussions of 'climate change' to local politics and attitudes, generating grassroots and community organization support, engaging vulnerable populations, engaging in peer-to-peer networking, and a focus on dialogue, learning, and collaboration, among others—a list of implementation strategies with a distinctly local flavor [3].

That said, it is also clear that cognizant federal agencies could—if so inclined—have acted to disincentivize, stagnate, or even prohibit use of elements of the regime to support adaption initiatives at the local scale. It is easy to imagine how a hostile Congress or Executive could use tools such as appropriations riders, budget cuts or reprograming, OMB regulatory reviews, or executive orders to limit agency discretion when it comes to use of environmental protection, resource management, or disaster response statutes in the context of climate adaptation at the local scale. This is reminder that the regime construct is not immutable, but rather an active, emergent, and sometimes contested policy space.

It should be emphasized that nearly all the Kresge case studies include reference to tensions between stakeholders, strategic disagreements among partners, and conflicts between champions and affected parties. None of the Kresge initiatives could be described as a bed of roses. This acknowledged, it is nevertheless the case that participants found ways to work past their differences. Given this context, it is interesting to note that the applied policy analysis and public administration literature is replete with references to factors believed to render climate adaptation difficult—perhaps impossible—to achieve. As cataloged by Biesbroek and others [26], these include (1) fragmentation among actors and agencies with a stake in an adaptive project, (2) institutional voids due to a lack of an established, authoritative adaptation policy, (3) the short time horizons of politicians, and (4) a lack of governmental motivation to begin activities to address climate change. It is beyond the scope of this paper to explore—or even explicate—any of these factors. And while we acknowledge the pervasive problematicity of factors such as these within the literature of public policy and government administration, we remind that none of these factors derailed efforts to design and implement adaptation initiatives in the 17 Kresge case localities.

None of the local adaptation initiatives were a simple act of government implemented by and through a lone agency or entity. As summarized in Table 1, all the Kresge initiatives were formulated and implemented by multiple partner organizations under a more or less collaborative model of interaction. It is our sense that the federal climate adaptation policy regime complex suggested through reanalysis of the Kresge case studies bears the hallmarks associated with theories of adaptive governance [27–32]. Adaptive governance is characterized by decentralized decision making, a reliance on procedural rationality, and a highly contextual application of science and technical information. Adaptive models of governance are often impelled by perceived stagnation under incumbent top-down patterns of governance controlled by centralized authorities relying on technically rationalized methods and metrics. As Milward and Provan explain, "government refers to the formal institutions of the state—the executive, legislative, and courts—and their monopoly of legitimate coercive power. Governance is a more inclusive term, concerned with creating the conditions for ordered rule and collective action, often including agents in the private and nonprofit sectors as well as within the public sector" [32]. p. 360. "Whereas government refers to political institutions, governance refers to processes of rule wherever they occur" [33], p. 3. The idea of governance, then, includes relationships between government and society, including means through which networks of private actors influence policy decisions and self-organize to mediate their own actions and behavior. As summarized by Ruhl and others, "the role of law and government in adaptive governance is to leave space for local innovation and private governance" [28], p. 1688.

In addition to its adaptive utilization of federal law and statutes, the federal climate adaptation policy regime we identified can be characterized as:

- Non-prescriptive: For the most part, the federal climate adaptation policy regime does not impose specific actions or outcomes on units of local government. With respect to climate adaptation, it mandates no methods, tools, or strategic orientations. Except for NFIP rate determinations, it does not designate enforceable actions or measures. Unlike the environmental protection, resource management, and disaster response statutes from which it is derived, it includes no standards, minimum requirements, or technological stipulations.
- "Situational" in its applicability [10,27]: Adaptation initiatives driven or enabled through the federal policy regime do not necessarily apply to all jurisdictions in the same way. Rather, contingencies and circumstances determine the degree to which governance models, experienced extreme events, or anticipated changes in climate parameters will impact resources or service streams addressed through a given policy or approach to governance.
- De-centered: Local scale adaptation initiatives are not coordinated among cognizant agencies or framed for consistency by means of an overarching strategy or vision. As already emphasized, adaptive governance by local entities is derivative to the original mission of nearly all regime components.
- Scope-limited application of existing policy tools: Local scale adaptation initiatives enabled by federal policy tend not to be synoptic or expansive in nature. Quite the

opposite. Limited by the scope of the authorizing statute or program, adaptation initiatives tend merely to operationalize a policy provision or modify a particular service stream or sphere of activity (e.g., planning process, permit renewal, grant review process). Local adaptation initiatives tend to be operationalized by means of familiar tools of local governance such as ordinances, permits, bond issues, easements and property buy-outs, utility fees, comprehensive plans, disaster mitigation plans, zoning, and community or municipal staff capacity-building.

The Kresge case studies and other research efforts reveal a significant amount of adaptation work being conducted by U.S. localities [2,3,34]. Insufficient attention has been paid to the pragmatic role that federal policy plays in community resilience through existing elements of federal climate adaptation policy regimes. For example, the Kresge case reanalysis suggests this role tends to be a 'bottom-up' effort to take advantage of available policy tools to accomplish local policy objectives as they emerge through adaptive governance. However, this does not preclude the possibility of using federal climate adaptation policy regimes in a 'top-down' effort to incentivize or drive community resilience, for example, through climate projection requirements set by FEMA for hazard mitigation plans.

6. Some Implications for the Ongoing Enterprise of Climate Services

Having characterized a federal climate adaptation policy regime and described how it plays out at the local scale, we turn our attention to what these insights might imply for the enterprise of climate services. We stated earlier that it would be valuable to identify types of policy with potential to improve local uptake of resilience measures, and then to apply this information to help guide the development and deployment of climate services. It is crucial then to note that the evidence from the Kresge case studies and this reanalysis points to many disassociated local scale decisions that are improving the resilience of specific components of individual communities. Taken individually, any one of these actions may seem like a stand-alone case, when in fact they can be understood as part of a larger pattern of adaptive governance focused on community resilience. This section first reviews what the evidence from the Kresge case studies and this reanalysis suggests about common assumptions in the literature about climate services. We then turn to a discussion of how the specific needs articulated by the Kresge case communities suggest augmentation of the path forward for climate services.

As conventionally conceived, the enterprise of climate services has arisen due to the confluence of two phenomena: (1) continued improvement in the predictive capacity, scaling, and applicability of climate data and information; and (2) the fact that improved information has not always translated into effective adaptation [35–40]. Owing to this gap, leading researchers, government program administrators, and political leaders have called for significant organizational and strategic overhauls to the ways in which national governments and international agencies produce and deliver climate information and services [41–43]. Indeed, it is not uncommon to hear that such change must be "paradigm shifting" or "transformational" in nature [29,44–47]. In the U.S., this perspective has led to calls for a national climate service, recently amplified through Executive Order 14008 which directs the Office of Science and Technology Policy, NOAA, and FEMA to study and report on ways to expand and improve the delivery of climate services to the American public [48].

The literature on climate services takes it as almost an article of faith that local scale adaptation is being impeded, constrained, or blocked entirely by mismatches and incongruities between available information and the perceived needs of local decision makers and stakeholders [44,46,49]. Our reanalysis of the Kresge case studies suggests that these perspectives are not fully consistent with the observed experience of local communities working to adapt to climate impacts. None of the Kresge case studies revealed the clear and stark bifurcation between knowledge producers and knowledge users that has become a fixture of the climate services literature, certainly nothing as dramatic or lurid as the so-called "valley of death" described by Buontempo and others [50]. Stakeholder interviews that inform the cases studies do not suggest a debilitating distraction due to tension between "answerable" and "unanswerable" questions [44]. None of the cases hinged upon the often-discussed dynamic wherein policy actors insist upon answers to questions that the scientific community is unable to provide [46,51]. There was no evidence of unrealistically "deterministic" views of future change or so-called "projection shopping". And none of the adaptation champions interviewed for the case studies seemed to harbor unrealistic expectations that action should be delayed because—given time—science would provide a much better characterization of climate impacts upon their specific "neck of the woods" [44].

As illustrated with Figure 1, all the policy initiatives chronicled in the Kresge case studies were framed in terms of some type of data acquisition or assessment process. In many cases, the assessment component was limited and informal. Other cases, such as El Paso, TX; Seattle, WA; and Miami-Dade County, FL involved comprehensive, formalized vulnerability assessments, including provision of tailored climate information products. In nearly all cases, actors seemed capable of seeking information, assessing its relevance, recognizing and appreciating uncertainties, and moving ahead accordingly. We found decision makers and stakeholders willing to "span boundaries" or "make do" with available information [52]. What might explain such a noticeable departure from the climate services orthodoxy?

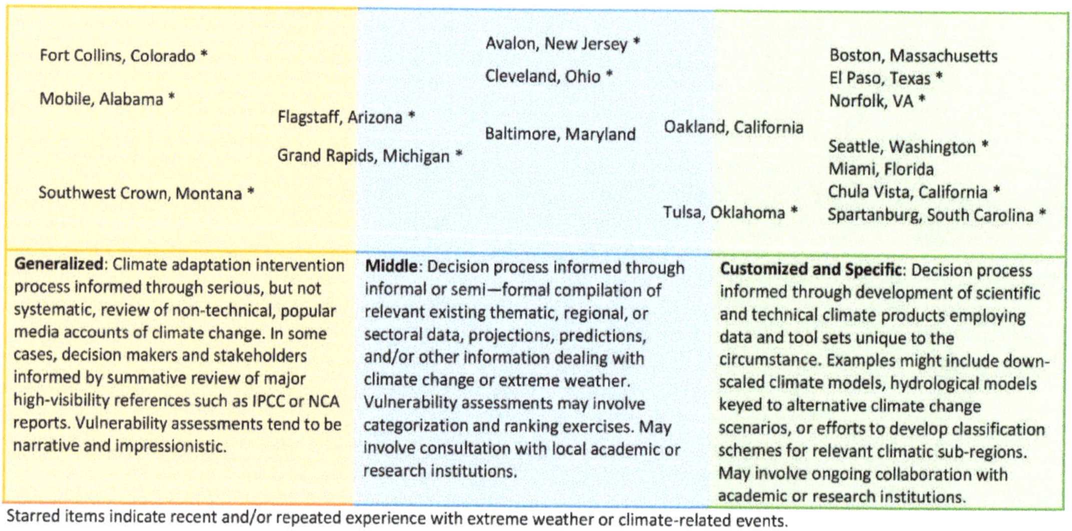

Figure 1. Conceptualizing Climate Change Vulnerability: Lived Experience, Sources of Information, and Locality-specific Analysis.

As we have already discussed, in most of the cases included in the Kresge study, climate adaptation or resilience actions were in some way associated with one or more federal policy mandates, none of which had an explicit textual focus on climate adaption or the idea of climate services. As per their statutory origins, regulatory actions undertaken by EPA, FEMA and other agencies must be grounded upon the "best available science", a condition that flows down to primacy agencies at the state and municipal level, and ultimately to regulated entities. In other words, local policy objects such as All-Hazard Plans, permit applications, and environmental impact assessments must be based on the best available science, but nothing more. This means that municipalities, public utilities, and other bodies of local governance are under no compunction to conduct original research, trade in "cutting edge" science, or somehow advance the state of the science in a particular field. It may also be significant that most of the localities covered through the Kresge case studies had experienced extreme weather events. As shown in Figure 1, all but 3 of the 17 case localities were impelled to some extent by recent stakeholder experience—sometimes repeated—with high-impact weather and climate events. Owing to this experience, decision makers and stakeholders may have been unusually motivated to act, adopting mental models animated by regulatory pragmatism and a perceived need to avoid known, experiential threats rather than the calculated output of rationalized, expert-produced, scenario-driven, and risk-optimized projections of future conditions [53].

As emphasized in a classic study by Rayner, Lach, and Ingram, the operational environment of municipalities, utilities, and other regulated entities can be complex, a-rational, programmatically oriented, and dependent upon craft skills and knowledge of localized systems [54]. Climate service developers need to understand and learn how to navigate this milieu; to become connoisseurs of practices and "rules" that explain things like "who and what sources of evidence to trust" and other factors that influence patterns of power and influence at the local scale [55], p. 42. It stands to reason that climate services provision may be especially helpful if conceptualized, framed, and delivered in a manner that is consistent with the concepts, parlance, operational environment, and/or institutional rationalities present in the parent activity.

In this regard, it would seem reasonable for climate service providers to adopt a regime perspective such as outlined in this paper to inventory and map federal policies that regulate resources and/or service streams subject to climatic disruption. This accomplished, would-be climate services providers can proceed in one—or both—of two directions: (1) work directly with communities to navigate applicable federal policy regimes to enable desired changes in local adaptation policy and governance, or (2) engage with federal mission agencies that administer existing regimes of environmental protection, resource management, or disaster response policy to learn how driving and/or enabling policies are being implemented to achieve climate resilience at the local scale, in order to help to expand the uptake and use of such practices. Whichever route is pursued, it seems to us that standard models of information production and provision may need to be modified or augmented in light of the adaptive governance dynamics and regimes described in this paper.

Building on the work of Meadow and colleagues [56], we suggest that co-productive methods such as Action Research (AR) or the Rapid Assessment Process (RAP) could be tailored to address and fit within the epistemic and operational confines of the regulatory context in which local scale initiatives of adaptive governance seem often to be couched. As described by Lather, these techniques are "openly ideological" in the sense that they are undertaken not to discover new knowledge, but rather to alleviate a known and bounded problem [57]. Under this conceptualization, climate products and services would be configured to "plug into" the practices, guidance, tools, and methods used within incumbent service streams. We worry that some co-productive models—such as Participatory Integrated Assessment, transdisciplinarity, or interaction with boundary organizations—may be too focused on open-ended research to be effective within extant federal climate adaptation policy regimes at the local scale.

Meadow and colleagues designate three primary "modes" of deliberative co-production: collegial, consultative, and collaborative. It may be that a distinct, fourth co-productive modality would help to assure better utilization of climate services at the local scale. Specifically, we are suggesting that there may be circumstances when climate service producers would benefit through something like an *apprenticeship* with local resource or service stream managers, enabling the "experts" at climate services or boundary organizations to learn and appreciate the nuance of place-specific regulatory processes, institutional rationalities, and operational environments at the nexus of federal policy and local governance. We think a similar dynamic could apply if climate service providers sought to work with federal regulatory and oversight agencies to identify procedural and/or functional policy objects that might be impacted by climatic change or extreme weather events.

In both cases, would-be climate service providers would need to become immersed in the culture and operational minutia of the service recipient in order to learn things like whether a particular action requires better information, why a contemplated activity may not need state-of-the-science inputs in order to proceed, and how such information would need to be configured in order to assure up-take through incumbent policy or governance modalities. Before attempting to engage in deliberative, co-productive activities, climate service providers need to "learn how, when, and where" it might be productive to build climate information into existing decision tools, best practices, and applications—or whether sufficient information already exists to support community action now [54], p. 224. We are suggesting that localized adaptations of federal policy may provide an especially rich culture for the inculcation of effective climate services, but that would-be providers of such services need to develop a robust, empathetic appreciation of these operational environments before attempting to engage in the co-productive role [58].

Due to the magnitude of some projected impacts, it is common to hear that climate change policies—both mitigation and adaptation—must be synoptic and transformational in character. Indeed, it has even been suggested that climate change could necessitate new forms of governance [59]. The Kresge case studies and this reanalysis suggest the viability of an alternative perspective [3]. As we have discussed, many of the case communities demonstrate that climate adaptation can be, indeed is being, addressed by collaborative bodies of governance under an extant legal regime using common policy tools. Appropriate application of climate services can help to fuel this emergent movement.

7. Analytical Limitations and Potential Research Needs

We believe it is significant that each of the 17 Kresge case localities exhibit a federal policy influence upon their climate adaptation initiatives. It is our position that the regime construct provides an important tool with which to characterize and evaluate the status of adaptation policy in the United States. However, we acknowledge several limitations in our line of analysis and suggest opportunities to advance research in this important arena.

Constraints of case selection: Our study focused on 17 communities that were able to undertake adaptive initiatives consistent with the scope of specified federal policies (although the federal role was not a factor considered in the original case selection). As with all case-based research, our observations could be an accident of case selection rather than evidence of a broader trend. For example, had the population of Kresge case communities included mostly small municipalities (population less than 300,000), we likely would have found fewer proactive climate adaptation-related initiatives due to limited resources alone. We do not believe that lesser peer learning opportunity or relative exclusion from national adaptation networks (which are also evident in small municipalities; will prove as significant a limitation in part because many of the Kresge case communities were motivated more by recent existential threats from weather or climate events than by a principled prioritization of climate change or adaptation to climate impacts—rendering such networks and peer learning largely beside the point [60]. Alternatively, a case population drawn from politically conservative localities might have exhibited ideological reluctance to build upon federal authorities rooted in environmental protection or resource management in particular, although perhaps less-so for those rooted in disaster response, and may consequently have resisted altogether the idea of addressing climate resilience through those authorities [3,61]. And finally, none of the Kresge case communities were animated by strong religious or faith-based motivations; another factor that could lead to differing approaches and policy outputs [62]. Moreover, it could be argued that had we focused on different localities, we might have found that federal policy actually served to constrain or even prevent adaptation activities envisioned by local actors. We doubt this argument because original case interviews and archival research did not suggest such a dynamic, but nevertheless acknowledge the logical possibility of alternative interpretations. A broader program of policy research could help to illuminate this topic and substantiate our observations.

Our research is U.S. specific: The Kresge case communities are all in the United States. And while tools such as the regime construct have international bona fides and have been used outside the U.S., it could be argued that the bulk of our evidence is unique to the distributed federalist nature of contemporary U.S. environmental protection, resource management, and disaster response policy and therefore of limited applicability. We note however, a literature base that includes influential studies assessing factors affecting climate adaptation policy initiatives in non-U.S. settings, including Western Europe, Scandinavia, Australia, Canada, Russia, and parts of South-Central Asia [63–67]. While none of this research builds on the regime perspective adapted for this paper, it does consider factors that are broadly relevant to our analysis such as the relationship between central and local units of governance; applicability of alternative policy tools; dominant regulatory perspectives; and perhaps most importantly, the role of climate-related data and information. This acknowledged, we nevertheless suggest that our finding of a national level climate adaptation policy regime complex with significant implications for local resilience activity is ripe for further international research.

Assuming effective outcomes: It could be argued that the ultimate effectiveness of the initiatives described in the case studies is unknown. Only time will tell if the actions undertaken by the Kresge case communities will materially reduce vulnerabilities or enhance adaptive capacity.

Promoting the adequacy of the status quo: Climate change is a highly contested and politicized topic. We have suggested that the Kresge case reanalysis points to the viability of a broadly incrementalist approach to climate change adaptation. This could be interpreted as a foil for ideological conservatism, as support for arguments that we do not need to address climate change in a concerted and aggressive manner. We reject the notion that an incrementalist strategic orientation suggests inaction or a passive acceptance of current rates of resilience uptake at the local level. But we acknowledge that some may argue—wrongly in our view—that this strategic orientation amounts to a defense of the status quo.

Policy design and the influence of regime layering: As we have already noted, the literature on policy regimes includes an important focus of the phenomena of layering and drift. There is ample recognition that policy design activities often occur upon the legacy of past decisions [68,69]. This means that new policy elements may or may not be fully consistent with the incumbent portfolio of policy tools. Our reanalysis of the Kresge case studies shows how climate impacts can be successfully layered onto existing environmental protection, resource management, and disaster response policies, but we did not address the question of whether new climate foci either augmented or degraded incumbent policy performance. We see this as a potentially fruitful area of research.

The role of states: Many federal policies are administered and enforced by state-level primacy agencies. Moreover, state constitutions typically delineate areas of allowed local governmental authority and activity. Further, there are several U.S. states that have taken aggressive policy action to promote resilience and adaptation to climate impacts [70]. While the Kresge case materials did not reveal or suggest determinate state roles in local climate decision making, there is clearly a basis in experience and literature to suggest non-trivial state-level involvement. Indeed, application of the regime lens could reveal unexpected and important patterns of state-level policy influence on local adaptation decision making and policy design. This is clearly an area in which focused research could prove illuminating.

Author Contributions: Formal analysis, C.H. and J.V.; Investigation, C.H. and J.V.; Methodology, C.H. and J.V.; Writing—original draft, C.H. and J.V. All authors have read and agreed to the published version of the manuscript.

Funding: This research received no external funding.

Institutional Review Board Statement: Not applicable.

Informed Consent Statement: Not applicable.

Data Availability Statement: Not applicable.

Conflicts of Interest: The authors declare no conflict of interest.

References

1. EcoAdapt. Climate Adaptation Knowledge Exchange. 2022. Available online: https://www.cakex.org/ (accessed on 24 February 2022).
2. Georgetown Climate Center (GCC). *Adaptation Clearinghouse*; Georgetown University: Washington, DC, USA, 2022; Available online: www.adaptationclearinghouse.org/about.html (accessed on 16 February 2022).
3. Vogel, J.; Carney, K.; Smith, J.; Herrick, C.; Stults, M.; O'Grady, M.; St Juliana, A.; Hosterman, H.; Giangola, L. *Climate Adaptation: The State of Practice in U.S. Communities*; The Kresge Foundation: Detroit, MI, USA, 2016.
4. Vogel, J.; Herrick, C.; Hosterman, H. Climate change adaptation: An assessment of accomplishments at the community level. In *Applied Policy Research: Concepts and Cases*; Springer, J.F., Haas, P., Porowski, A., Eds.; Routledge: New York, NY, USA, 2017.
5. Intergovernmental Panel on Climate Change (IPCC). *Climate Change 2007: Impacts, Adaptation and Vulnerability. Contribution of Working Group II to the Fourth Assessment Report of the Intergovernmental Panel on Climate Change*; Cambridge University Press: Cambridge, UK, 2007.
6. Rittle, H.; Webber, M. Dilemmas in a general theory of planning. *Policy Sci.* **1973**, *4*, 155–169. [CrossRef]
7. Herrick, C. A review of the U.S. invasive species policy mix: Questioning the prospect of an integrated regime. *Environ. Policy Gov.* **2019**, *29*, 262–279. [CrossRef]
8. May, P.; Jochim, A. Policy regime perspectives: Policies, Politics, and Governing. *Policy Stud. J.* **2013**, *41*, 426–452. [CrossRef]
9. Rayner, J.; Howlett, M. Introduction: Understanding integrated policy strategies and their evolution. *Policy Soc.* **2009**, *28*, 99–109. [CrossRef]
10. Jordan, A.; Lenschow, A. Environmental policy integration: A state of the art review. *Environ. Policy Gov.* **2010**, *20*, 147–158. [CrossRef]
11. Nilsson, M.; Eckerberg, K. (Eds.) *Environmental Policy Integration in Practice*; Earthscan: London, UK, 2007.
12. Underdal, A. Integrated marine policy. What? Why? How? *Mar. Policy* **1980**, *4*, 159–169. [CrossRef]
13. Congressional Research Service (CRS). *Public Health Service Agencies: Overview and Funding (FY2016-FY2018)*; Congressional Research Service: Washington, DC, USA, 2017.
14. Keohane, R.; Victor, D. *The Regime Complex for Climate Change. Discussion Paper 2010-33*; Harvard Project on International Climate Agreements: Cambridge, MA, USA, 2010.
15. Thelen, K. How institutions evolve: Insights from comparative historical analysis. In *Comparative Historical Analysis in the Social Sciences*; Mahoney, J., Rueschmeyer, D., Eds.; Cambridge University Press: New York, NY, USA, 2003.
16. Mahoney, J.; Thelen, K. (Eds.) *Advances in Comparative-Historical Analysis*; Cambridge University Press: New York, NY, USA, 2015.
17. Brunner, R.D. Science and the Climate Change Regime. *Policy Sci.* **2001**, *34*, 1–33. Available online: http://www.jstor.org/stable/4532520 (accessed on 10 February 2022). [CrossRef]
18. Bierbaum, R.; Smith, J.; Lee, A.; Blair, M.; Carter, F.; Flemming, P.; Ruffo, S.; Stults, M.; McNeely, S.; Wasley, E.; et al. A Comprehensive Review of Climate Adaptation in the United States: More Than before, But Less Than Is Needed. *Mitig. Adapt. Strateg. Glob. Chang.* **2013**, *18*, 361–406. Available online: http://link.springer.com/article/10.1007%2Fs11027-012-9423-1#page-2 (accessed on 3 March 2022). [CrossRef]
19. Melillo, J.; Richmond, T.; Yohe, G. (Eds.) *Climate Change Impacts in the United States: The Third National Climate Assessment*; U.S. Global Change Research Program, U.S. Government Printing Office: Washington, DC, USA, 2017. [CrossRef]
20. Hansen, L.; Gregg, R.; Arroyo, V.; Ellsworth, S.; Jackson, L.; Snover, A. *The State of Adaptation in the United States: An Overview. A Report for the John D. and Catherine T*; MacArthur Foundation: Chicago, IL, USA; EcoAdapt: Paris, France, 2013.
21. Kerwin, C. *Rulemaking: How Government Agencies Write Law and Make Policy*; Congressional Quarterly Press: Washington, DC, USA, 1994.
22. Dewey, S.; Hanson, L.; Horan, C.; Jacobs, W.; Goho, S.; Opportunities to Address Climate Change in the Farm Bill. Harvard Law School: Emmett Environmental Law and Policy Clinic. 2017. Available online: http://www.farmbilllaw.org/report/climate-change/ (accessed on 12 March 2022).
23. Dailey, A. Integrating Climate Change Adaptation into the Superfund Program. In *Climate Change Adaptation Webinar*; United States Environmental Protection Agency: Washington, DC, USA, 2015; Available online: https://clu-in.org/conf/tio/climatechange_040115/slides/Dailey.pdf (accessed on 12 March 2022).
24. Executive Order (EO) 13514. *Federal Leadership in Environmental, Energy, and Economic Performance*; Executive Office of the President: Washington, DC, USA, 2009.
25. Executive Order (EO) 14008. *Tackling the Climate Crisis at Home and Abroad*; Executive Office of the President: Washington, DC, USA, 2021.
26. Biesbroek, R.; Termeer, C.; Kabat, P.; Klostermann, J. Institutional governance barriers for the development and implementation of climate adaptation strategies. In Proceedings of the Working Paper for the International Human Dimensions Programme Conference "Earth System Governance: People, Places, and the Planet", Amsterdam, The Netherlands, 2–4 December 2009.

27. Kauffman, N.; Hill, K. Climate change, adaptation planning and institutional integration: A literature review and framework. *Sustainability* **2021**, *13*, 10708. [CrossRef]
28. Ruhl, J.; Cosens, B.; Soininen, N.; Gunderson, L. Designing Law to Enable Adaptive Governance of Modern Wicked Problems. *Vanderbilt Law Rev.* **2020**, *73*, 1687. Available online: https://scholarship.law.vanderbilt.edu/faculty-publications/1189 (accessed on 27 April 2022).
29. Granberg, M.; Bosomworth, K.; Moloney, S.; Kristianssen, A.-C.; Funfgeld, H. Can regional-scale governance and planning support transformative adaptation? A study of two places. *Sustainability* **2019**, *11*, 6978. [CrossRef]
30. Brunner, R.; Lynch, A.H. *Adaptive Governance and Climate Change*; American Meteorological Society: Boston, MA, USA, 2010.
31. Brunner, R.; Steelman, T.; Coe-Juell, L.; Cromley, C.; Edwards, C.; Tucker, D. *Adaptive Governance: Integrating Science, Policy, and Decision Making*; Columbia University Press: New York, NY, USA, 2005.
32. Milward, H.B.; Provan, K. Governing the hollow state. *J. Public Adm. Res. Theory* **2000**, *10*, 359–379. [CrossRef]
33. Bevir, M. *Governance: A Very Short Introduction*; Oxford University Press: Oxford, UK, 2012.
34. Carmin, J.; Nadkarni, N.; Rhie, C. *Progress and Challenges in Urban Climate Adaptation Planning: Results of a Global Survey*; Massachusetts Institute of Technology: Cambridge, MA, USA, 2012.
35. Vincent, K.; Daly, M.; Scannell, C.; Leathes, B. What can climate services learn from theory and practice of co-production? *Clim. Serv.* **2018**, *12*, 48–58. [CrossRef]
36. Ismail-Zadeh, A.; Cutter, S.; Takeuchi, K. Forging a paradigm shift in disaster science. *Nat. Hazards* **2017**, *86*, 969–988. [CrossRef]
37. Perrels, A.; Le, T.-T.; Cortekar, J.; Hoa, E.; Stegmaier, P. How much unnoticed merit is there in climate services? *Clim. Serv.* **2012**, *17*, 100153. [CrossRef]
38. Vaughan, C.; Dessai, S. Climate services for society: Origins, institutional arrangements, and design elements for an evaluation framework. *WIREs Clim. Chang.* **2014**, *5*, 587–603. [CrossRef]
39. Dilling, L.; Lemos, M. Creating usable science: Opportunities and constraints for climate knowledge use and their implications for science policy. *Glob. Environ. Chang.* **2011**, *21*, 680–689. [CrossRef]
40. Hecht, A. Meeting the challenge of climate service in the 1980s. *Bull. Am. Meteorol. Soc.* **1984**, *65*, 365–366. [CrossRef]
41. WMO. Global Framework for Climate Services. 2022. Available online: https://gfcs.wmo.int/node/1276 (accessed on 14 February 2022).
42. Lucio, F.D.F.; Grasso, V. The global framework for climate services (GFCS). *Clim. Serv.* **2016**, *2–3*, 52–53. [CrossRef]
43. Hewitt, C.; Mason, S.; Walland, D. Commentary: The global framework for climate services. *Nat. Clim. Chang.* **2012**, *2*, 831–832. [CrossRef]
44. Jacobs, K.; Street, R.B. The next generation of climate services. *Clim. Serv.* **2020**, *20*, 100199. [CrossRef]
45. Kundzewicz, Z.; Forland, E.; Piniewski, M. Challenges for developing climate services—Poland and Norway. *Clim. Serv.* **2017**, *8*, 17–25. [CrossRef]
46. Buizer, J.; Jacobs, K.; Cash, D. Making short-term climate forecasts useful: Linking science and action. *Proc. Natl. Acad. Sci. USA* **2016**, *113*, 4597–4602. [CrossRef]
47. Rickards, L. Transformation is adaptation. *Nat. Clim. Chang.* **2013**, *3*, 690. [CrossRef]
48. Office of Science and Technology Policy, National Oceanic and Atmospheric Administration, and Federal Emergency Management Agency. Opportunities for Expanding and Improving Climate Information and Services for the Public—A Report to the National Climate Task Force. 2021. Available online: https://downloads.globalchange.gov/reports/eo-14008-211-d-report.pdf (accessed on 10 April 2022).
49. Cortekar, J.; Bender, S.; Brune, M.; Groth, M. Why climate change adaptation in cities needs customized and flexible climate services. *Clim. Serv.* **2016**, *4*, 42–51. [CrossRef]
50. Buontempo, C.; Hewitt, C.; Doblas-Reyes, F.; Dessai, S. Climate service development, delivery and use in Europe at monthly to inter-annual timescales. *Clim. Risk Management* **2014**, *6*, 1–5. [CrossRef]
51. McNie, E. Delivering climate services: Organizational strategies and approaches for producing useful climate-science information. *Weather. Clim. Soc.* **2013**, *5*, 14–26. [CrossRef]
52. Guston, D. Boundary organizations in environmental policy and science: An introduction. *Sci. Technol. Hum. Values* **2001**, *26*, 399–408. [CrossRef]
53. Herrick, C. Self and place constructs in climate change vulnerability assessments: Gaps and recommendations. *Sustainability* **2021**, *13*, 2990. [CrossRef]
54. Rayner, S.; Lach, D.; Ingram, H. Weather forecasts are for wimps: Why water resource managers do not use climate forecasts. *Clim. Chang.* **2005**, *69*, 197–227. [CrossRef]
55. Cairney, P. *The Politics of Evidence-Based Decision Making*; Palgrave Pivot: London, UK, 2016.
56. Meadow, A.; Ferguson, D.; Guido, Z.; Horangic AOwen, G.; Wall, T. Moving toward the deliberate coproduction of climate science knowledge. *Weather. Clim. Soc.* **2015**, *7*, 179–191. [CrossRef]
57. Lather, P. Issues of validity in openly ideological research: Between a rock and a soft place. *Interchange* **1986**, *17*, 63–84. [CrossRef]
58. Herrick, C.; Pratt, J. Sustainability in the water sector: Enabling lasting change through leadership and cultural transformation. *Nat. Cult.* **2012**, *7*, 285–314. [CrossRef]
59. Di Paola, M.; Jamieson, D. Climate change and the challenges to democracy. *Univ. Miami Law Rev.* **2018**, *72*, 369.

60. Lioubimtseva, E. The role of inclusion in climate vulnerability assessment and equitable adaptation goals in small American municipalities. *Discov. Sustain.* **2022**, *3*, 3. [CrossRef]
61. Stiles, W. All adaptations are local. *Issues Sci. Technol.* **2014**, *30*, 57–64.
62. Bhagwat, S.; Economou, A.; Thornton, T. The idea of climate change as a belief system: Why climate activism resembles a religious movement. *GAIA* **2016**, *25*, 94–98. [CrossRef]
63. Bulkeley, H.; Kern, K. Local government and the governing of climate change in Germany and the UK. *Urban Stud.* **2006**, *43*, 2237–2259. [CrossRef]
64. Ayers, J.; Forsyth, T. Community-based adaptation to climate change. *Environ. Sci. Policy Sustain. Dev.* **2009**, *51*, 22–31. [CrossRef]
65. Soares, M.B.; Alexander, M.; Dessai, S. Sectoral use of climate information in Europe: A synoptic overview. *Clim. Serv.* **2018**, *9*, 5–20. [CrossRef]
66. Nachmany, M.; Byrnes, R.; Surminski, S. *Policy Brief: National Laws and Policies on Climate Change Adaptation: A Global Review*; Grantham Research Institute on Climate Chane and the Environment: London, UK; London School of Economics and Political Science: London, UK; University of Leeds: Leeds, UK, 2019.
67. Swart, R.; Celliers, L.; Collard, M.; Garcia Prats, A.; Huang-Lachmann, J.T.; Llario Sempere, F.; de Jong, F.; Manez Costa, M.; Martinez, G.; Pulido Velazquez, M.; et al. Reframing climate services to support municipal and regional planning. *Clim. Serv.* **2021**, *22*, 100227. [CrossRef]
68. Schneider, A.; Ingram, H. *Policy Design for Democracy*; University Press of Kansas: Lawrence, KS, USA, 1997.
69. Howlett, M.; Mukherjee, I. Policy Design and Non-Design: A Continuum of Formulation Modalities. In *Routledge Handbook of Policy Design*; Taylor and Francis: London, UK, 2018; Available online: https://ink.library.smu.edu.sg/soss_research/2760 (accessed on 21 January 2022).
70. Institute for Sustainable Communities (ISC). Regional Collaboratives for Climate Change—A State of the Art. 2019. Available online: https://us.sustain.org/wp-content/uploads/2019/04/Regional-Collaboratives-for-Climate-Change-FINAL-1-pdf (accessed on 15 June 2022).

MDPI
St. Alban-Anlage 66
4052 Basel
Switzerland
Tel. +41 61 683 77 34
Fax +41 61 302 89 18
www.mdpi.com

Sustainability Editorial Office
E-mail: sustainability@mdpi.com
www.mdpi.com/journal/sustainability

www.ingramcontent.com/pod-product-compliance
Lightning Source LLC
Chambersburg PA
CBHW061510210926
51850CB00008B/1871